Concise Textbook of Equine Clinical Practice Book 1

This concise, practical text covers the essential information veterinary students, new graduates and practitioners need to succeed in equine practice, focussing on lameness. Written for an international readership, the book conveys the core information in an easily digestible, precise form with extensive use of bullet-points, lists, diagrams, protocols and extensive illustration (over 650 full colour, high quality photographs).

Part of a five-book series that extracts and updates key information from Munroe's *Equine Surgery, Reproduction and Medicine, Second Edition*, the book distils best practice in a logical straightforward clinical-based approach. It details clinical anatomy, physical clinical examination techniques, diagnostic techniques and normal parameters, emphasising the things regularly available to general practitioners with minimal information of advanced techniques. The clinical information is split into anatomy-based sections.

Ideal for veterinary students on clinical placements with horses as well as for practitioners needing a quick reference 'on the ground'.

Concise Textbook of Equine Clinical Practice Book 1

Lameness

Antonio Cruz

Graham Munroe

Michael Schramme

Edited by

Graham Munroe

CRC Press
Taylor & Francis Group
Boca Raton London New York

CRC Press is an imprint of the
Taylor & Francis Group, an **informa** business

First edition published 2024
by CRC Press
6000 Broken Sound Parkway NW, Suite 300, Boca Raton, FL 33487-2742

and by CRC Press
4 Park Square, Milton Park, Abingdon, Oxon, OX14 4RN

CRC Press is an imprint of Taylor & Francis Group, LLC

Library of Congress Cataloging-in-Publication Data

Names: Cruz, Antonio M., author. | Munroe, Graham A., author, editor. | Schramme, Michael, author. | Equine clinical medicine, surgery and reproduction. 2nd edition.

Title: Concise textbook of equine clinical practice. Book 1, Lameness / Antonio Cruz, Graham Munroe, Michael Schramme ; edited by Graham Munroe.

Other titles: Lameness
Description: First edition. | Boca Raton, FL : CRC Press, 2023- | Abridgement of: Equine clinical medicine, surgery and reproduction / edited by Graham Munroe. 2nd edition. 2020. | Includes bibliographical references and index. | Summary: "This concise, practical text covers the essential information veterinary students need to succeed in Equine practice, focussing on lameness. Written for an international readership, the book conveys the core information in an easily digestible, precise form with extensive use of bullet-points, lists, protocols and extensive illustration (over 650 full colour, high quality photographs and radiographs). Part of a fur-book series that extracts key information from Munroe's Equine Surgery, Reproduction and Medicine, the book distils best practice in a logical clinical-based approach. The spiralbound format allows the book to lie open during practice"--

Provided by publisher.
Identifiers: LCCN 2022053992 (print) | LCCN 2022053993 (ebook) | ISBN 9781032438863 (hardback) | ISBN 9781032066141 (paperback) | ISBN 9781003369226 (ebook)
Subjects: MESH: Horse Diseases | Lameness, Animal | Horses | Handbook Classification: LCC SF959.L25 (print) | LCC SF959.L25 (ebook) | NLM SF 959.L25 | DDC 636.1/089758--dc23/eng/20230201

LC record available at https://lccn.loc.gov/2022053992
LC ebook record available at https://lccn.loc.gov/2022053993

ISBN: 978-1-032-43886-3 (hbk)
ISBN: 978-1-032-06614-1 (pbk)
ISBN: 978-1-003-36922-6 (ebk)

DOI: 10.1201/9781003369226

Typeset in Sabon
by Evolution Design & Digital

Contents

Preface

There is a vast array of clinical equine veterinary information available for the under- and postgraduate veterinarian and veterinary nurse to peruse. This is contained in textbooks, both general and specialised, and increasingly online at websites of varying quality and trustworthiness. It is easy for the veterinary student or nurse, recent graduate, and busy general or equine practitioner to become overwhelmed and confused by this diverse range of information. Often what is required, particularly in the clinical situation, is a distillation of the essential knowledge and best-practice required to treat the horse in the most suitable way. This concise, practical text is designed to fulfil that need.

This book focuses on lameness, which is the most common group of equine clinical problems affecting all types of horse, pony and donkey. It is part of a five-book series, which together covers all the areas of equine clinical practice. The information is extracted and updated from *Equine Clinical Medicine, Surgery and Reproduction (2nd Edition)* that was published in 2020. It is written for an international readership and is designed to convey the core, best-practice information in an easily digested, quick reference form using bullet-points, lists, tables, flow-charts, diagrams, protocols, and extensive illustrations and photographs.

The locomotor system is split into sections on approach to lameness, the foal, muscle, soft-tissue injuries, and anatomical regions, and each is approached in the same logical straightforward clinical-based way. There are details of relevant clinical anatomy, physical clinical examination techniques, normal parameters, aetiology/pathophysiology, clinical examination findings, differential diagnosis, diagnostic techniques, management and treatment, and prognosis. The emphasis is on information tailored to general equine clinicians with just enough on advanced techniques to make the practitioner aware of what is available elsewhere.

This series of books is intended to be used on a day-to-day basis in clinical practice by student and graduate veterinarians and nurses. The small size and spiral binding format allow them to lie open on a surface near to the patient, readily available to the veterinary student or practitioner while looking at, or treating, a clinical case.

About the Authors

Antonio Cruz is double boarded in Equine Surgery and Sports Medicine. He is currently a faculty member at the Justus Liebig Universität Giessen in Germany, having spent most of his career in North America. He has spent 25 years in academic practice in different capacities, having been tenured faculty (Associate Professor) at the University of Guelph (Canada) for several years. He has also worked at the Universities of Saskatchewan, Minnesota, Prince Edward Island and Bern in Switzerland. He also spent 7 years in private practice establishing a leading referral surgical and sports medicine facility in Vancouver, Canada. He holds several postgraduate degrees. He has supervised many graduate students, residents and interns and published over 60 articles and many book chapters. He is a regular speaker at international meetings. His main clinical focus is equine orthopaedics, and he is actively involved in research in the area of gait analysis and equine surgery.

Graham Munroe qualified from the University of Bristol with honours in 1979. He spent 9 years in equine practice in Wendover, Newmarket, Arundel and Oxfordshire, and a stud season in New Zealand. He gained a certificate in equine orthopaedics and a diploma in equine stud medicine from the RCVS while in practice. He joined Glasgow University Veterinary School in 1988 as a lecturer and then moved to Edinburgh Veterinary School as a senior lecturer in large animal surgery from 1994 to 1997. He obtained FRCVS in 1994 and DipECVS in 1997 by examination, and was awarded a PhD in 1994 for a study in neonatal ophthalmology. He has been visiting equine surgeon at the University of Cambridge Veterinary School, University of Bristol Veterinary School and Helsingborg Hospital, Sweden, and was team veterinary surgeon for the British Driving Teams from 1994 to 2001, the British Dressage Team from 2001 to 2002 and the British Vaulting Team in 2002. He was also FEI veterinary delegate at the Athens Olympics in 2004. He currently works in private referral surgical practice, mainly in orthopaedics. He has published over 60 papers and book chapters.

Michael Schramme qualified from the Rijksuniversiteit Gent, Belgium, in 1985. He has since worked as an equine surgeon at the University of Ghent, the Royal Veterinary College, the Animal Health Trust, Cornell University, North Carolina State University and the Ecole Nationale Vétérinaire de Lyon in France, where he is currently a full professor in equine surgery and orthopaedics. He is a Diplomate of the European and American Colleges of Veterinary Surgeons and of the European College of Veterinary Sports Medicine and Rehabilitation, and an associate of the European College of Veterinary Diagnostic Imaging. He is a past President of the European College of Veterinary Surgeons and the European Society of Veterinary Orthopaedics and Traumatology. He has an interest in all aspects of equine surgery, lameness, and diagnostic imaging with special emphasis on MRI.

List of Abbreviations Book 1

AL-DDFT	accessory ligament of the deep digital flexor tendon
ALD	angular limb deformity
AST	aspartate aminotransferase
CBCT	cone-beam computerised tomography
CDE	common digital extensor
CK	creatine kinase
CNS	central nervous system
Co	coccygeal vertebrae
CR	computed radiography
CSA	cross-sectional surface area
CT	computed tomography
DBLPN	deep branch of the lateral plantar nerve
DCP	dynamic compression plates
DDFT	deep digital flexor tendon
DDR	direct digital radiography
DDSL	deep distal sesamoidean ligaments
DE	digestible energy
DFTS	digital flexor tendon sheath
DIP	distal interphalangeal
DIT	distal intertarsal
DJD	degenerative joint disease
DOD	developmental orthopaedic disease
DP	dorsopalmar/plantar
DR	digital radiography
DSL	distal sesamoidean ligament
DSP	dorsal spinous process
ECR	extensor carpi radialis
EDTA	ethylenediamintetraacetic acid
FAD	flavin adenine dinucleotide
FFD	film focal distance
FT	fibularis (peroneus) tertius
GA	general anaesthesia
GI	gastrointestinal
HYPP	hyperkalaemic periodic paralysis
i/m	intramuscular(ly)
i/v	intravenous(ly)
ICL	inferior check ligament
IRAP	interleukin-1 receptor antagonist protein
IRU	increased radiopharmaceutical uptake
ISL	intersesamoidean ligament
LaDE	lateral digital extensor
LDE	long digital extensor
LDET	long digital extensor tendon
LDF	lateral digital flexor
LPL	long plantar ligament

MC	metacarpal
MCP	metacarpophalangeal
MDF	medial digital flexor
MDP	methyl diphosphonate
MIC	minimum inhibitory concentration
MIZ	maximum injury zone
MPICL	medial palmar intercarpal ligament
MRI	magnetic resonance imaging
MSC	mesenchymal stems cells
MT	metacarpal
NSAID	non-steroidal anti-inflammatory drug
NSC	non-structural carbohydrates
OA	osteoarthritis
OAAM	occipitoatlantoaxial malformation
OCD	osteochondrosis
OCLL	osseous cyst-like lesions
ODSL	oblique distal sesamoidean ligament
p/o	per os, orally
P1	first phalanx
P2	second phalanx
P3	third phalanx
PAL	palmar/plantar annular ligament
PCR	polymerase chain reaction
PIP	proximal interphalangeal
PIT	proximal intertarsal
POD	plantar osteochondral disease
PPID	pituitary pars intermedia dysfunction
PRP	platelet-rich plasma
PS	proximal scutum
PSB	proximal sesamoid bone
PSLD	proximal suspensory desmitis
PSSM	polysaccharide storage myopathy
PT	peroneus tertius
PTH	parathyroid hormone
RER	recurrent exertional rhabdomyolysis
SDFT	superficial digital flexor tendon
SDSL	straight distal sesamoidean ligament
SI	sacroiliac
SL	suspensory ligament
STIR	short tau inversion recovery
TC	tarsocrural
TCT	tibialis cranialis tendon
TENS	transcutaneous electrical nerve stimulation
TMT	tarso metatarsal
WBC	white blood cells

Approach to the Lame Horse

Introduction

- Lameness is common in all types of horse.
- defined as an alteration in the animal's normal stance and/or mode of progression:
  - caused by pain or neural or mechanical dysfunction.
- ideally logical sequence to approach:
  - define which limb or limbs are involved.
  - find the exact site of pain.
- other techniques such as radiography or advanced imaging are then used to:
  - determine a specific pathological process and make a diagnosis.
- once a provisional or accurate diagnosis is made:
  - management plan can be formulated, and a prognosis given to the owner.
- many factors may alter this sequence:
  - e.g. environment where the examination is taking place.
  - financial considerations.
  - impending competition.

History

- accurate history is essential and requires careful questioning.
- **Signalment of the case:**
  - age, sex, breed and use of the horse can suggest certain conditions.
- length of ownership, training pattern, type and amount of work, and previous lameness problems should be noted.
- **Present lameness:**
  - **when** (duration of lameness) and **where** (at pasture or in work) first noted.
  - lameness improved or worsened with rest or exercise.
  - limb affected or what are they noticing when the horse is ridden.

- traumatic episodes associated with onset.
- other associations (e.g. recent change in shoeing for nail-bind).
- regions of heat, swellings or other clinical signs noted by the owner.
- any treatment, rest or shoeing tried by the owner before the examination.

Clinical examination

At rest

- visual inspection standing square on a level surface, from both sides, front and behind.
  - overall conformation
  - signs of asymmetry of the muscles and bones.
  - swelling   ○ sites of trauma
  - foot conformation and stance (Figs. 1.1 and 1.2).
  - conformation affects the way a horse moves at all gaits:
    - may predispose to lameness. (Fig. 1.3).
  - foot conformation is important in the incidence of foot lameness (Figs. 1.4–1.8):
  - evaluation of pastern and foot angle (dorsopalmar/plantar hoof balance).
  - mediolateral foot balance
  - foot symmetry/shape/size.
  - conformation determines the load distribution of the structures in the foot and elsewhere within the limb.
  - hoof horn defects (e.g. hoof cracks) should be noted (Fig. 1.9).
  - posture or stance may indicate:
    - horse's response to acute or chronic pain in one or more limbs (Fig. 1.10).
    - specific loss of function of part of the locomotor system (Fig. 1.11).

DOI: 10.1201/9781003369226-1

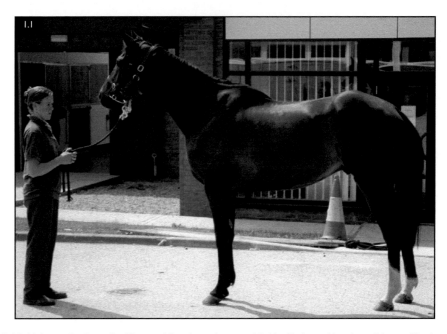

FIG. 1.1 Initial examination of a Thoroughbred racehorse with hindlimb and back problems. The horse is stood squarely on a hard surface and is examined from each side, behind and in front for stance, swellings, muscle wastage and conformation.

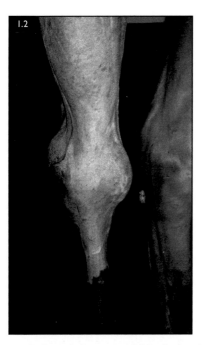

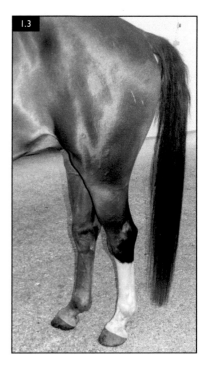

FIG. 1.2 A distended tarsocrural joint in the right hindlimb of a horse with lameness, which was localised to the joint by intra-articular analgesia.

FIG. 1.3 A Swedish Warmblood horse with straight hindlimb conformation.

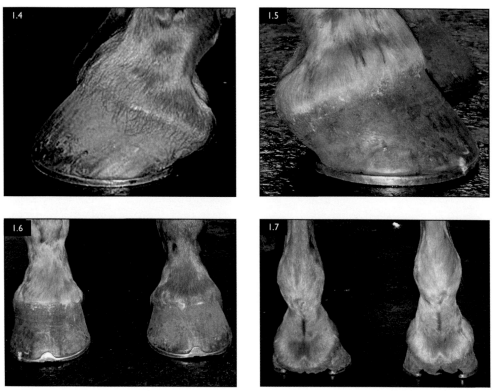

FIGS. 1.4–1.7 Observation of both forefeet at rest from both sides (1.4, 1.5), the front (1.6) and behind (1.7) for conformation. Note the contracted right forefoot and the long toe/low heel conformation of both forefeet. A diagnosis of chronic palmar hoof pain of navicular bone origin was made after the results of diagnostic analgesia, radiography and MRI.

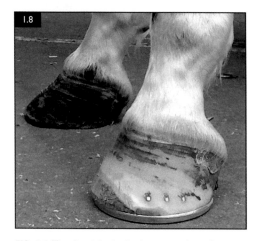

FIG. 1.8 The front feet of a horse undergoing corrective trimming and shoeing. Note the right foot has a long toe with a broken-back foot pastern axis. The left foot has had the toe trimmed back to re-establish a straight foot pastern axis.

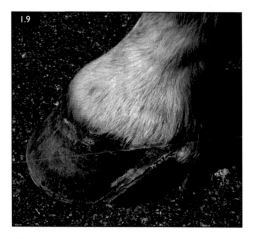

FIG. 1.9 This horse suffered a severe trauma to the coronary band and hoof of this foot several years ago. It is now left with a permanent injury to the coronary band, which is leading to a crack in the quarter of the hoof.

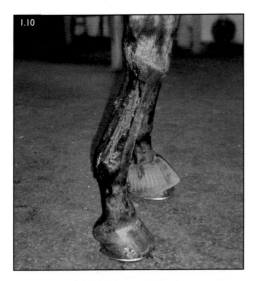

FIG. 1.10 Observation of the forelimb stance of a horse from the right side. Note the acquired carpal contracture of the right forelimb in this case of a chronic proximal suspensory ligament insertional desmitis of over 2-years' duration.

FIG. 1.11 Dropped elbow with flexed carpus and fetlock in a horse exhibiting temporary radial nerve paresis post a general anaesthesia in right lateral recumbency. Note this posture is also seen in olecranon fractures.

- palpation of the affected limb(s) is essential:
  - knowledge of normal palpable anatomy is a prerequisite.
  - comparison with contralateral limb may be useful.
    - **beware bilateral conditions e.g. osteochondrosis dissecans.**
  - palpate sequentially regardless of the suspected lame limb suggested by the owner:
    - all the limbs, both weight bearing and held off the ground.
    - neck and back.
    - unless an obvious reason merits immediate inspection (e.g. a recent wound).
    - look for cardinal signs of inflammation of heat, swelling and pain.
    - assess normal reaction to gentle digital pressure versus that of pain response.
      - careful repeat examination of the area.
      - palpation in the same area of the contralateral limb.

- examination of the foot:
  - ground bearing surface of the foot inspected carefully for any abnormalities.
  - hoof testers can be used to assess:
    - solar reaction.
    - application across the frog and heels may reveal sites of pain.
  - shoe inspected for type, wear and nail placement.
    - shoe should be removed.
    - foot inspected further for evidence of a foot-related condition.
  - digital pulse to the feet should be palpated.
    - bounding pulses commonly felt in:
      - acute laminitis
      - subsolar abscesses
      - bruising (Fig. 1.12).
      - pedal bone fractures.

At exercise

- lameness is generally assessed in a straight line on a level, hard, even surface:
  - from in front (for forelimbs) (Fig. 1.13).
  - from behind (for hindlimbs) (Fig. 1.14).

FIG. 1.12 This horse presented with an acute-onset, moderate lameness of a forelimb, which was localised to the foot by hoof tester reaction, and a strong digital pulse. After removal of the shoe, clear areas of fresh subsolar and white line haemorrhage are visible at the toe underneath where the shoe was placed.

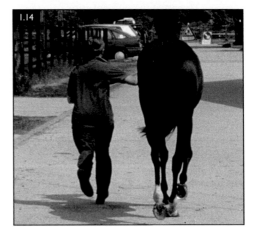

FIGS. 1.13, 1.14 Trotting a horse towards the examining veterinarian on a firm level gravel surface (1.13). Note the loose way the horse is led to allow any movement of the head to be clearly seen. Horses that are lame should always be examined from in front, behind and the side (1.14).

- at the walk and trot, in-hand.
- movement of the whole horse and individual limbs should be evaluated:
  - assess foot placement and breakover.
  - examination from the side helps reveal:
    - shortened stride length and lowered foot flight arc in a lame limb.
  - lameness that is subtle or involves multiple limbs:
    - best assessed on a circle on the lunge at the trot.
  - comparison between movement on a soft and hard surface is helpful:
    - may increase the degree of lameness in some cases (Fig. 1.15).
  - if the lameness occurs in certain situations, examine the horse at:
    - higher speeds, such as canter and gallop.
    - when ridden/driven.
- symmetrical movement is the normal situation in the sound horse:
  - appreciating changes in this is key to identifying lameness and limb(s) involved.
  - weight-bearing forelimb lameness:
    - 'head nod' – head being raised when the lame limb strikes the ground.
  - hindlimbs:
    - 'hip hike' in the lame hindlimb.
      - pelvis 'hikes' upward when the lame hindlimb hits the ground.
      - moves downward when the sound limb hits the ground.
      - easier for some clinicians to see a downward movement of the pelvis:
        - on the side of the lame limb as it leaves the ground.

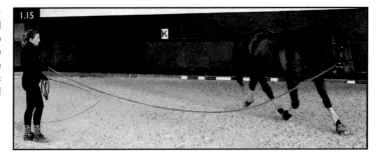

FIG. 1.15 Lungeing exercise on a soft and/or hard surface can be useful to give a further baseline lameness to record before embarking on diagnostic analgesia as part of a full lameness examination.

- o unilateral hindlimb lameness may appear as an ipsilateral forelimb lameness:
  - ◆ due to horse shifting its weight forward in compensation.
  - o digital recording of a lame horse in motion, on a smart phone:
    - ◆ later evaluation in slow motion, can help visualise a lameness further.
- lameness is graded to indicate severity, commonly on a scale of 0–5 or 0–10:
  - o 0 is sound and 5 out of 5 or 10 out of 10 is non-weight bearing.

Manipulative tests

- aim to exacerbate temporarily the degree of lameness and help identify source.
- **Flexion tests** should ideally be performed on all pairs of limbs for comparison:
  - o lastly on the lame limb.
  - o time and force required to carry out flexion tests are personal but should be consistent.
    - ◆ generally, 45–60 seconds with mild force is used.
  - o horse is trotted away from, and back to, the examiner immediately after the test for 20–30 metres.
  - o persistent increase in lameness over baseline is positive and should be graded.
  - o **not appropriate to carry out flexion tests on severely lame horses.**
- forelimb flexion tests may be divided into:
  - o full forelimb flexion test aims to flex all the joints by supporting the limb at the toe with the cannon and radius parallel to the ground.
  - o distal forelimb flexion test (Fig. 1.16) aims to flex the foot and fetlock by supporting the limb at the toe with the carpus at around 90° and with

minimum flexion of the elbow and shoulder joints.
  - o carpal flexion test aims to flex mainly the carpus by holding the cannon parallel to the ground and allowing the distal joints to remain unflexed.
  - o proximal forelimb flexion test flexes mainly the shoulder joint and involves holding the limb at the radius and pulling the entire limb caudally and slightly proximally.
- hindlimb flexion tests can be divided into:
  - o hindlimb flexion tests are less specific than in the forelimb.
    - ◆ presence of reciprocal apparatus in hindlimb.
  - o full flexion test, the limb is supported at the toe with the cannon and tibia parallel to the ground (Fig. 1.17).
  - o distal hindlimb flexion test consists of flexing the foot and fetlock by supporting the limb at the toe with the

FIG. 1.16 Right forelimb distal limb flexion test. Note how the carpus is kept as minimally flexed as possible.

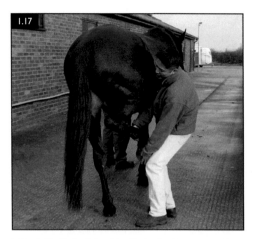

FIG. 1.17 Full right hindlimb flexion test. Note how all the joints are flexed, with the limb directly underneath the horse.

cannon perpendicular to the ground and the hock and stifle at 90°.
- ○ proximal hindlimb flexion test (hock or 'spavin' test) involves supporting the limb at the cannon parallel to the ground and held fully flexed at the hock and stifle.
- extension tests may also be carried out:
  - ○ placement of a reverse heel wedge under a foot is used to evaluate suspected caudal hoof pain.
- evaluation of lameness after:
  - ○ direct digital pressure to a painful region.
  - ○ hoof tester application to a suspicious region of the solar surface of the foot.

Diagnosis

Diagnostic analgesia

- perineural nerve blocks, local infiltration and intrasynovial joint/sheath/bursa blocks:
  - ○ used to abolish lameness temporarily in the limb being investigated.
  - ○ help isolate site of pain.
- mechanical or neurological types of lameness are not suitable for this approach.
- **lameness severe (grade 4 or 5/5) and localising signs of pain or anatomical abnormalities:**
  - ○ **subjected to diagnostic imaging before any diagnostic analgesia techniques.**

- Mepivacaine is the most common local anaesthetic agent used:
  - ○ Lidocaine and Prilocaine are used occasionally.
  - ○ differ in time of onset/duration of analgesia, and degree of inflammatory reaction they cause.
- selection of which blocks are carried out is based on the previous clinical examination:
  - ○ no apparent signs for a particular region are identified:
    - ♦ start as distally as possible and work sequentially proximally with perineural and intrasynovial analgesia techniques. (Fig. 1.18).
- partial or complete response to diagnostic analgesia warrants further investigation of that region by appropriate diagnostic imaging.
- if a contralateral limb lameness is revealed:
  - ○ examine this limb with diagnostic analgesia, starting distally as before.

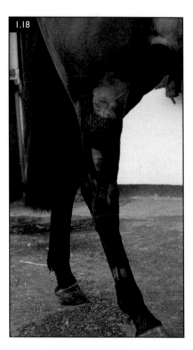

FIG. 1.18 This horse has had sequential regional and intra-articular analgesia carried out up to the level of the stifle joints in an attempt to localise the source of the hindlimb lameness.

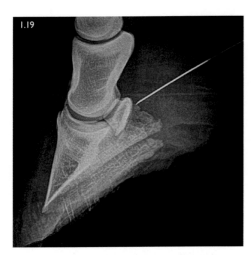

FIG. 1.19 Lateromedial radiograph of the foot confirming the correct placement of a needle into the navicular bursa before the local anaesthetic is injected.

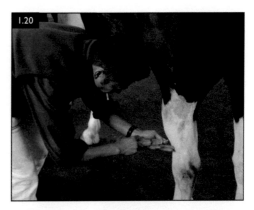

FIG. 1.20 Left hindlimb superficial and deep peroneal (fibular) nerves are being injected with local anaesthetic as part of analgesia of the tarsus and entire distal hindlimb. This nerve block is routinely combined with a tibial nerve block.

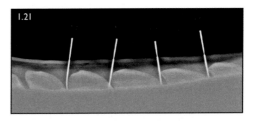

FIG. 1.21 Lateromedial radiograph of the dorsal spinous processes of a horse with back pain showing needles in the interspinous spaces prior to the injection of local anaesthetic.

- o imaging is then carried out after the site of lameness in this limb is confirmed.
- sites for perineural nerve blocks should be cleaned and can be clipped if identification of anatomical landmarks is difficult or the coat is dirty.
- **all intrasynovial injections require full aseptic precautions, after clipping the hair.**
- physical restraint and consideration of personnel safety are advisable in most horses.
  - o placement of a nose twitch, if the horse tolerates this.
  - o assistant lifting a contralateral limb for forelimb blocks or ipsilateral forelimb for hindlimb blocks may be useful.
- chemical restraint is required for intractable animals:
  - o intravenous acepromazine or shorter-acting alpha-2 agonists (e.g. Xylazine).
    - ♦ dosage dependent on the individual's behaviour.
  - o re-examine horse after the sedation has worn off, usually after 20–45 minutes.
- horses are usually re-examined 5–30 minutes after blocking (depends on block used):
  - o trot in a straight line and/or on the lunge.

- o compromise between allowing enough time to anaesthetise the nerve or joint and minimising the spread of the drug to other anatomical sites.
- o intrasynovial blocks in larger joints (scapulohumeral and stifle):
  - ♦ re-examined after 30–60 minutes by some clinicians.
- o some perineural nerve blocks may be tested for efficacy by loss of skin sensation prior to re-examination.
- o interpretation of the results based on knowledge of which region the block desensitises (Table 1.1).

Synovial fluid collection and analysis

- sites for diagnostic intrasynovial (joint/sheath/bursa) analgesia may also be used for synovial fluid collection for analysis.

TABLE 1.1 More commonly used perineural nerve blocks, local infiltration and intrasynovial blocks for the forelimb and hindlimb.

	FORELIMB	HINDLIMB
Perineural blocks	Palmar digital Abaxial sesamoid Low 4-point High 4-point (subcarpal) Lateral palmar Median/ulnar/musculocutaneous	Plantar digital Abaxial sesamoid Low 6-point Deep branch of the lateral plantar nerve High 6-point Tibial/superficial and deep fibular (peroneal) (**Fig. 1.20**)
Local infiltration	Painful exostoses including 'splints' Origin of the suspensory ligament	Painful exostoses including 'splints' Origin of the suspensory ligament Dorsal spinous processes (**Fig. 1.21**) Sacroiliac region
Intrasynovial blocks	Navicular bursa (**Fig. 1.19**) Distal interphalangeal joint Proximal interphalangeal joint Metacarpophalangeal joint Digital sheath Intercarpal joint (communicates with the carpometacarpal joint) Antebrachiocarpal joint Carpal sheath Humeroradial joint Bicipital bursa Scapulohumeral joint	Navicular bursa Distal interphalangeal joint Proximal interphalangeal joint Metatarsophalangeal joint Digital sheath Tarsometatarsal joint Central tarsal joint Tarsocrural joint Tarsal sheath Femoropatellar joint Medial femorotibial joint Lateral femorotibial joint Coxofemoral joint

- main indication is a suspicion of a synovial septic process.
- **clipping the hair and aseptic preparation of the collection site are mandatory.**
- aspirated fluid should be placed in EDTA and plain tubes for analysis.
- visual assessment of synovial fluid:
  - normal joint fluid is slightly viscous and has a clear, straw colour.
  - less viscous in inflamed joints (decreased hyaluronan), similar colour to normal.
  - varying degrees of haemorrhage may be present depending on:
    - sampling technique or the presence of a haemarthrosis.
  - septic fluid is less viscous and often cloudy and discoloured (Fig. 1.22).
- Parameters routinely measured include cytology and total protein concentration:

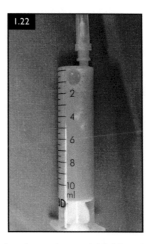

FIG. 1.22 A syringe of synovial fluid aspirated from the distended metacarpophalangeal joint of a very lame Thoroughbred yearling that had sustained a wound in the region of the fetlock 48 hours earlier. Note the very turbid and discoloured synovial fluid that had a WBC count of >100 × 10^9/1.

TABLE 1.2 Synovial fluid cytology reference ranges.

	APPEARANCE	TOTAL WBCs (×10⁹/l)	NEUTROPHILS	TOTAL PROTEIN (g/l)
Normal	Clear Straw coloured	≤0.5	<10%	<20
Sepsis	Turbid, degenerate	15–150	>90%	30–60
Osteoarthritis	Pale yellow	≤1.0	10–15%	<25
Osteochondrosis	Pale yellow	0.5–1.0	10–30%	<25
Acute trauma (e.g. intra-articular fracture)	Serosanguineous	3–10	<10%	<30

- o cytology is useful for identifying/ monitoring sepsis or post-injection reactions.
- o total protein concentrations normally tend to be higher in the larger joints.
- o if sepsis is suspected:
  - ♦ bacterial culture is essential – often difficult.
  - ♦ positive results are increased by placing the fluid directly in blood culture medium immediately after collection.
  - ♦ Polymerase chain reaction (PCR) analysis useful where cytology is equivocal:
    - – reveal bacterial DNA in suspected samples.
- o parameters of normal synovial fluid and certain conditions are shown in Table 1.2.

Radiography

- most important diagnostic imaging modality used for the musculoskeletal system.
- basic settings that determine the quality of the image include:
  - o total number of X-rays produced (mAs).
  - o ability of the X-ray to penetrate the region of interest (kV).
  - o film focal distance (FFD).
- Digital radiography (DR) has now become standard in equine practice:
  - o two basic types: computed radiography (CR) and direct digital radiography (DDR or simply DR).

- o CR uses crystals, that can be photo-stimulated within a flat panel plate inside cassettes, which go through a processor that converts the exposed crystals into a greyscale image.
- o DR uses an electrical photoconductor within the panel to convert X-ray photons into electrical charges which are immediately converted into a greyscale on-screen image.
- o digital systems provide many advantages to the equine clinician:
  - ♦ better image quality and diagnostic capability.
  - ♦ ability to manipulate images after they are obtained to highlight different tissues and structures.
  - ♦ increased portability and on-site image availability.
  - ♦ digital storage and sharing of images.
- o generator of X-rays can be a mobile, semi-mobile or fixed gantry machine.
- multiple projections are necessary since a 2-D image of a 3-D structure is produced (Figs. 1.23, 1.24).
- plain radiography refers to a standard radiograph and is used routinely (Fig. 1.25).
- contrast radiography (Fig. 1.26) refers to the placement of a metallic probe or radiodense contrast material (e.g. non-ionic, water-soluble compounds such as iohexol) into a specific region to highlight certain pathology:
  - o useful for confirming synovial wound penetration.
  - o injected into tendon sheaths, bursae and fistulous tracts.

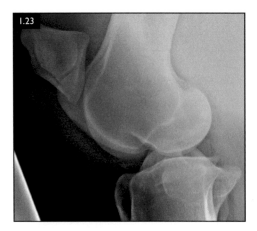

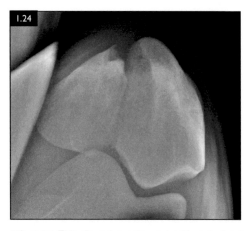

FIG. 1.23 Lateromedial radiograph of the stifle of a horse with an acute and severe lameness. There was swelling of the femoropatellar joint and patella. It is not clear what has happened in this horse.

FIG. 1.24 This flexed cranioproximal/craniodistal oblique (skyline) view of the stifle of the horse in 1.23 clearly shows a parasagittal fracture of the patella.

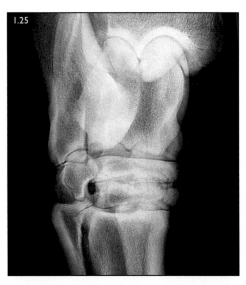

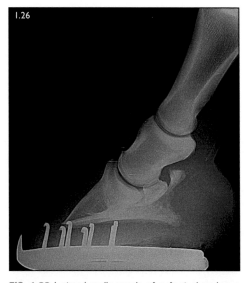

FIG. 1.25 Standard dorsolateral/plantaromedial oblique radiographic projection of a tarsus illustrating degenerative joint disease of the small tarsal joints ('bone spavin').

FIG. 1.26 Lateral radiograph of a foot showing a positive contrast study of the navicular bursa. Contrast medium was injected into the bursa to ascertain its integrity following a puncture of the solar surface of the foot. No contrast is seen moving from the bursa into the solar penetration tract.

- X-rays are a radiation hazard, and radiation safety is essential to protect personnel and to comply with health and safety regulations:
  - using long-handled cassette holders where possible.
  - minimum number of personnel present during radiographic examination.
  - collimation of primary X-ray beam to reduce scatter.
  - wearing of lead-lined protective gowns, gloves and thyroid shields (Fig. 1.27).

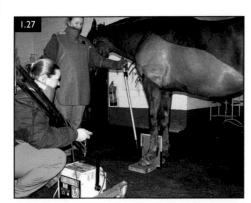

FIG. 1.27 A lateral radiograph of the foot is being taken. Note the radiation safety measures and that the foot is raised off the ground to assist in correct positioning.

- film badges worn at all radiographic examinations and checked regularly to monitor personnel X-ray total exposure.
- radiographic interpretation (radiology) requires extensive knowledge of the normal structure and appearance of the body on radiographs, possible normal variations, and the range of pathological changes that can be detected.

Ultrasonography

- emission of high-frequency sound waves by electrically stimulated crystals in a transducer or probe that are transmitted through the region of interest.
- attenuated by different tissues and reflected back to the transducer as echoes:
  - electronically processed to provide 2-D greyscale real-time image.
  - represents the acoustic impedance of the tissues scanned.
- different frequencies determine the detail and depth of the image acquired:
  - higher the frequency (MHz):
    - better the resolution (detail) but lower the penetration (depth).
- Linear probes ranging from 5 to 12 MHz are now commonly available for most machines and are sufficient for most orthopaedic examinations.
- **hypoechoic** denotes a decreased echogenicity of the tissue (darker image).
- **anechoic** denotes no echogenicity (i.e. fluid [black image]).
- **hyperechoic** denotes an increased echogenicity (brighter image).
- ultrasonography is a safe, non-invasive imaging technique commonly used to image:
  - soft-tissue structures (e.g. tendons, ligaments) [Figs. 1.28, 1.29].
  - joint capsules
  - synovial linings [Figs. 1.30, 1.31].
  - localised soft-tissue swellings [Fig. 1.32].
  - muscles, nerves and blood vessels.
- transverse and longitudinal images are essential to allow complete examination.

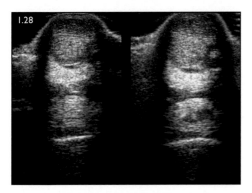

FIG. 1.28 Transverse ultrasonogram of the upper palmar metacarpus of a racehorse with acute-onset lameness and soft-tissue swelling of the area. Note the core lesion in the right upper mid-body of the suspensory ligament.

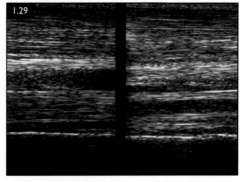

FIG. 1.29 Sagittal longitudinal ultrasonogram of the same horse demonstrating the lesion in the right limb and confirming the length of ligament that is injured.

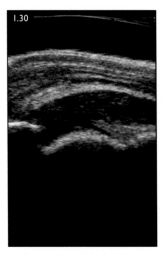

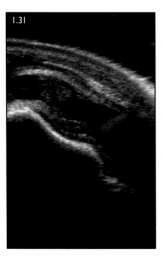

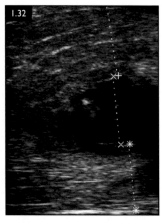

FIGS. 1.30, 1.31 Sagittal (1.30) and oblique (1.31) ultrasonograms of the dorsal fetlock region of a young Thoroughbred with sepsis of the joint. Note the distended dorsal metacarpophalangeal joint filled with hyperechoic joint fluid and the hyperplasia of the dorsal synovial membrane.

FIG. 1.32 Ultrasonogram of an injection abscess in the neck of a horse following vaccination. Note the oval shaped hypoechogenic abscess (between the crosses) deep within the muscles of the neck.

- monitoring of the healing process is routinely carried out with ultrasonography:
  - provides information to help recommend an appropriate rehabilitation.
- useful for imaging bone and joint contours:
  - ilium for ilial wing fractures.
  - articular cartilage, e.g. femoropatellar joint for osteochondrosis.
- guide biopsy or injection techniques such as:
  - intra-articular facet joints of the neck vertebrae.
- accurate interpretation of images is required, since artefact production (e.g. through probe contact and positioning) is common.

Nuclear imaging (gamma scintigraphy or 'bone scanning')

- intravenous injection of a radioactive substance that is distributed throughout the horse.
- gamma camera is placed alongside the horse and the energy emitted from radioactive decay of the substance is recorded, processed by a computer and an image pattern produced for interpretation.

- Technetium (^{99m}Tc) - radioactive substance used in equine musculoskeletal nuclear medicine.
  - bound to methyl diphosphonate (MDP) as a carrier.
- bone scans are usually obtained at 3 hours post injection.
- skeletal structures that are actively remodelling, both normally and abnormally, bind more ^{99m}Tc–MDP than surrounding bone.
- abnormal intensity of increased uptake ('hot spot') in a particular site may indicate pathology within that osseous structure (Fig. 1.33).
- regions not easily radiographed and of large bulk (e.g. back and pelvis) are amenable to bone scanning (Fig. 1.34).
- pathology, such as stress fractures or bone/joint remodelling lend themselves to the technique.
- scanning is carried out in sedated, standing horses.

Thermography

- surface temperature of an object can be measured and illustrated using a thermographic camera.
- circulatory pattern and blood flow in an area dictate the thermal pattern seen, and

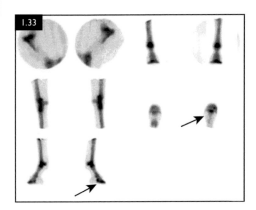

FIG. 1.33 Bone scintigraphy results of a scan of the forelimb of a horse with pathology involving the right navicular bone. (The arrows show increased radiopharmaceutical uptake in the navicular bones.)

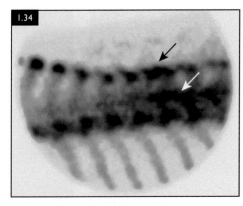

FIG. 1.34 Scintogram of the thoracic region of the back of a horse showing increased uptake in the dorsal spinal processes (black arrow) (this can be insignificant in some horses) and also the caudal thoracic intervertebral facet joints (white arrow). (Photo courtesy Alex Font)

this forms the basis for thermographic interpretation.
- non-invasive and can detect superficial inflammation.
- used by some clinicians in the diagnosis of certain types of lameness:
  ○ little serious scientific evidence confirming its efficacy.
  ○ results do not always correlate with those of other diagnostic techniques.
  ○ at the present time its use is controversial.

Magnetic resonance imaging (MRI)

- at present, MRI in the horse is mainly confined to the limbs (up to and including the carpus and tarsus usually) and head only.
  ○ inability to position other regions of the horse into human MRI machines.
- performed under general anaesthesia when using high-field machines.
- most equine examinations are now performed using low-field machines in sedated, standing horses (Fig. 1.35).
- MRI interpretation is a specialist field and generally multiple views at different settings are taken and viewed.
- most useful for evaluating conditions of structures within the hoof, although

recently, fetlocks, proximal cannon, carpus/tarsus and even stifles have been examined.
- MRI has become essential in the understanding of entheseopathies and intraosseous bone pathologies such as subchondral bone injuries, osteochondral lesions and the origin and insertion points of ligaments on to bone.

Computed tomography (CT)

- CT scanning involves the use of advanced X-ray technology, radiation detectors and a computer system and operating console.
- CT images represent 2-D or 3-D representations (depending on the software) of tissue volume. Cross-sectional images are obtained as required.
- depending on the type of machine, general anaesthesia may be required to allow acquisition:
  ○ CT scanners are now available that allow examination of the head and neck routinely in standing horses.
  ○ Robotic CT machines are being developed to allow limbs (up to distal radius or tibia) to be scanned in the standing horse.
- acquisition and interpretation of CT images is a specialist field.

FIG. 1.35 Standing MRI being carried out on the distal right forelimb of a horse. (Photo courtesy Alex Font)

- MRI is generally most useful for soft tissue and CT for bone pathology.
- cervical spinal cord CT scans in standing horses have recently been acquired and can reveal sites of spinal cord and nerve compression.
- CT can be used as an adjunct to preoperative surgical planning of complex fractures, for the assessment of joints for cartilage damage in osteoarthritis (OA) (with arthrography), and for the detection of subchondral bone anomalies and osseous cysts.

Laboratory tests

- muscle damage may be indicated from measuring serum enzymes from a heparinised blood or serum sample:
  - ○ Creatinine phosphokinase (CPK).
  - ○ Aspartate aminotransferase (AST).
- an exercise test can be performed in helping to diagnose metabolic muscle disease such as chronic exertional rhabdomyolysis. (See page 251)

Serology

- high antibody titres to the tick-borne organism *Borrelia burgdorferi* may be seen in Lyme disease.
- *Brucella* titres may be useful in the diagnosis of fistulous withers, vertebral osteomyelitis or unexplained neck pain.

Muscle biopsy

- useful for diagnosing specific muscle disorders. (See page 252)

Electromyography (EMG)

- diagnosis of neuromuscular problems such as specific areas of muscle atrophy or fasciculations. (See page 252)

Gait analysis

- objective kinetic and kinematic measurements from lame horses (using force plates and motion capture systems, respectively) have been the scientific 'gold standard' for lameness detection and evaluation.
- these methodologies were only available in research institutes or large clinical practices and involved large purpose-built facilities.
- portable systems using different combinations of horse-mounted accelerometers and gyroscopes are now becoming increasingly affordable and reliable.
  - ○ clinicians can now use these tools in everyday lameness work-ups.
- do not replace the '**clinician's eye**'.
  - ○ do provide objective assessments of lameness that are particularly useful when:
    - ◆ evaluating responses to analgesia.

- repeated evaluations after long periods of time are used in assessing response to treatments.
  - these systems measure asymmetry of movement between left and right sides of the body, and differences in motion of contralateral body segments and head are calculated.
    - accuracy and reliability of such systems are dependent on:
      - hardware (number of sensors and frequency of readings).
    - software algorithms that interpret the raw data and produce on-screen visual aids for the clinician to read.

Management of lameness

Arthroscopy, tenoscopy and bursoscopy

- endoscopic surgery ('key-hole' surgery) has revolutionised the treatment of synovial structure diseases in the horse.
- key advantages are:
  - direct visualisation of most synovial structures in the horse.
    - diagnostic information about cartilage, joint capsule, menisci, tendons and certain ligaments.
  - surgical tool to remove chip fractures and osteochondral fragments and debride and lavage septic joints.
  - small 'stab' incisions reduce trauma, lead to better cosmesis and allow earlier return to function.
  - specialised, expensive instrumentation and a high surgeon skill factor.
  - not always possible to evaluate entire joint surfaces, so case selection is important.
- tenoscopy and bursoscopy are developments from arthroscopy, using the same techniques to evaluate tendon sheaths and bursae.
  - digital tendon sheath, the tarsal and carpal sheaths.
  - navicular, calcaneal and intertubercular (bicipital) bursae (Fig. 1.36).
- use of these techniques has allowed new conditions to be diagnosed, e.g.
  - desmitis of the intercarpal ligaments.
  - cartilage lesions of the medial femoral condyle.
  - longitudinal tears of the flexor tendons in the digital sheath.
  - meniscal and cruciate ligament injuries in the stifle joint.

Physiotherapy

- important part of the rehabilitation of lame horses.
- wide variety and combinations of techniques are utilised, depending on lameness treated.
- **Massage:**
  - promotes muscle relaxation and good circulation.
  - useful for relieving focal spasm in longissimus dorsi and neck musculature.
- **Muscle ('Faradic') stimulation.**
  - Transcutaneous electrical nerve stimulation (TENS) can stimulate muscle groups.
    - used for neurogenic atrophy cases (e.g. supraspinatus and infraspinatus atrophy following suprascapular nerve damage ['Sweeney']).
    - improving muscle tone and mass to atrophied muscles in:
      - chronic back problems (after pain has been resolved).
      - limb disuse.
        - after fracture repair or chronic poor/non-weight-bearing lameness.
    - used as a diagnostic tool.

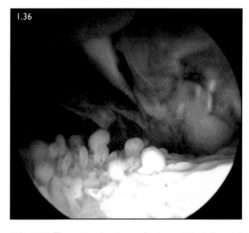

FIG. 1.36 Tenoscopic view of an acutely inflamed carpal sheath showing considerable synovial proliferation, haemorrhage and fibrin clots, and in the background damaged fibres of the deep digital flexor tendon.

- Controlled exercise
  - simply walking out in-hand through to schooling over poles or using a treadmill.
  - allows graduated increase in strength and coordination.
  - timed to coincide with the natural healing processes of recovering tissues.
- Swimming
  - exercises the cardiovascular system while reducing load on the limbs and allowing muscle groups to work.
  - carried out in a specially designed swimming pool environment, with or without a treadmill, or even in the sea.

Therapeutic ultrasound

- utilises high-frequency sound waves to promote tissue healing:
  - exact mechanisms of action are unknown.
  - may have biomechanical effects that produce positive healing processes in damaged soft tissue.

Therapeutic laser

- low-intensity lasers may influence the local circulation, trigger cell proliferation and provide analgesia.
- exact mechanisms within the tissue are unknown.
- used to aid the healing of wounds, superficial flexor tendonitis and in the treatment of OA.
- few peer-reviewed publications on its use, mode of action or effectiveness.

Magnetic and electromagnetic therapy

- used to assist fracture healing and treat back problems due to the finding that bone has piezoelectrical properties.
- suggested that pulsed electromagnetic fields may stimulate bone healing and provide analgesia.
- there is no objective data as to their efficacy.

Extracorporeal shock-wave therapy

- high acoustic wave impulses are generated either by:
  - focused, where the waves are generated electrohydraulically, piezoelectrically or electromagnetically and converge on a small point.
  - radially, where the waves are generated pneumatically and expose surrounding tissues and targeted tissue under treatment.
- may increase regional blood flow, have direct cellular effects, activate osteogenic factors, and have analgesic properties.
- conditions treated are:
  - sore shins.
  - insertional desmopathies (proximal suspensory desmitis, suspensory branch insertions, avulsion fractures of proximal attachment of suspensory ligament).
  - impinging dorsal spinous processes.
  - other conditions reported to be treated include tibial stress fractures, incomplete proximal phalangeal fractures, OA of the distal hock joints, superficial and deep digital flexor tendonitis.
  - treatment of angular limb deformities by retarding growth on the convex side of the deformity.
  - many of the conditions where shock-wave therapy has been used have no or minimal scientific evidence basis and as such it should be prescribed with this in mind.

First-aid treatment of the fracture patient

- Immediate fracture support with first-aid measures is essential to allow the best possibility for appropriate treatment if this is achievable.
  - stabilises the limb helping to relieve pain.
  - minimises further bony and soft-tissue damage.
  - prevents further contamination if open and renders the horse safer to travel.

- Initial approach:
  - brief, but thorough, examination of the whole horse to ascertain that there are no life-threatening injuries present (attended to first).
  - horses can (but not always) be extremely distressed and severely lame with fractures of the limb.
    - physical (e.g. a twitch) or chemical restraint may be needed before application of any splint support.
    - chemical restraint used with care since ataxia and collapse may occur in a compromised horse; small doses are advisable.
    - systemic analgesics such as NSAIDs can be given.
  - antibiotics should be prescribed if an open fracture is present.
  - check horse's tetanus vaccination status.
- First-aid fracture support is best achieved with the use of bandages and splints:
  - casts are difficult to apply properly in the standing horse.
    - do not accommodate post-injury swelling.
    - must be removed and reapplied to readjust them.
    - time consuming and unnecessarily expensive.
- splint should ideally immobilise the joints proximal and distal to the fracture:
  - quick and easy to apply:
    - Wood (45 mm × 20 mm) and/or PVC guttering cut lengthways to give a U-shape are suitable splint materials.
    - length of splint depends on size of horse and the injury.
    - padding at the proximal and distal ends of the splint before application lessens the incidence of pressure points.
  - commercial splints (e.g. Kimsey splint) can be used with certain fracture types.
- sufficient bandaging materials for a half- or full-length Robert Jones bandage should be available depending on the fracture.
- divide the fore- and hindlimbs into regions for appropriate splint application.

Forelimbs

Fractures of the distal metacarpus and the proximal and middle phalanges

- transverse or oblique fracture is suspected:
  - align dorsal cortices of third metacarpus and phalanges to minimise a fulcrum effect of the fracture on loading.
  - assistant holds the limb off the ground by the forearm, so the distal limb is vertical.
  - one or two bandage layers are applied to the distal limb and the splint is applied dorsally.
  - further one or two layers are applied over this to protect the splint and further stabilise the limb.
  - heavy, tightly applied taping from toe to carpus prevents the splint loosening.
  - heel wedge is sometimes a useful addition.
- fracture in the sagittal plane is suspected:
  - two splints on the lateral and medial aspects of the distal limb are applied over a half-limb Robert Jones bandage, with the limb weight bearing.

Fractures from the mid-metacarpus to distal radius

- full Robert Jones bandage applied from toe to elbow in normal standing position.
- splints are then taped tightly to the lateral and caudal aspects of the bandaged limb.
- proximal ends of the splints must be padded to prevent rubbing.

Fractures of the mid and proximal radius

- minimise abduction of the limb due to the lateral musculature of the forearm.
- Robert Jones bandage is applied from the ground to the elbow.
- splint is tightly applied laterally, extending from the foot to the mid-scapula level.
- proximal end of the splint must be padded as previously.

Fractures of the ulna, humerus and scapula

- fractures cannot be splinted and supported.
- their location disables the triceps muscles, which affects ambulation of the horse.

- ○ Robert Jones bandage is applied from the foot to the elbow.
- ○ splint is applied using tape caudally from the fetlock to the elbow to fix the carpus in extension and allow the horse to move more easily.

Hindlimbs

- the reciprocal apparatus of the hindlimb presents problems with splinting and the splint can be less well tolerated than in the forelimb.
- further bandaging is often necessary after ambulation due to loosening of the splint.

Fractures of the distal metatarsus and below

- as for the forelimb, but the splint is placed on the plantar rather than dorsal aspect.
- assistant should hold the limb above the hock so that the distal limb is vertical for application.

FIG. 1.37 A full-limb splinted Robert Jones bandage used in the hindlimb of a horse for the conservative treatment of a medial condylar fracture of the distal third metatarsus. Note the lateral and plantar splints.

Fractures of the mid and proximal metatarsus

- Robert Jones bandage is applied from toe up to and including the calcaneus, with the limb weight bearing.
- splints are tightly applied caudally and laterally to the level of the calcaneus (Fig. 1.37).

Fractures of the tibia and tarsus

- splint must counteract the medial force of the lateral musculature of the tibia and the destabilising effect of stifle flexion by the reciprocal apparatus.
- full Robert Jones bandage is applied from the toe to the proximal tibia.
- splint is tightly applied to the lateral aspect of the limb and extends to the tuber coxae.
  - ○ light steel rod (12 mm), shaped to form a loop proximally, can be used as a splint but this is rarely available.
  - ○ long, thin wooden splint can be used instead.

Fractures of the stifle, femur and pelvis

- fractures are not amenable to external coaptation.
- if possible the horse should be cross-tied to minimise further damage.

Fracture treatment options

- fractures can potentially occur in any of the bones of the musculoskeletal system, and they are individually discussed in the various anatomical sections.
- range of treatments currently available and they can be divided into conservative and surgical treatments.
- goal of treatment is to restore function to the affected limb so that the horse can either return to full work, become a breeding animal or retire to pasture pain-free.

Conservative

Box rest

- all types of fracture require box rest whether treated by conservative or surgical means.
- those that are amenable to box rest and bandaging as a sole treatment for a full return to function include:
  - certain splint fractures.
  - incomplete non-displaced fractures of long bones, such as the radius and tibia.
  - some incomplete fractures of P1.
- decision for this form of treatment alone must be based on the individual case.
- length of time required for complete healing will depend on the type of fracture and any complicating factors encountered, but it is often 10–12 weeks.
- some cases may need cross-tying to prevent lying down and getting up.
  - temperament of a horse is extremely important to the final outcome.

External coaptation

Splints
- generally used for first-aid treatment of fractures.
- used in fracture treatment as an adjunct after internal fixation immediately postoperatively or when a cast has just been removed.

Casts
- impregnated fibreglass casting materials are used in the horse (Figs. 1.38, 1.39).
- used as a primary treatment only for a limited number of fracture types.
  - half-limb cast for an incompletely fractured pastern or a foot cast for a pedal bone wing fracture.
- most commonly used to protect internal fixations during anaesthetic recovery or until good fracture healing is well underway, e.g. a lateral condylar third metacarpal fracture.
- proper application of a cast requires:
  - careful preparation and positioning of the limb.
  - cleaning of the foot and preplacement of suitable wound dressings.
  - application of a stockinette lining.
  - appropriate padding with orthopaedic felt and general limb padding prior to placement of the cast material.
- frequent cast changes due to loosening of the cast, and cast-related soft-tissue problems such as pressure sores, are common and expensive.
  - increases risks if carried out under general anaesthesia.
  - long-term cast use can lead to:
    - DJD due to immobilisation of joints.
    - laxity of the soft-tissue structures around the joints.

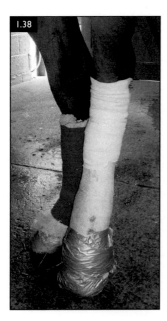
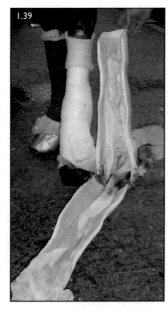

FIGS. 1.38, 1.39. Hindlimb half-limb cast used for the post-anaesthetic recovery of a horse that had undergone a surgical repair of a hindlimb first phalanx fracture (1.38). Note the layers of the cast, including the yellow cast foam, after it has been split prior to removal (1.39).

◆ disuse osteoporosis if the treated limb remains non-weight bearing for an excessive period of time.

- daily cast checks are necessary to prevent sores developing:
  ○ especially dorsoproximally at the cannon, fetlock and coronary band for a half-limb cast (Fig. 1.40).
- casts should ideally be changed 7–10 days after surgery.
  ○ allows for a decrease in soft-tissue swelling.
  ○ then at least every 4 weeks in adults and every 7–10 days in foals depending on the case.

Surgical

- various basic types of surgical fracture fixation techniques are used in the horse (Table 1.3).
- specific types of fixation used are described within the individual fracture templates.

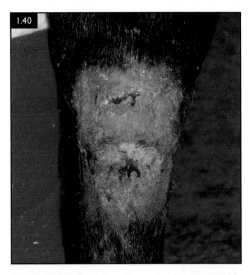

FIG. 1.40 Healing dorsal proximal cannon skin pressure sores after the cast has been removed. The area was inadequately padded, and the cast was not checked regularly enough.

TABLE 1.3 Types of fixation, definitions and use.

TYPE OF FIXATION	DEFINITION	USE
External fixation		
Transfixation casting	2 or 3 4–6 mm positive profile pins in MC3/MT3 and fibreglass cast	Comminuted fracture distal to pins within P1, P2, distal MC3/MT3
External fixator	Small animal versions in foals Positive profile pins and external side bars (PMMA) External skeletal fixation device (ESFD)	Severely comminuted fractures of P1, P2, distal MC3/MT3
Internal fixation		
Intramedullary implants	Steinmann pins	Foals for medullary stack pinning of humeral fractures, certain olecranon fractures and femoral capital physeal fractures
	Interlocking nails	Foals with humeral and femoral fractures
Cerclage wire (Fig.1.41)		Tension band wiring mid-body proximal sesamoid fractures Rostral fractures of lower and upper jaws Certain olecranon fractures in foals
Screws and plates (Figs. 1.42–1.44)	AO/ASIF system of lag screw fixation and DCP (dynamic compression plates) 4.5 and 5.5 mm screws	P1, P2. P3, MC3/MT3 fractures Olecranon fractures
	Locking compression plates (LCP)	Stronger and more versatile for long bone fractures
	Dynamic condylar screw (DCS) Dynamic hip screw (DHS)	Some other long bone fractures

- Some simple fractures of MC3/MT3 and P1 are now repaired routinely under standing sedation and regional analgesia avoiding risks of general anaesthesia and recovery.

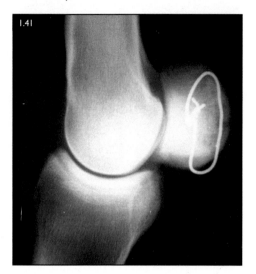

FIG. 1.41 Postoperative lateromedial radiograph of a proximal sesamoid bone mid-body fracture repaired by a cerclage wire.

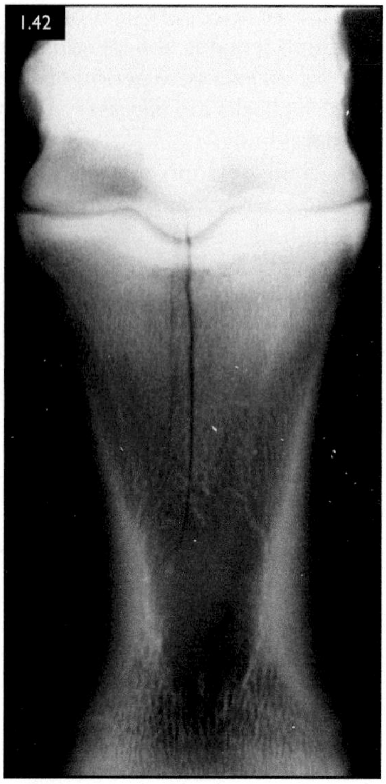

FIG. 1.42 Dorsopalmar radiograph of the first phalanx of a Thoroughbred racehorse that has sustained a non-displaced midline sagittal fracture during training. Note the two fracture lines that correspond to the dorsal and palmar cortices, the starting point at the sagittal groove and the way the fracture turns and exits through the lateral cortex of the mid-pastern.

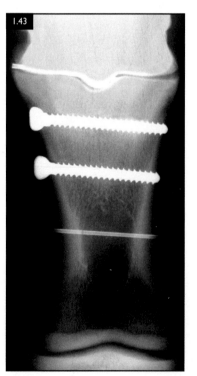

FIG. 1.43 Intraoperative radiograph showing placement of two lag screws from lateral to medial compressing the fracture line. Note the hypodermic needle markers at the level of the metacarpophalangeal joint and distal fracture line.

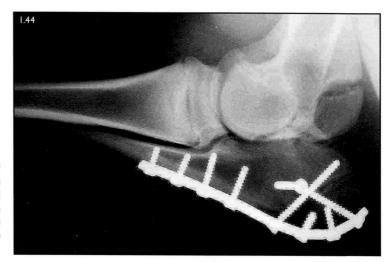

FIG. 1.44 Lateral radiograph of the repair of an olecranon fracture by the tension-band principle using a bone plate applied to the caudal surface of the bone.

Complications of fracture treatment

- most common and serious complication is infection at the fracture site and within the surrounding soft tissues:
  - heat, local pain and swelling are present.
  - discharging sinus and/or wound breakdown is common.
  - infective osteitis and osteomyelitis may be seen on postoperative radiographs.
  - poor or non-weight bearing increases strain on other limbs.
  - delays fracture healing.
  - significantly increases costs.
  - chances of contamination are increased in:
    - open fractures and/or where the soft tissues are badly damaged.
  - mixed bacterial populations are usually present and include anaerobes.
- other complications include:
  - refracture through the original fracture plane due to:
    - premature implant removal or delayed healing.
  - implant failure during anaesthetic recovery can be catastrophic.
    - use of external coaptation and assisted recovery with head and tail ropes can help prevent this.
  - delayed healing can occur for a variety of other reasons where the healing environment is less than optimal (e.g. movement at the fracture repair site).
  - overloading of the opposite limb after fracture repair can lead to:
    - angular limb and hyperextension of the fetlock deformities in foals.
    - laminitis and suspensory ligament breakdown in the adult horse.
  - owner should be made aware of these potential complications prior to fracture repair since they may lead to euthanasia on humane and/or financial grounds.

The Foal and Developing Animal

CONGENITAL MUSCULOSKELETAL ABNORMALITIES

- abnormalities present at birth, either structural and/or functional.
  - result from abnormalities in embryogenesis or intrauterine factors.
- may be genetic or environmental in origin, but often no definitive cause is identified.

Congenital angular limb abnormalities (ALDs)

(See page 37 for Acquired angular limb deformities.)

Definition/overview

- present at birth
- uni- or bilateral
- forelimbs are more commonly affected.
- usually seen in larger, faster growing breeds such as Thoroughbreds.
- common, especially at the carpus (typically valgus) and fetlock (typically varus) joints.
  - valgus is a deviation in alignment of the limb which is lateral to its long axis from the point of deviation. (Fig. 2.1)
  - varus is a deviation in alignment of the limb which is medial to its long axis.
- coincidental limb rotation which exacerbates the appearance of the ALD is common.
  - rotational deformities are not correctable.
  - resolve as the foal increases in size and its chest and pelvis become broader.

Aetiology/pathophysiology

- multifactorial and not clearly understood, but possible cited factors include:

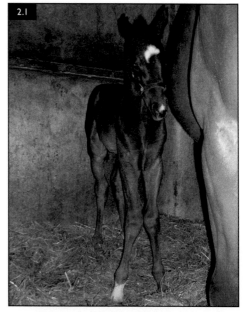

FIG. 2.1 Bilateral congenital carpal valgus angular limb deformity in a neonatal foal.

  - intrauterine malpositioning.
  - overnutrition of the mare in the last third of pregnancy.
  - joint laxity.
  - incomplete/defective ossification of the cuboidal bones of carpus and tarsus.
  - other hereditary (poor conformation), nutritional and hormonal influences.
- joint laxity is common at birth:
  - often appears as an angular or hyperextension deformity.
  - usually resolves spontaneously within a few days following gentle exercise:
    - encourages strengthening of muscles, ligaments, tendons and periarticular structures.

DOI: 10.1201/9781003369226-2

Clinical presentation

- examine all four limbs for deformity in all planes:
  - particularly perpendicular to the frontal plane through the limb.
  - manipulation of the affected limbs will reveal:
    - gross instability in cases of joint laxity, but not in other cases.
  - several limbs may be involved.
  - lameness is not usually a feature (Figs. 2.2, 2.3).

Differential diagnosis

- Acquired ALDs.

Diagnosis

- clinical history of a deformity present at birth, or shortly thereafter, and the clinical findings are diagnostic.

- radiographs of the affected area will confirm the exact nature of the problem:
  - lateromedial and dorsopalmar/plantar views with long plates:
    - allows visualisation of the long bones proximal and distal to deformity.
  - lines drawn on the radiographs through the mid-points of the long bones proximal and distal to the affected area should intersect within the affected joint.
    - allow the angle of deformity to be measured.
  - some cases, there are hypoplastic bones present within the joint (Fig. 2.4).
  - stressed views may be useful where joint laxity is suspected.

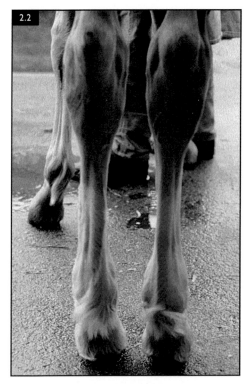

FIG. 2.2 2-week-old Clydesdale foal with a left hindlimb fetlock varus that has been present since birth. Manipulation revealed marked medio-lateral joint instability in the left fetlock compared with the right.

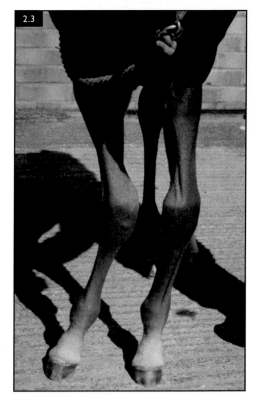

FIG. 2.3 Neonatal dysmature Thoroughbred foal with a 'windswept' appearance due to a right carpal valgus and a left carpal varus. There was carpal joint instability in all planes on manipulation and carpal bone hypoplasia on radiography.

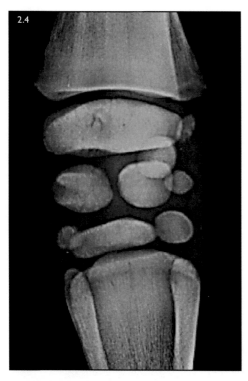

2.4

FIG. 2.4 Dorsopalmar radiograph of the left fore-limb carpus of a neonatal foal showing evidence of carpal bone hypoplasia. Note the rounded profile of the individual carpal bones with apparently increased space between them due to incomplete ossification of the cartilage surrounding the centres of ossification.

Management

- in most newborn foals that demonstrate mild to moderate ALDs, particularly of the carpus or tarsus, no treatment is required other than restricted exercise and corrective foot trimming:
  - remove any excessive lateral [valgus] or medial [varus] hoof growth at 2–4-week intervals.
  - gradual improvement with time.
- cases with cuboidal bone hypoplasia, use of a tube cast is generally advisable.
- foals with moderate or more severe fetlock varus/valgus:
  - more proactive approach is required.
  - window of opportunity for limb correction is small (i.e. up to 12 weeks of age).

- foals with more extreme deformities of carpus or tarsus should be managed by surgery.

Defective ossification of the cuboidal bones of the carpus and tarsus

- defects in the ossification process can occur in any young foal:
  - premature and dysmature foals are most affected.
  - foetal growth retardation due to placental disease, severe metabolic or parasitic disease in the mare, twin pregnancy or poor mare nutrition may be involved.
- carpal valgus is the most common presentation at birth.
  - either remains static or
  - worsens over first 2 weeks due to exercise deforming soft cartilage structures.
- clinical examination reveals no pain or swelling in the affected joint initially:
  - usually no lameness.
  - increased range of movement within the joint may be present:
    - both dorsopalmarly/dorsoplantarly and mediolaterally.
    - particularly if joint laxity is also present.
- radiographs can reveal one or more abnormal carpal bones:
  - not possible to fully estimate amount of cartilage precursor damage (Fig. 2.4).
  - full ossification occurs at about 30 days and allows a more accurate prognosis.
  - secondary degenerative joint disease (DJD) is a possible sequela.
- treatment involves the use of a tube cast:
  - support of the limb in correct axial alignment.
  - prevents crushing of cartilaginous precursors of the ossified cuboidal bones.
  - **monitor casts, as skin damage in these foals from pressure sores is a high risk:**
    - change the cast every 3–4 days.
- cuboidal bone defective ossification can also lead to collapse of the third and/or central tarsal bone(s).
  - usually bilateral

- excessive flexion of the hocks ('curby' conformation).
  - characteristic bunny-hop gait (Fig. 2.5).
  - tarsal valgus may also be present.
  - radiography reveals wedging of the central and third tarsal bones (Fig. 2.6).
  - treatment similar to cases of joint laxity to allow ossification to progress.
    - DJD is a possible sequela.

Congenital flexural limb deformities

Definition/overview

- present at birth or become evident very shortly afterwards
- unknown aetiology.

- possible causes include:
  - intrauterine malpositioning
  - genetic factors.
  - toxic insults during embryonic life
  - neuromuscular disorders.
  - influenza virus infecting pregnant mares
  - dams fed goitrogenic diets.
- not 'contracted tendons' but relative shortening of musculotendinous unit in relation to bony structures.

Clinical examination

- careful palpation and manipulation of the affected joint(s) in a weight-bearing and non-weight-bearing position.
- mainly affects the fetlock joint of newborn foals and, less commonly, the distal interphalangeal (DIP) joint.

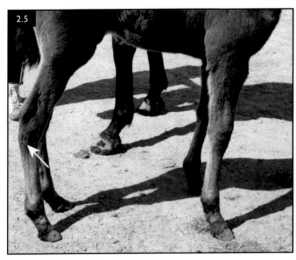

FIG. 2.5 Three-week-old Warmblood foal with partial collapse of the small tarsal bones in a dorsal plantar plane, leading to a 'curby hock' appearance (arrow). (Plantar aspect from point of hock to plantar fetlock is not straight.)

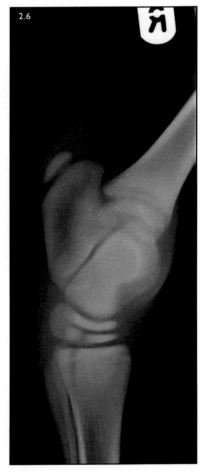

FIG. 2.6 Premature foal with gross hypoplasia of the tarsal bones with obvious wedging of the central and third tarsal bones.

- hindlimbs are affected as, or more, commonly than the forelimbs.
- palpation of the flexor tendons, suspensory ligament and inferior check ligament may further indicate those structures involved.

Diagnosis

- radiographic examination of affected joints is useful to determine the presence of specific bone or joint abnormalities, which may affect the prognosis of the case.

Treatment

- usually effectively managed using a proprietary splinting system or by creating a splint using either half-tubing PVC or from fibreglass casting material.
- splint is taped dorsally over a well-bandaged limb with the limb in maximum extension:
  - incorporate wiring at the toe when the DIP joint is involved.
- alternatively, immobilise the distal limb, in extension, with a light fibreglass cast.
- both systems should be reapplied every few days (3–5) to avoid pressure sores.
  - at each replacement, the limb may be further extended.
- most cases straighten within a few days with resulting normal limb conformation.

Congenital flexural deformity of the distal interphalangeal joint

- uncommon condition that presents as varying degrees of DIP joint flexion (Fig. 2.7).
  - foal walks on its toe and the heel does not contact the ground.
  - hoof shape does not alter in the early stages, unlike the acquired forms:
    - increased toe wear does occur.
- uni- or bilateral
- newborn or very young foal.
  - can be associated with other flexural deformities:
    - metacarpophalangeal (MCP) joint.
- increased tension on palpation of inferior check ligament and deep digital flexor tendon (DDFT) often present.

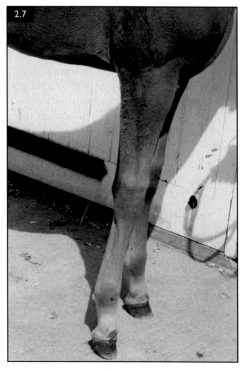

FIG. 2.7 Young Thoroughbred foal with a right forelimb congenital flexural deformity of the distal interphalangeal joint. Note the typical 'ballerina' foal stance on the toe tip in the right fore.

- treated by a combination of:
  - regular gentle exercise on hard, even surfaces.
  - specific exercises, such as walking up inclines.
  - passive manipulation.
  - analgesics (beware of NSAID medication in foals) help reduce pain which is involved in some cases.
  - Oxytetracycline may be useful (2–4 g in 500 ml saline slowly i/v once or twice in the first 48 hours post-partum or even, in some cases, later).
    - method of action is unclear.
    - kidney function should be checked prior to using this treatment.
  - corrective foot trimming consisting of mild heel rasping.
  - short toe extension with a glue-on plastic shoe or hoof composite over the toe:
    - prevents excessive toe wear.

- helps stretch affected periarticular soft tissue structures.
  - **caution when using toe extensions that do not support the solar surface of the foot as significant risk of dorsal hoof wall fracture.**
  - ○ splinting or casting with incorporation of the foot up to the carpus or tarsus:
    - 7–14 days leads to a relaxation of the muscle–tendon unit.
    - cast pressure sores can be a problem.
  - ○ unresponsive cases, inferior check ligament desmotomy, but seldom necessary.

Congenital flexural deformity of the metacarpo/metatarsophalangeal joint (Fig. 2.8)

- may be uni- or bilateral and often involves the distal joints as well.

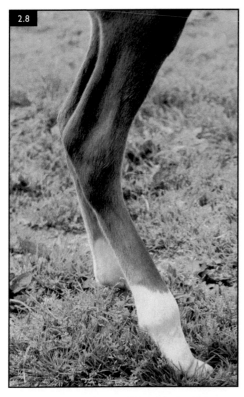

FIG. 2.8 Newborn Thoroughbred foal with a congenital flexural deformity of the right hind metatarsophalangeal joint.

- foal tends to knuckle over at the fetlock.
  - ○ may even walk on the dorsal aspect of the fetlock in severe cases.
- most common congenital flexural deformity and hindlimbs more commonly affected.
- treatment varies with the degree of the deformity:
  - ○ mild cases may resolve with exercise, bandage support and the use of NSAIDs.
  - ○ more severe cases often involve the DIP joint as well and require extension splinting or casting as described previously.
  - ○ Oxytetracycline may also be used in such cases as previously described.
  - ○ surgical intervention with inferior check ligament desmotomy may be required in unresponsive or severe cases, where the distal joints are also involved.

Congenital flexural deformity of the carpal joint (Fig. 2.9)

- usually bilateral and quite common.
- affected foals are either able to stand but 'buckling' forward at the carpus or, in severe cases, unable to stand.
- many cases exhibit normal distal limb conformation.

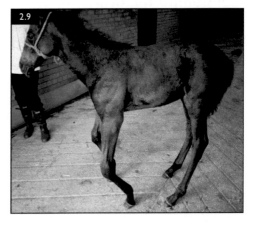

FIG. 2.9 Older suckling Thoroughbred foal with history of a bilateral carpal flexural deformity at birth that was not corrected. The foal is shown at 9 weeks of age with bilateral severe carpal flexural and secondary varus angular limb deformities.

- careful examination of these foals is necessary since this presentation may be part of the 'contracted foal syndrome'.
- in most severe cases, the foal is unable to assume the 'diving position' necessary for birth, leading to dystocia, and possible delivery by caesarean section.
- foals unable to straighten their forelimbs rely on forearm muscular contraction to stand:
  - if allowed too much exercise, foal's forearm musculature will become fatigued and show marked tremors or spasm.
- initially, many foals need assistance to suckle as they find standing unassisted difficult.
- unilateral cases can rapidly develop overuse limb injuries in the contralateral limb.
- mild cases may resolve with:
  - cautious gentle exercise thereby avoiding over-fatigue of the foal.
  - Oxytetracycline.
- treatment with proprietary splints or a half tube PVC or fibreglass dorsal splint (extending just proximal and distal to the carpus), with heavy bandaging:
  - usually effective in straightening the limb within a few days.
  - severe cases, or where splinting is ineffective, surgical release techniques are available.
- if the foal's carpus can be passively extended into or near to its normal position:
  - prognosis for normal limb conformation is good.
- grave prognosis indicated:
  - failure to resolve the condition with splinting.
  - foals with 'contracted foal syndrome'.
  - severe bilateral cases leading to dystocia.
  - poor response to surgical release:
  - euthanasia may be indicated in these circumstances.

Congenital hyperextension of newborn foals ('flaccid tendons')

- unknown aetiology and pathogenesis:
  - may be a physiological variant or a temporary failure of the agonist/antagonist muscle balance.
- possible higher incidence in Thoroughbreds and heavy horse breeds.
- common in premature (<320 days) or dysmature foals.
  - more severe if there is accompanying systemic illness or inadequate exercise.
  - flaccidity of the flexor tendons is often accompanied by periarticular ligament laxity and joint instability.
  - both hindlimbs, or all four limbs, are affected:
    - MCP/MTP and/or interphalangeal joints are usually affected.
    - often dropping of the fetlock.
    - plantar/palmar aspect of pastern and fetlock contacts the ground in severe cases (Fig. 2.10).
- many cases resolve spontaneously as muscle tone/ligament strength improves post-partum:
  - encouraged by careful exercise.
  - corrective foot trimming of the heel is useful:
    - provides flat weight-bearing surface and eliminates 'rocker heel' effect.

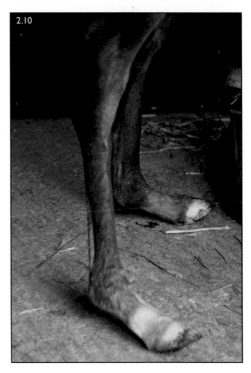

FIG. 2.10 Neonatal foal with severe bilateral hyperextension of the distal limb joints (flaccid tendons).

- light protective bandage to protect the heel/pastern/fetlock from trauma:
  - ◆ **too much support exacerbates the condition.**
- in unresolving or severe cases, heel extension shoes should be used (Fig. 2.11).
- surgical management as a salvage procedure has been described but rarely necessary or used.

Rupture of the common digital extensor tendon (CDE) (Fig. 2.12)

- relatively uncommon, usually present in combination with other musculoskeletal defects and affects both forelimbs:
  - carpal and metacarpophalangeal flexural deformities.
  - hypoplasia of cuboidal bones and underdeveloped pectoral muscles.

- may be inheritable and higher incidence has been reported in Arabs and Quarter Horses.
  - almost always follows carpal contracture and may be a direct physical consequence in Thoroughbred foals.
- swelling of the tendon sheath over the dorsolateral surface of the carpus:
  - bilateral cases are common.
  - affected foals adopt a slightly bowlegged, over at the knee stance.
  - may knuckle forward at the fetlock when walking.
  - palpation of the swelling may reveal tendon rupture.
- radiography and ultrasonography can confirm the presence of hypoplastic carpal bones and partial/complete tendon rupture.

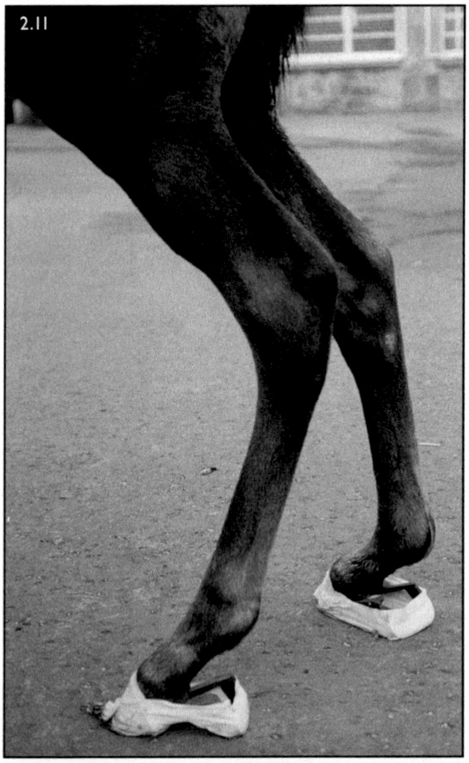

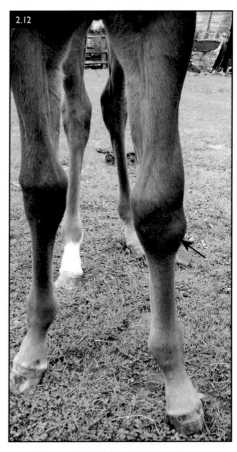

FIG. 2.11 Older Thoroughbred suckling foal with a history of bilateral hindlimb distal joint hyperextension that did not respond to conservative treatment and was eventually treated with heel extension shoes.

FIG. 2.12 Two-week-old foal that presented with an acute onset of a dorsolateral distal radius fluid-filled swelling (arrow) due to rupture of the CDE tendon. This was not associated in this foal with a flexural deformity.

- rupture of the CDE tendon heals spontaneously, in most foals, without specific treatment.
  - where it is secondary to, or accompanied by, other orthopaedic problems, appropriate management of these is indicated.
- prognosis is good provided concurrent flexural deformities are correctable.

Contracted foal/ limb syndrome.

- variety of combinations of congenital appendicular and axial contractures and curvatures that are uncommon, but well recognised. (Fig. 2.13):
  - Arthrogryposis.
  - torticollis or 'wry neck'.
  - scoliosis, lordosis and kyphosis of the thoracolumbar spine.
  - varying degrees of flexural deformity involving the carpus, MCP/MTP joints and, sometimes, the tarsus.
  - asymmetric formation of the cranium or 'wry nose'.

  - attenuation (thinning of the ventral abdominal wall or even visceral eventration) is present in some cases of severe scoliosis.
- may be due to unfavourable, restrictive uterine conditions.
- no evidence of a genetic aetiology has been described.
- if the deformities are mild, affected foals can develop normally given time and adequate nursing care.
- surgical correction may be attempted in more severe cases, but if multiple abnormalities are present, euthanasia is indicated.

Polydactyly

- rare condition of more digits than normal in a limb.
- dominant hereditary transmission with incomplete penetrance has been postulated.
- **teratological** where duplication is distal to the fetlock joint:
  - results in two separate digits articulating with a third metacarpal bone, which may or may not be divided distally (Fig. 2.14).

FIG. 2.13 A foal suffering with contracted foal syndrome was euthanased immediately after birth. There is a wry nose, neck torticollis, distal limb hyperextension and carpal flexural deformity visible in this view.

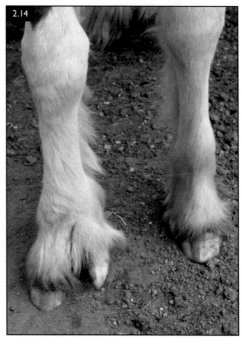

FIG. 2.14 Teratological polydactyly in the right forelimb of a Cob foal.

- more common **atavistic** form involves an extra digit on the medial forelimb:
  - fully articulates with a fully developed second metacarpal bone.
  - managed successfully by surgical resection of the extra digit at its base.
- usually unilateral in a forelimb:
  - bilateral forelimb, and cases affecting all four limbs, also described.

Lateral luxation of the patella

- rare condition in most breeds, but commonly seen in miniature Shetlands and other small breeds:
  - hereditary mode of transmission (autosomal recessive gene) in Shetland ponies.
  - usually due to hypoplasia of the lateral trochlear ridge of the femur (Fig. 2.15).
- uni- or bilateral.
  - bilaterally affected foals typically present with a 'squatting' stance, with hips, stifles and hocks in extreme flexion (Fig. 2.16).
- lateromedial, caudocranial and flexed skyline radiographs confirm the condition (Fig. 2.17).
- surgical repair can be attempted in unilateral cases where underlying bony development is adequate.
- prognosis is guarded in unilateral cases and guarded to poor in bilateral cases:
  - miniature breeds can cope quite well with milder bilateral deformities.

FIG. 2.16 Typical squatting stance of a foal with bilateral patellar luxation.

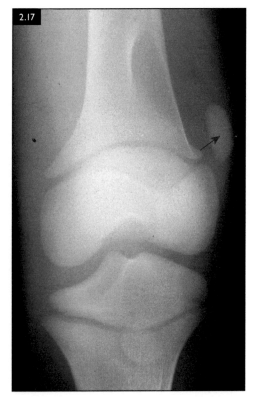

FIG. 2.17 Craniocaudal radiograph of the stifle joint of the foal in 1.79 showing the patella luxated laterally outside the plane of the femur (arrow).

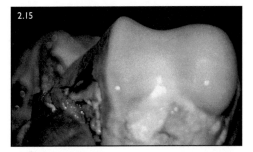

FIG. 2.15 Post-mortem specimen of the distal femur of a foal that presented with bilateral patellar luxation. Note the rather flattened hypoplastic lateral trochlear ridge (on the left) and shallow trochlear groove.

Osteochondrosis

- part of a group of orthopaedic conditions affecting young, growing horses, termed 'developmental orthopaedic diseases' (DOD):
  - includes physitis, angular limb and flexural limb deformities, and cervical vertebral malformation.
- defined as a 'failure of normal endochondral ossification':
  - disturbance of normal differentiation of cells in growing cartilage at the end of the long bones in the articular–epiphyseal growth plate.
  - leads to retention of cartilage or dyschondroplasia.
  - these changes can either repair or develop into a clinical entity and lead to lameness.
- manifestations of this disturbance in the horse are multiple:
  - necrosis of the affected cartilage may lead to cartilage fibrillation and fissuring at various depths.
  - osteochondral fragments ('joint mice') may detach to float free in the joint.
  - inflammation of the joint in response to these changes is termed 'osteochondritis'.
  - dissecting flaps of cartilage are termed 'osteochondritis dissecans' (Fig. 2.18).
  - subchondral bone cysts of the distal femoral condyle were regarded as a manifestation of osteochondrosis (OCD), but most clinicians now consider it a separate condition.
- aetiology of OCD is not clear, but it is multifactorial with various trigger factors:
  - rapid growth rate in fast-growing horses such as Thoroughbreds, Standardbreds and Warmbloods:
    - possibly due to bone growth outstripping blood supply.
  - genetic predisposition may exist in these breeds.
  - dietary imbalances and mismanagement are major factors:
    - excessive carbohydrate and protein intake.
    - excess phosphorus, calcium and zinc and insufficient copper.
  - trauma is strongly implicated as a factor (whether primary or secondary is still controversial).

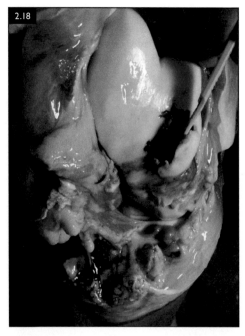

FIG. 2.18 Large dissecting flap lesion of the mid-lateral trochlear ridge of the distal femur of a 6-month-old Warmblood foal, which represents the most common form of an OCD lesion in the stifle joint.

- stage of development when the articular–epiphyseal cartilage complex is vulnerable to damage is unknown at the present.
- more commonly encountered in the hindlimb joints:
  - tarsocrural joint (Warmbloods and Standardbreds) (Fig. 2.19).
  - femoropatellar joint (Warmbloods and Thoroughbreds) (Fig. 2.20).
  - MCP/MTP joints (Thoroughbreds and Standardbreds).
  - less common in other breeds but does occur.
  - bilateral lesions are common but multiple different joint lesions are not.
  - variable lameness and joint distension (Fig. 2.20) are usually present.
  - radiography (Fig. 2.21) and ultrasonography (Figs. 2.22, 2.23) for imaging lesions.
    - examination of the contralateral limb is always advisable.

- treatment may be conservative or surgical (arthroscopy) depending on the type of lesion and age of the horse (Fig. 2.24).

- prognosis depends on the location, type and severity of the lesion and age of the horse.
- clinical signs, treatment and prognosis of OCD in each joint are discussed in the relevant sections on the forelimbs and hindlimbs.

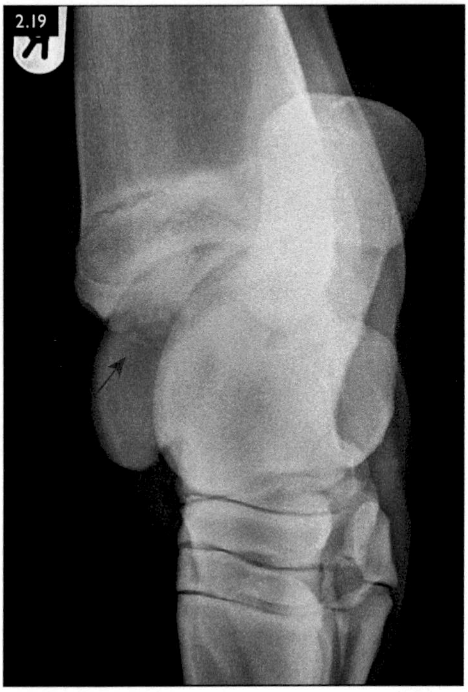

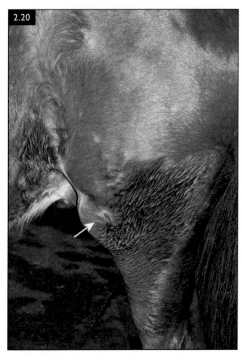

FIG. 2.19 Dorsomedial/plantarolateral oblique radiograph of the hock of a young Warmblood horse presenting with tarsocrural joint distension and lameness localised to this joint. There is an OCD lesion of the distal intermediate ridge of the tibia (arrow), which is multi-fragmented.

FIG. 2.20 Six-month-old Thoroughbred foal with marked bilateral stifle joint distension and hindlimb lameness. Note the enlarged femoropatellar joint in the left hindlimb just below the fold of the flank (arrow).

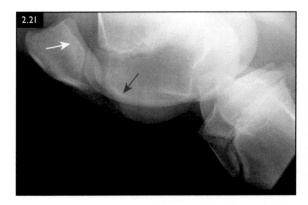

FIG. 2.21 Lateral radiograph of the stifle joint of a weanling Warmblood with femoropatellar joint distension. An upper third lateral trochlear ridge OCD lesion with loss of the bony outline of the ridge and subchondral bone lysis (red arrow) and, more unusually, a subpatellar OCD lesion (white arrow) are visible.

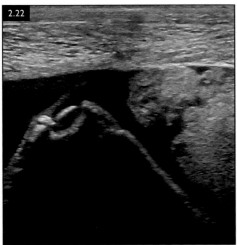

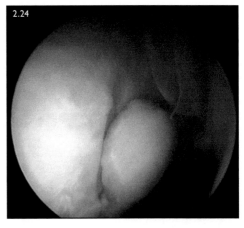

FIG. 2.24 Arthroscopic image of an osteochondral flap lesion of OCD off the lateral trochlear ridge within the femoropatellar joint.

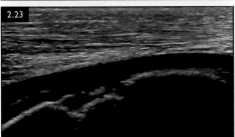

FIGS. 2.22, 2.23 Ultrasound scans of the stifle joint of the same horse as in 2.21. (2.22) Transverse scan showing femoropatellar joint distension, synovial proliferation and clear disruption of the subchondral bone with a possible cartilage flap overlying the defect. (2.23) Longitudinal scan confirming the defect in the subchondral bone and disruption in the cartilage layer.

ACQUIRED MUSCULOSKELETAL ABNORMALITIES

Acquired angular limb deformities

Definition/overview
- usually manifest as a lateral or medial deviation to the long axis of the limb in the frontal or transverse plane, commonly seen at the carpus, fetlock and tarsus.
- lateral (valgus) and medial (varus) deviations occur.
- coincidental postural or rotational deformities, especially with carpal valgus (lateral rotation), are common.

Aetiology/pathophysiology
- multifactorial and includes one or more of:
  - genetics
  - fast growth rate and/or excessive body weight.
  - over- or undernutrition
  - mineral and trace element imbalances.
  - excessive exercise, internal or external trauma and contralateral limb lameness.
    - asymmetric or imbalanced longitudinal bone growth may

occur from a distal physis or epiphysis due to overload of the physis on one side.
- decreased growth on that side and greater growth from opposite side.
♦ direct trauma to the physis can lead to asymmetric damage and growth.

Clinical presentation

- development of ALD is in two main periods of age:
  ○ birth to 6–8 weeks
  ○ 6 weeks to 9 months.
- all breeds are affected, but it is particularly common in fast-growing larger breeds.
- more common in the forelimb and can be uni- or bilateral
- no sex predisposition.
- other types of DOD may be present.
- examine the foal at rest and walking:
  ○ site(s) and degree of angulation of affected joint(s) are assessed in the forelimbs from in front and in the hindlimbs from behind. (Figs. 2.25, 2.26).
  ○ stand the foal as square as possible:
    ♦ foot directly under the upper part of the limb.
    ♦ concurrent rotation from the chest must be considered:
      - evaluate forelimb by viewing limbs perpendicular to frontal plane of limb.
  ○ all limbs should be palpated and manipulated in non-weight-bearing position for:
    ♦ joint instability, growth plate swelling and heat/pain/swelling suggestive of external trauma.
    ♦ bony swelling associated with the growth plate is quite common in longer-standing cases.
  ○ lameness examination is essential, including assessing foot placement:
    ♦ **lameness is significant because foals with ALDs are not normally lame.**
    ♦ trauma-induced cases usually present acutely with lameness and varying instability.

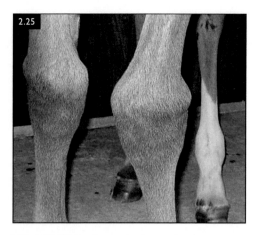

FIG. 2.25 Bilateral acquired carpal valgus angular limb deformity. In this detailed view of the left forelimb, note the severe valgus deformity and enlarged distal radial physis.

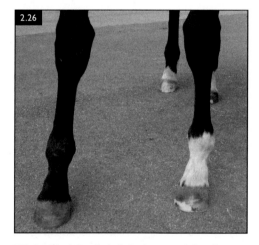

FIG. 2.26 Left forelimb fetlock varus deformity.

Differential diagnosis

- concurrent flexural deformities and/or OCD
- traumatic injuries.

Diagnosis

- clinical history and findings are diagnostic.
- specific problems and their subsequent management are confirmed by radiography:
  ○ see pages in congenital ALDs.

- objective assessment of the angulation of the deformity is achieved by:
  - ◆ measuring the angle of deviation on the radiographs.
  - ◆ normal range of angulation of carpus and tarsus expected to be <5°:
    - – mild/moderate angulation (5–10°).
    - – severe angulation (>10°) (Fig. 2.27).
  - ◆ degree of angulation is more critical to the prognosis in the fetlock joint.

- concurrent malformation or injury to the distal physis and/or epiphysis and to bones making up the articulation should also be evaluated (Figs. 2.28, 2.29):
  - ◆ affects the treatment selected and the prognosis.
    - – Salter–Harris type V or VI injury to the physis may lead to premature closure and loss of major growth potential on the affected side.

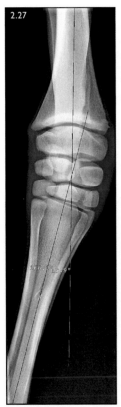

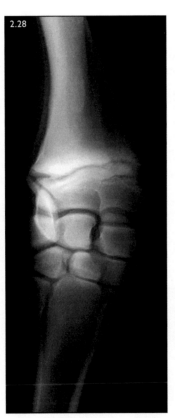

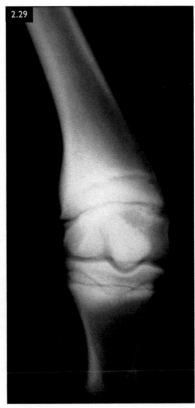

FIG. 2.27 Dorsopalmar radiograph of a foal with moderately severe carpal valgus. Lines drawn between the mid-points of the long bones proximal and distal to the carpus intersect at the level of the carpal bones and the degree of deformity is measured as 13.9°.

FIG. 2.28 Dorsopalmar radiograph of the right forelimb of a foal with severe carpal valgus deformity showing the degree of angulation, wedging of the epiphysis, ectasia of the physis and metaphyseal flaring often noted in such cases.

FIG. 2.29 Severe varus angular limb deformity of the fetlock joint. Note the wedging of the epiphysis and severe tipping of the physis in the distal third metacarpus.

Management

Conservative

- more effective in early cases, and exercise restriction is crucial:
  - box rest and daily in-hand short walks, or small-yard rest (not a paddock).
  - corrective foot trimming to alter the mediolateral foot balance:
    - ◆ lower lateral wall for carpal valgus
    - ◆ lower medial wall for fetlock varus
    - ◆ decreasing toe length and squaring off encourages symmetrical breakover.
  - glue-on shoes or composite application to the hoof for more advanced cases.
    - ◆ medial extension for carpal valgus/ lateral extension for fetlock varus.
    - ◆ contraction of the hoof capsule can occur if fitted for long periods (Fig. 2.30).
- diet of the foal and lactating broodmare will require manipulation:
  - reduce excessive protein and decrease growth rate of foal.
  - assess the diet for any deficiencies or excesses of trace elements and minerals.
    - ◆ balance these using supplements.
- monitor these cases frequently (every 7–10 days).
  - follow their progression.

- respond with different treatments if the foal is not steadily improving.
- shock-wave therapy to retard growth on the convex side of various ALDs has been described but is controversial:
  - three to five sessions at weekly intervals were used concurrently with the use of corrective farriery techniques, with comparable results to surgical correction.

Surgical

- used when conservative treatment is unsuccessful, the condition is worsening, or the deviation is severe at the outset.
- aimed at manipulating growth acceleration and/or growth retardation of the physis and epiphysis.
- reported surgical techniques include:
  - hemicircumferential periosteal transection and elevation (periosteal stripping).
  - temporary transphyseal bridging using a single screw:
    - ◆ foal must be monitored after surgery as there is an overcorrection risk:
      - – postoperative radiographic monitoring is recommended.

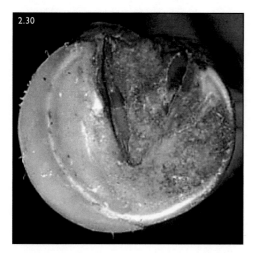

FIG. 2.30 Solar view of a foal's foot after composite has been moulded and attached to the medial wall to create an extension for the treatment of a carpal valgus.

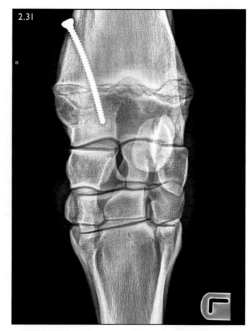

FIGURE 2.31 Dorsopalmar radiograph of the carpus of a foal 2 months after a transphyseal screw has been inserted laterally for the treatment of a carpal valgus.

- ◆ restricted exercise is maintained.
- ◆ optimal time for screw removal is just before the limb is completely straight, as further straightening will occur for a short time after implant removal (Fig. 2.31).

Prognosis

- good if cases are managed early and regularly monitored.
- guarded if there is a poor initial response to conservative treatment, the condition is severe from the outset, concurrent problems are present (e.g. contralateral limb lameness) or if there is rapid deterioration.

Acquired flexural limb deformities

Definition/overview

- acquired (developmental) deviation of the limb in a sagittal plane leading to persistent hyperflexion of a joint(s) region.
- congenital cases (see page 28)
- condition is not due to contracted tendons, but to a relative shortening of a tendon unit in relation to bony structures.

Aetiology/pathophysiology

- multifactorial:
  - ○ excessive feeding in predisposed individuals leads to excess and rapid growth.
  - ○ genetic predisposition.
  - ○ these factors plus mineral/trace element imbalances can lead to DOD (physeal dysplasia, OCD) and pain, which alter the stance and loading of limbs.
    - ◆ initiate contraction and shortening of the musculoskeletal unit.
- during rapid bone growth the accompanying lengthening of the tendinous unit is limited due to the accessory ligaments, and a discrepancy may occur leading to the deformity.
- true tendon contraction can occur post severe tendon injuries in adults.
- chronic non-weight bearing in very painful limbs in any age of animal can lead to flexural deformities, especially acquired carpal contracture.
  - ○ e.g. conservatively managed displaced ulnar fractures.

Clinical presentation

- develops after birth through to 2 years old, most commonly in forelimbs:
  - ○ DIP (coffin) joint between 3 and 18 months.
  - ○ MCP (fetlock) flexural deformity between 9 and 18 months.
- no sex predisposition and unilateral and bilateral cases are seen.
- DIP joint deformity clinical signs include:
  - ○ prominent bulge at the coronary band.
  - ○ 'dishing' of the dorsal hoof wall.
  - ○ increase in heel length and eventually a 'boxy' foot ('club foot') appearance.
    - ◆ Stage I DIP joint deformity (Fig. 2.32).
    - ◆ more severe Stage II DIP joint deformity:
      - – heel does not contact the ground.
      - – dorsal hoof wall goes beyond the vertical (weight on toe) (Fig. 2.33).

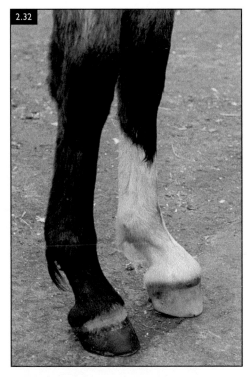

FIG. 2.32 A yearling pony with an acquired flexural deformity of the left forelimb distal interphalangeal joint. Note the upright, boxy and contracted left forelimb foot, with the heel raised just off the ground. Stage I deformity.

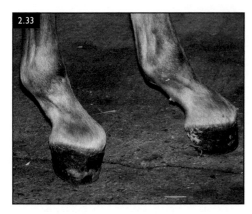

FIG. 2.33 More severe case of bilateral acquired distal interphalangeal joint flexural deformity, with a Stage II deformity in the right forelimb.

- o excessive wear at the toe of the hoof may allow infections to establish and further decrease weight bearing on the limb.
- o palpation of the AL-DDFT (inferior check ligament), both weight and non-weight bearing, may reveal increased tension.
- MCP deformity is less common and presents after a period of compensatory growth.
  - o upright conformation with intermittent dorsal knuckling in early or mild cases (Fig. 2.34).

- o persistent in advanced cases and 'knuckling over' and stumbling are sometimes seen at the walk.
- o palpation of the AL-DDFT may reveal some DIP joint component.

Differential diagnosis

- none.

Diagnosis

- clinical history and findings are diagnostic.
- radiography of the foot of the affected limb may reveal bone remodelling at the tip of the pedal bone in severe cases of DIP joint deformity:
  - o may affect the long-term prognosis in some cases.
  - o infrequently, persistent infection in the hoof at the toe can be associated with damage and infection in the pedal bone.

Management
Distal interphalangeal flexural deformities

Conservative
- severity of the problem and the age of presentation determine the management options:
  - o controlled exercise programme on hard even surfaces (walk up small inclines).

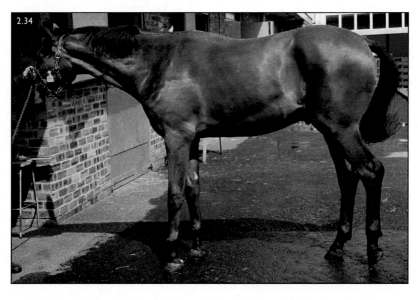

FIG. 2.34 Yearling Thoroughbred with rapid onset of bilateral flexural deformity of the metacarpophalangeal joints of both forelimbs. Note the very good condition and size of the animal and the upright forelimb conformation through the fetlock. The left forelimb is intermittently knuckling forward.

- o placing the foal and mare into a small pen with an all-weather or dry surface for controlled but regular exercise is beneficial.
- o judicious use of NSAIDs to control any pain improves the benefit of exercise.
- o intravenous injection of Oxytetracycline in younger foals (see page 29).
- o regular foot trimming (every 10 days) is important to lower the heels.
- o application of a plastic, glue-on toe extension shoe (changed every 10–14 days):
  - ◆ prevent further 'boxiness' developing.
  - ◆ placing a hoof composite cap over the toe is a very effective alternative.
  - ◆ both methods protect the toe and dorsal hoof capsule from bruising during exercise and act as a lever arm (Fig. 2.35).
  - ◆ use of a built-up heel, which is then gradually reduced, is also advocated by some clinicians.
- o dietary restriction to lower the carbohydrate and protein intake for the mare and foal.
- o early weaning helps control feed intake and decrease growth rate.

Surgical
- desmotomy of the AL-DDFT (inferior check ligament desmotomy) is recommended if no improvement occurs in a few weeks, depending on the severity of the deformity (Fig. 2.36).
- very severe cases may require DDFT tenotomy although this is a salvage procedure only.
- postoperative management involves continuing with the conservative management techniques outlined above.
  - o not uncommon to have a cosmetic blemish at the incision sites.

Metacarpophalangeal flexural deformities

Conservative
- very frustrating to deal with as neither conservative nor surgical treatment is particularly effective:
  - o conservative treatment is usually only effective in mild deformities.

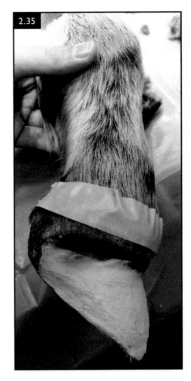

FIG. 2.35 Fitting a composite plastic toe extension to the foot of a foal with flexural deformity of the distal interphalangeal joint.

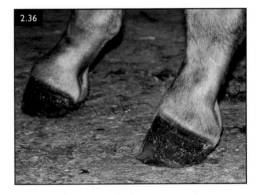

FIG. 2.36 Postoperative results of surgical treatment of the horse in 2.33.

- o controlled exercise programme (walking on level ground with secure footing) and small-area free exercise are less helpful.
- o feet should be balanced in both planes.
- o some reports of success using heel wedges in early cases.

- application of a toe extension combined with a vertical bar ('fetlock brace') shoe can be useful (Fig. 2.37).
- dietary restriction to lower the carbohydrate and protein intake and decrease the growth rate is essential:
  - many of these cases are in a period of excessive compensatory growth.
  - ensure correct trace element and mineral intake.
- pain control will help the animal to weight bear more normally on the affected limb(s).
  - treat any specific orthopaedic disease.
  - careful use of NSAIDs as analgesics.

Surgical
- frequently unsuccessful in correcting this deformity.
- usual practice to section both the AL-DDFT and the accessory ligament of the superficial digital flexor tendon ('superior check ligament').
  - combined with aggressive management, including forced exercise on the limb and hopping exercises, or use of a fetlock brace.

FIG. 2.37 A fetlock brace shoe individually created to fit the yearling in 2.34. Note it is in place, with the strap that forms the brace part of the shoe tightened around the fetlock, to stop it knuckling forward when the animal is weight bearing.

Prognosis
- good for Stage I and guarded for Stage II DIP joint flexural deformities treated before 24 months of age.
- MCP flexural deformities carry a guarded to very poor prognosis, depending on presenting severity.
- factors that worsen the prognosis include delay in appropriate treatment and any concurrent orthopaedic problems in the affected limb.
- most horses with acquired carpal contracture have a chronic unresponsive or untreated lameness, and it is uncommon to be able to restore such horses to athletic soundness.

Subchondral bone cysts

Overview
- most recorded at the weight-bearing surface of the medial femoral condyle.
- found less frequently at the distal epiphyses of the proximal phalanx, third metacarpus/metatarsus, proximal epiphyses of the middle phalanx, radius and tibia, glenoid of the scapula, articular margin of the third phalanx and, occasionally, the carpal and tarsal bones.
  - often termed '**osseous cyst-like lesions**' at these sites.
- aetiology/pathophysiology is not completely understood but two main theories:
  - **OCD** – focal failure of endochondral ossification.
    - thickened plug of cartilage invaginates to form a bone cyst.
    - necrosis/collapse, secondary inflammatory response plus trauma important factors.
  - **Traumatic** – more widely held view.
    - trauma at the weight-bearing point of a joint leads to microfissure formation in the articular surface.
    - subsequent ingress of synovial fluid and cyst development.
- development of a cystic lining and production of inflammatory mediators results in a synovitis and, typically, an intermittent exercise-induced lameness.
- cysts have been recorded in older horses associated with osteoarthritis or articular/periarticular injury.

Clinical presentation

- more common in younger horses (<4 years) but can present in older animals.
- mainly faster growing large breeds are affected, especially Thoroughbreds and Warmbloods.
- lameness is usually seen when athletic work begins:
  - usually a subtle, insidious or intermittent lameness that improves with rest only to recur once back in work.
  - acute-onset lameness is less common.
  - mild joint distension may be present, but often difficult to identify.
  - flexion tests may be slightly positive.
  - intra-articular analgesia is often required to identify the site of pain.

Differential diagnosis

- subchondral lucencies primarily associated with advanced DJD
- bone abscess.

Diagnosis

- full lameness examination and systematic diagnostic analgesia is required to assess the relevance of any cystic structure within a joint.
- full radiographic series for the affected joint should be obtained (Fig. 2.38):
  - certain views are more useful in depicting cysts at different locations.
  - can be bilateral, so the contralateral limb should always be radiographed.
  - positive contrast arthrograms may also be useful in certain instances.
  - cysts vary in size from small indentations to a large radiolucent lucency depending on their stage of development.
  - often oval or rounded in shape, usually confluent with the articular margin, and may have a sclerotic peripheral margin (see Fig. 2.39).
- ultrasonography may be useful in some cases to evaluate the cartilage contour of the joint.
- bone scintigraphy results are not consistent.
- MRI and CT can help locate and define more accurately cysts of the distal limb and may be a useful presurgical intervention.

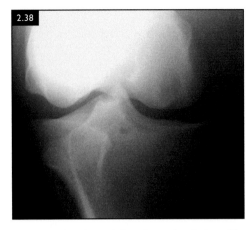

FIG. 2.38 Caudocranial radiograph of the stifle showing the femorotibial joint in a 4-year-old pony with lameness localised to this joint by intra-articular analgesia. Note the large subchondral bone cyst in the medial condyle of the distal femur and the large connection between the cyst and joint.

Management

Conservative

- box or small-paddock rest for up to 6 months.
  - temporarily helpful in cysts located elsewhere than medial femoral condyle.
  - NSAIDs to control the pain and inflammation in the joint(s).
- various intra-articular medications have been used, including corticosteroids.
  - generally limited success and often only temporary resolution in lameness.
- long-term oral joint supplementation with chondroitin sulphate and glucosamine.

Surgical (Fig. 2.39)

- debridement of the cyst via arthroscopy.
  - enlargement of the cyst and worsening of lameness have been observed.
  - subchondral bone forage is currently contraindicated.
  - cancellous bone graft and stem cells can be placed after debridement.
- injection of corticosteroids (usually triamcinolone) into cystic lining, under arthroscopic or ultrasonographic guidance.
  - may be successful in some cases, but in others, improvement may be temporary and subsequent cystic enucleation is required.

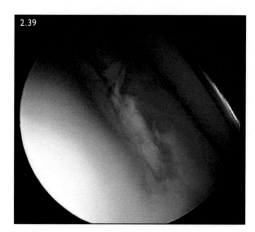

FIG. 2.39 An arthroscopic view of the medial femoral condyle of a 2-year-old Thoroughbred racehorse with a subchondral bone cyst. Note the flattening of the condyle and an obvious defect in the cartilage. Probing with a needle allowed entrance into a large cyst, which was treated by injection of corticosteroids into the cyst cavity.

- cortical screw placed across the cyst under arthroscopic or radiographic control.

Prognosis

- depends on the joint affected, the severity of the lesion, associated DJD, intended use of the horse and the treatment option selected.
- see page 183 for subchondral bone cysts of the medial femoral condyle.
- too few numbers of cysts treated in other locations are available to make definitive statements but usually a guarded prognosis is given.
- accompanying DJD and bilateral cysts worsens the prognosis.

Physitis (epiphysitis, physeal dysplasia)

Definition/overview

- form of DOD involving a disturbance of endochondral ossification at the peripheral or axial region of the metaphyseal growth plate.
- most common at distal radius/tibia and distal third metacarpus/metatarsus of foals and young horses.
- mainly encountered in fast-growing breeds, such as the Thoroughbred.

Aetiology/pathophysiology

- aetiology is currently unclear:
  - compression trauma to the growth plate due to excessive exercise or overuse of one limb during the most active growth phases of young horses.
  - more common where nutritional (lush pasture, high concentrates) and supplement excesses or deficiencies are implicated.
  - some clinicians believe it may be another manifestation of OCD in the horse.

Clinical presentation

- most common in distal third metacarpus/metatarsus from 3–6 months and distal radius from 8 months to 2 years (Fig. 2.40).
- uncommonly in the distal tibia (from 9–18 months).
- affected physes are:
  - usually enlarged, particularly on the medial aspect.
  - firm, warm and painful to palpation.
- flexion of adjacent joints may be painful.
- concurrent angular and/or flexural deformities may be present in severe cases.
- lameness can be overt or, more usually, presents as a slight gait stiffness.

Differential diagnosis

- Infectious physitis
- Salter–Harris type V and type VI growth plate injuries.

Diagnosis

- clinical history and findings are suggestive.
- radiography is necessary to confirm a diagnosis:
  - lateromedial and dorsopalmar/plantar views are most useful (Fig. 2.41).
  - metaphysis of the bone will appear widened and asymmetrical:
    - ♦ flaring and sclerosis of the metaphysis adjacent to the physis.
  - physis appears irregular and widened.
  - ALDs may also be present along with overlying soft tissue swelling.

Management

- restriction of exercise helps to reduce further physeal trauma:

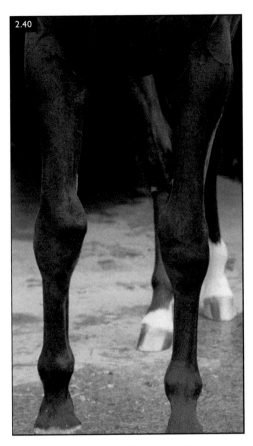

FIG. 2.40 Yearling Warmblood colt with enlarged physis of both distal radii. The animal was being overfed and growing rapidly. This is a typical site for physitis.

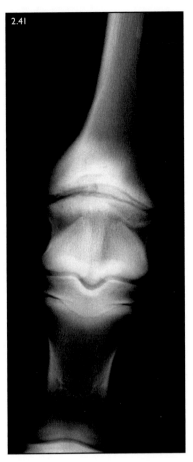

FIG. 2.41 Dorsopalmar radiograph of a weanling foal with physitis showing changes within the physis and metaphyseal flaring and subchondral bone sclerosis.

- o may include box rest (in severe cases) to small-yard rest.
- o initially from 2 weeks to 2 months depending on the severity of the condition.
- o NSAIDs should be used if the horse is very painful or lame to provide analgesia.
- o correction of any nutritional and supplement excesses and/or deficiencies:
  - ♦ reduction in energy content (to decrease body weight).
  - ♦ assess and correct calcium, phosphorus, copper and zinc levels.
  - ♦ corrective foot trimming to balance all four feet and correct any angular and/or flexural deformity present.

Prognosis

- good, providing it does not lead to severe permanent conformational defects due to premature closure of growth plates.
- self-limiting disease process that ceases when skeletal maturity is attained.

The Foot

Clinical presentation

- foot problems are the most common cause of lameness in the forelimbs.
  - generally, less common in the hindlimb.
- most common presenting signs for diseases affecting the horse's foot are:
  - lameness, usually caused by a focus of pain.
    - ◆ much less frequently, by a mechanical change in function.
  - change in appearance, shape, or size in chronic cases.
  - signalment and history are frequently variable and non-specific.
    - ◆ knowledge of the horse's shoeing routine may offer important clues.
      - – when last shod, how frequent, any recent farriery change.

Examination of the foot with the horse at rest involves:

- visual inspection to identify:
  - gross lesions such as hoof cracks or hoof wall avulsions.
  - size of the foot in relation to the size of the horse.
  - shape of each foot and symmetry between the feet.
  - flares and circumferential patterns of the growth rings.
- palpation to identify:
  - excessive warmth.
  - soft areas in the sole.
  - moistness, particularly at the hairline.
- paring of the sole (**best with shoe not in place**) to expose:
  - bruises.
  - defects in the white line or sole.
- application of hoof testers to identify withdrawal response due to a focus of pain:
  - solar inflammation, bruising or abscess.

- inflammation, infection or fracture of the distal phalanx.
- manipulation of the digit by flexion, extension, and rotation to determine:
  - restriction in range of motion.
  - pain resulting in a withdrawal response on forced flexion of the digit.
  - pain resulting in a withdrawal response on a board extension test of the digit.

Examination of lameness in horses with foot pain involves:

- at walk with particular attention to the horse's foot fall.
  - normal front foot lands slightly heel-before-toe with lateromedial symmetry.
    - ◆ pain in the navicular region land toe first.
    - ◆ pain in the toe region demonstrate exaggerated heel-before-toe landing.
      - – horses with laminitis, seedy toe, keratoma of the toe.
    - ◆ asymmetric pain land preferentially on side of the foot away from the lesion.
      - – palmar process fracture of distal phalanx, quittor, injury of an ossified ungular cartilage.
- trotting in a straight line while observing the horse from the side and the front:
  - identify the head nod associated with foot lameness.
  - bilateral foot pain frequently presents with a stiff and short-striding gait.
- trotting in circles on firm and soft ground is useful in the detection of foot lameness:
  - head nod generally exacerbated with lame foot on the inside circle on firm ground.
    - ◆ compared to trotting on a straight line or trotting in circles on sand.

DOI: 10.1201/9781003369226-3

- bilateral foot pain that does not show a head nod on a straight-line trot:
  - generally, show an obvious head nod on both circles on firm ground.
  - corresponds to exacerbated foot lameness on the inside of the circle.
  - **some injuries that cause foot pain paradoxically may cause exacerbation of lameness when the foot is on the outside of the circle.**
    - lateral collateral ligament of the distal interphalangeal (DIP) joint.
    - lateral lobe of the deep digital flexor tendon (DDFT).
    - medial palmar process of the distal phalanx.
    - medial ossified ungular cartilage of the foot.
- examination in straight line trot following stress tests on the digit (1 minute each).
  - may exacerbate lameness caused by pain in following anatomical areas:
  - flexion test of the digit:
    - distal and proximal interphalangeal joints.
    - fetlock joint.
    - navicular bone or the DDFT.
  - extension with a board or wooden wedge:
    - podotrochlear apparatus (navicular bone, ligaments and bursa, DDFT)
  - heel elevation test (wooden wedge):
    - distal part and insertion of the DDFT.
    - collapsed heels.
  - elevation of the lateral or medial quarter with a wooden wedge:
    - elevated part of the foot:
      - bruising
      - eccentric keratoma.
      - bone injury of the elevated palmar process or ossified cartilage of the foot.
      - asymmetric osteoarthritis (OA) lesions of DIP joint.
    - lower part of the foot (injured collateral ligament of the DIP joint).

Ancillary diagnostic tests

Localisation of pain

(Figs. 3.1a, 3.1b)

- physical examination fails to determine the source of the pain causing lameness.

- localisation of pain to the foot using regional and/or intrasynovial analgesia:
  - local anaesthetic deposited in one synovial structure can diffuse into an adjacent synovial structure or around an adjacent nerve.
  - specificity of pain localisation is still better with anaesthesia of the navicular bursa or DIP joint compared with:
    - palmar digital or abaxial sesamoid nerve block.
    - improved accuracy by limiting amount of anaesthetic used and observing response in relation to time.

Imaging of the foot

- indicated when physical examination and diagnostic regional analgesia do not provide a conclusive diagnosis.
- **Radiography** is the first-line imaging technique:
  - examination typically consists of five radiographic projections:
    - lateromedial.
    - dorsopalmar.
    - 45° dorsoproximal–palmarodistal oblique (upright pedal).
    - 60° dorsoproximal–palmarodistal oblique.
    - Palmaroproximal–palmarodistal oblique (skyline).
  - additional or fewer views can be used depending on the target structure of interest:
    - distal phalanx, hoof capsule, navicular bone.
    - specific additional oblique views may be performed as necessary.
  - radiography is particularly suitable for imaging of:
    - osseous abnormalities like fractures.
    - advanced degenerative changes causing bone lysis and/or bone proliferation.
    - subtle bone margin changes like osteophytes, entheseophytes and small chip fractures.
    - position of the distal phalanx relative to the hoof capsule.

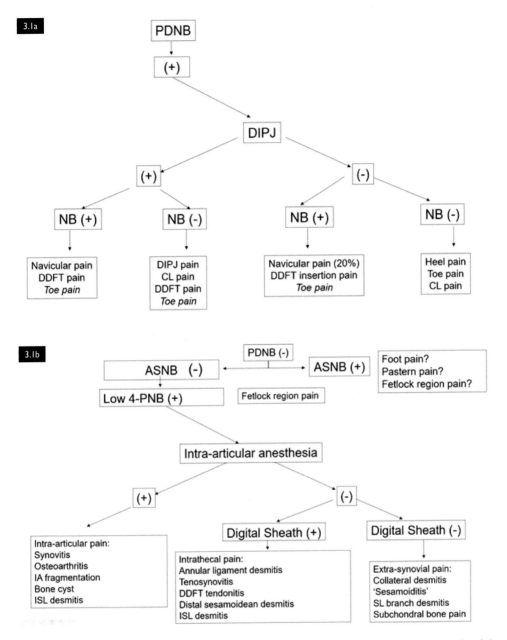

FIGS. 3.1A, 3.1B Flow diagrams showing the decision tree involved in the differential diagnosis of the localisation of pain causing foot lameness. PDNB = palmar digital nerve block; DIPJ = distal interphalangeal joint; NB = navicular bursa; DDFT = deep digital flexor tendon; CL = collateral ligament; ASNB = abaxial sesamoid nerve block; Low 4-PNB = low four point nerve block; IA = intra-articular fragment; ISL = intersesamoidean ligament; SL = suspensory ligament.

- o with severe lameness, radiography should be performed before diagnostic analgesia to prevent exacerbation of a traumatic injury.
- **Ultrasonographic examination of the foot** is difficult because of the presence of the hoof capsule and a strongly user-dependent technique.
  - o rarely offers a conclusive diagnosis but can be used to image:
    - ♦ DDFT in the region of the pastern.
    - ♦ proximal 1/3 of the collateral ligaments of the DIP joint.
    - ♦ dorsal joint recess of the DIP joint.
- **Scintigraphy** may be able to detect increased radiopharmaceutical uptake in the foot in the absence of radiographic abnormalities:
  - o use has declined because of its poor spatial resolution and the advent of superior imaging techniques.
- **CT** provides cross-sectional images and has excellent spatial resolution with fine structural detail, especially for bone.
  - o greatly improved the ability to evaluate osseous structures in the foot but images can also be windowed for soft tissue imaging.
- **MRI** provides cross-sectional images but has superior soft tissue contrast and definition than CT and can show abnormal water content in tissues, especially in bone.
  - o vastly improved the ability to make an early diagnosis of previously unidentified pathology within the bones, joints, and soft tissues of the foot.

Management

- treatment of bone, joint and soft tissue injuries and diseases of the foot is similar to that of other parts of the limb, but different shoeing techniques can be used to assist other treatments by:
  - o protecting the foot.
  - o changing the balance of the foot.
  - o changing foot motion during the stride to prevent interference or enhance animation.
  - o supporting the position of the foot on the ground.
  - o reducing the stress on an injured structure.
- specialised structure of the foot's integument is composed of cornified epidermis, dermis, and subcutaneous tissue.
  - o stratum corneum of the hoof epidermis forms the hoof capsule and this alters:
    - ♦ surgical access to deeper structures (compromise between exposure and stability of the hoof).
    - ♦ management of defects in the integument of the foot (cannot be sutured in comparable manner to skin).

DISEASES OF THE FEET

Diseases and injuries of the integument of the foot

- foot wounds take many forms:
  - certain common features in the way they heal and are treated.
- abrasions to the coronary band and iatrogenic hoof wall avulsions heal similarly to:
  - partial thickness skin wounds by epithelialisation from remnants of germinal epithelium distributed across the surface of the wound.
- full-thickness defects of the coronary band, wall, sole, or frog follow:
  - four phases of wound healing but without retraction or contraction of wound margins.
- superficial wounds, like abscesses, are usually satisfactorily treated with topical antibiotics or antiseptics.
  - **systemic antibiotics are seldom warranted.**
- wounds that extend deep to the dermis require systemic antibiotics:
  - plus topical antimicrobials until a healthy layer of granulation is present.
  - regional perfusion of the distal limb with antibiotics via a superficial vein:
    - ♦ achieves antibiotic concentrations in tissues that persist above the MIC for various bacteria for over 24 hours.
- bandaging for foot injuries requires three layers, of which a primary layer or surface dressing should be:
  - adherent if debridement is required.
  - non-adherent if epithelialisation and fibroplasia are to be encouraged.
- secondary, or padding layer (not always used), helps protect and support the foot.
- tertiary layer holds the other layers in place - **avoid excessive pressure.**

Abscesses

Definition/overview

- focal accumulation of purulent exudate between the germinal and keratinised epithelium of the hoof.

Aetiology/pathophysiology

- cause is not usually specifically identifiable:
  - bacterial access to underlying tissues through small defects in the hoof capsule.
  - puncture wounds.
  - hoof cracks.
  - pre-existing disease (e.g. laminitis) may lead to recurrent abscess formation.
  - predisposition of poor hoof structure (e.g. dropped or thin soles) to bruises:
    - ♦ secondarily infected (suppurative bruises).
- particularly painful because of rapid increase in pressure which causes extension of the abscess along the path of least resistance:
  - under the sole or frog by separation of the hoof capsule from the germinal layer of the epithelium (double sole).
  - proximally deep to the hoof wall causes separation at the coronary band ('gravel'/ 'grit').
  - through germinal layers of the integument and dermis affecting deeper structures (septic osteitis).

Clinical presentation

- acute, severe lameness in one limb, sometimes evident after exercise or turnout. (4–5/5).
- variable occurrence:
  - distal limb swelling and/or cellulitis.
  - swelling or a discharging sinus ('gravel') at the coronary band.
  - gas pocket at the level of the solar epidermis on radiographs.

Differential diagnosis

- any disease of the foot associated with acute onset of severe lameness in a single limb:
  - fracture of the distal phalanx or navicular bone.
  - sepsis of a deep digital structure following a street nail.
  - severe bruising.
  - severe injury to a ligament or tendon within the foot.

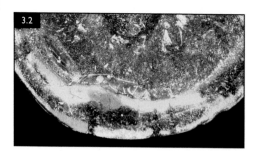

FIG. 3.2 Abscess drainage immediately inside the white line after exposure with a hoof knife.

Diagnosis

- increased warmth in the affected foot.
- increased digital pulse.
- marked withdrawal to application of hoof testers (**mostly but not always**).
- dark discolouration of the horn on exploration of the sole, indicating a track or injury site (Fig. 3.2).
- absence of a track after exploration:
  - **excessive paring of the foot is contraindicated.**
  - 24-hour poulticing may cause the abscess to drain spontaneously, often at the coronary band (Fig. 3.3).
  - further paring of the softened horn post poulticing may be successful.
- still no positive findings with persistent pain:
  - abaxial sesamoid perineural analgesia to localise pain to foot.
  - radiographs of foot to detect gas pockets and rule out other differentials.

Management

- drainage is the primary treatment:
  - use regional nerve blocks and/or sedation in difficult horses.
  - small hole, approximately 1 cm in diameter, will suffice.
  - abscesses adjacent to the white line drained through a small notch in the distal wall.
    - ♦ deficits in the wall are managed more easily than a hole in the sole.
  - lameness should resolve quickly.
- antiseptic dressing (few days).
  - repeated poultices or foot soaking not usually indicated.
- abscess wound should be dry within a few days (Fig. 3.4).

FIG. 3.3 Probe demonstrates the tract created by an abscess that spontaneously drained at the coronary band.

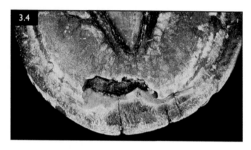

FIG. 3.4 Cornified sole of the horse in 3.2, 4–5 days after abscess drainage and dressing with povidone–iodine solution.

- shod with a full pad or plate to protect sizeable solar defects.
- tetanus prophylaxis in horses without a recent history of vaccination.
- systemic antibiotics not required unless the abscess has extended deep to the dermis.
- radiography should be performed if clinical signs fail to improve, or the abscess recurs.

Prognosis

- good for simple abscesses.
- recurrent abscessation likely if predisposing cause exists.
- varies with the involvement of deeper structures.

Bruising

Definition/overview

- extravasation of blood from ruptured blood vessels into the surrounding tissues caused by blunt trauma.

Aetiology/pathophysiology

- single forceful contusion.
- recurrent lower-grade trauma associated with conformational or pathological predisposition, such as flat feet or laminitis.
- pressure on the sole from poorly fitting shoes (e.g. corns in the angles of the sole).
- mostly in the solar or parietal lamellae, rarely at the coronary band or in the frog.
- occasionally become infected through microfractures in the hoof capsule and form an abscess.

Clinical presentation

- usually associated with lameness:
  o acute or chronic, mild or severe, unilateral or bilateral.
  o assess foot carefully for predisposing causes such as poor conformation.
- occasionally an incidental finding when a horse is being trimmed or shod.
- characteristic red discolouration caused by extravasation of blood into the tissues.
  o haemorrhage penetrates a variable distance into the substance of the horn.
  o appears stippled to reflect the tubular nature of the horn.
  o may not be visible until the affected hoof reaches the surface through normal growth.

Differential diagnosis

- any other cause of foot lameness.

Diagnosis

- withdrawal response to the application of hoof testers.
- identify blood-stained horn only when the foot has been cleaned and pared (Fig. 3.5) or the stained horn has grown out sufficiently superficial.
- bruising from the coronary band is visible in unpigmented horn of the wall:
  o horizontal red stripes, but these are seldom clinically significant.
- chronic bruising may lead to pedal osteitis:
  o resorption of bone from the margins of the distal phalanx.

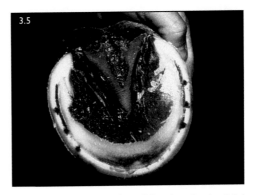

FIG. 3.5 Extensive bruising of the dorsal sole distal to the dorsal solar margin of the distal phalanx.

Management

- isolated stone bruises:
  o rest and NSAIDs.
  o icing affected foot for 24 hours.
- partially paring the surface of the horn over the bruise to relieve pressure from the area as it heals.
- **opening of the bruised area with exposure of the solar lamellae must be avoided.**
- recurrent bruising associated with flat soles requires protection of the sole:
  o seated-out wide-web shoes.
  o full leather or soft silicone pads.
- bruising of the solar lamellae associated with rotation of the distal phalanx in laminitis requires:
  o long-term corrective trimming and shoeing.
  o restore normal alignment of the distal phalanx and eliminate sole pressure.

Prognosis

- full recovery for isolated stone bruises and corns.
- guarded for recovery without recurrence if persistent primary cause (e.g. laminitis, flat feet).

Thrush

Definition/overview

- degenerative keratolytic condition of the frog and adjacent sulci.

Aetiology/pathophysiology

- probable anaerobic bacterial infection, but the specific aetiological agent has not been identified.
- horses standing in moist unhygienic conditions or subjected to poor daily foot management.
- common in moist frogs under a full pad.
- horses with deep and/or narrow sulci are more prone to developing thrush than those with wide and shallow sulci (Fig. 3.6).
- usually starts superficially on the surface of the frog, particularly in central and paracuneal sulci:
  - extends deeper into the stratum corneum of the frog.
  - occasionally reaching the germinal layers of the epidermis and the dermis.

Clinical presentation

- characteristic foul odour and discharge from the paracuneal sulci.
- ragged frog surface.
- often not lame, but deeper damage, especially between the bulbs of the heel, can be painful (Fig. 3.6).

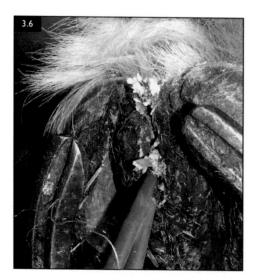

FIG. 3.6 A foot with severe thrush infection in the central sulci of the frog and midline of the bulbs of the heel.

Differential diagnosis

- Canker:
  - proliferative condition affecting the basal layers of the frog.
  - may extend to any part of the hoof.
  - unresponsive to remedies that cure thrush.

Diagnosis

- characteristic odour and appearance of the frog.
- degenerative change of the superficial layers of the frog.
  - confined to the frog and paracuneal sulci.
- thrush cases are less likely to be lame.

Management

- debridement of ragged and undermined frog.
- transfer the horse to hygienic and dry conditions.
- improve daily foot management with cleaning of the paracuneal sulci:
  - application of topical antiseptics and astringents.
  - povidone–iodine, 2% tincture of iodine, or proprietary preparations of formalin (not aldehyde).
- corrective trimming and shoeing for poor foot conformation.

Prognosis

- usually, good.
- recurrence is common:
  - narrow frogs and deep sulci conformation.
  - moist unhygienic stable surfaces.

Canker

Definition/overview

- defined histologically as a chronic proliferative pododermatitis.

Aetiology/pathophysiology

- aetiology and pathogenesis of canker are unknown:
  - anaerobic bacterial infection affecting the germinal layer of the epithelium has been hypothesised in the literature.

- o spirochaetes have been demonstrated to be present in biopsy samples.
- o bovine papilloma viral DNA identified in 100% of canker lesions in one study.
- o may represent one or more disease processes or aetiological agents.
- horses standing for prolonged periods on a wet ground surface that is contaminated with faeces are more prone.
- individual or breed-related genetic predisposition has been suggested.
- classically starts in the central or paracuneal (collateral) sulci and then rapidly spreads to the crura of the frog.
  - o may extend to sole or bulbs of the heel, and occasionally to walls of the hoof.

Clinical presentation

- usually chronic development, over several weeks to months.
- affects one or more feet.
- characteristic proliferative appearance from which extend finger-like projections:
  - o usually associated with a yellow, creamy exudate.
- typically located on the frog but may extend to involve any part of the hoof.
- palpation of the affected area is painful.
- lameness often initially mild but progressively increases as the disease worsens.

Differential diagnosis

- Thrush.

Diagnosis

- physical appearance of the foot (Fig. 3.7).
- biopsy and histopathology are required for confirmation.

Management

- trimming of the foot prior to debridement to ensure:
  - o maximal width between the heels to maximise drainage.
  - o limit crevices in the sulci where the causative agent can linger.
- surgical debridement consists of excision of the proliferative tissue:
  - o preserve as much of the germinal layer of the epidermis as possible.
  - o healing then occurs as a partial thickness wound:

- ◆ epithelialisation without the formation of fibroplasia.
- o usually satisfactorily performed in the standing sedated horse using an abaxial sesamoid nerve block, but difficult horses may require general anaesthesia.
- o remove the surface tissue with a hoof knife, scalpel, or laser.
  - ◆ intricate debridement is best performed with a rongeur.
- o aggressive debridement that removes significant amounts of the germinal epithelium results in:
  - ◆ healing by fibroplasia/ epithelialisation.
  - ◆ delays in healing without improving results.
- o alternatively, following gross debridement, sterile maggots can be used for more detailed debridement.
- o tourniquet is usually unnecessary unless more extensive debridement is to be performed.
- many topical medications have been used under foot bandages in treatment of canker:
  - o 2–5% suspension of metronidazole in saline or a paste made from crushed metronidazole tablets and saline applied as a wet-to-dry dressing after debridement.
  - o in refractory cases, 0.05% enrofloxacin or clindamycin in Tricide (Molecular Therapeutics LLC), applied either as a wet-to-dry dressing or as a dry-to-dry dressing after the dressing has been soaked and dried.

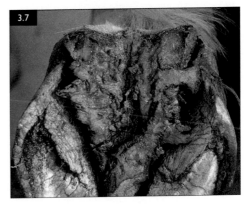

FIG. 3.7 Canker. Extensive moist proliferation of the epidermis of the crura and sulci of the frog.

- other topical dressings include:
  - saturated with 10% benzoyl peroxide in acetone.
  - chloramphenicol and fungicides.
  - topical cisplatin in a cream (potential role of bovine papilloma virus).
- systemic use of antibiotics with an anaerobic spectrum as an adjunct to local therapy:
  - metronidazole or penicillin.
- systemic or topical steroids as an adjunct to therapy.
- should be dramatic improvement in uncomplicated cases within 2 weeks and complete resolution within 4 weeks.
- in complicated cases, where there is a reservoir of the causative agent in a protected location in either the hoof or the environment, recurrence is common.

Prognosis

- response to treatment is highly variable.
- prognosis is always guarded.

Laminitis

Definition/overview

- disease in which a series of pathophysiological events cause injury to the dermal and epidermal lamellae of the foot (Fig. 3.8).
  - weakens the lamellar attachments of the distal phalanx to the hoof capsule.
- defined as acute, subacute or chronic:

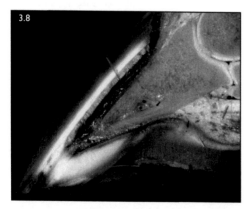

3.8

FIG. 3.8 Cross-section of a foot with acute laminitis before it has displaced. Note that the dorsal epidermal lamellae have been damaged and started to pull away.

- acute laminitis occurs within 72 hours of onset of clinical signs.
  - not accompanied by displacement of the distal phalanx.
- subacute laminitis is present for more than 72 hours duration.
  - not accompanied by displacement.
- chronic laminitis occurs once the distal phalanx has displaced in relation to the hoof capsule (regardless of duration).

Aetiology/pathophysiology

- several known pathophysiological entry pathways exist in the pathogenesis of acute laminitis:

1. **Sepsis-related laminitis:**
   - associated with diseases such as grain overload, colic, colitis, pleuropneumonia and metritis that are frequently accompanied by clinical signs of endotoxaemia.
   - expression of inflammatory mediators in the lamellae during the prodromal stages of the disease consistent with a systemic inflammatory response.

2. **Endocrinopathic laminitis:**
   - associated with pituitary pars intermedia dysfunction (PPID) and insulin resistance.
   - pasture-associated laminitis is associated with grazing on grasses with high non-structural carbohydrates:
     - may lead to weight gain and insulin resistance.
     - others develop laminitis without evidence of insulin resistance.

3. **Supporting limb laminitis:**
   - following prolonged weight bearing on one limb.
   - regular loading/unloading is needed to move arterial blood through the lamellar capillary beds.

4. **Corticosteroid-associated laminitis:**
   - anecdotally associated with systemic or intra-articular steroid administration.

5. **Traumatic laminitis:**
   - associated with prolonged exercise on hard ground (road founder).

- pathophysiology of chronic laminitis follows on from displacement of the distal

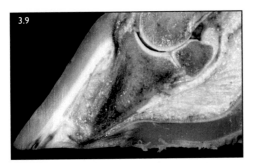

FIG. 3.9 Cross-section of a foot post laminitis rotation. Note the wedge of damaged tissue dorsally underneath the hoof wall and the beginning of bending of the tubules at the coronary band.

phalanx in relation to the hoof capsule caused by the stress of weight bearing and movement in the presence of severe tissue necrosis within the lamellae.

- o necrotic tissue can subsequently be repaired and in severe cases this forms a wedge of hyperplastic epidermal tissue between the parietal dermis and the hoof capsule (Fig. 3.9).
- o hoof wall grows faster at the heels than at the toe.
- o lamellar wedge and the abnormal growth give rise to the characteristically deformed slipper hoof, concave dorsal wall, high heels, and flattened sole.

Clinical presentation

- acute bilateral forelimb lameness, but any combination of feet is possible (Fig. 3.10).

- o characterised by an extremely stiff, short-striding gait with marked heel-before-toe landing of the foot:
  - ♦ forelimbs are placed out in front of the horse.
  - ♦ hindlimbs are thrust underneath the horse's body.
- o more severely affected horses are reluctant to move or to pick up a foot that is contralateral to an affected limb.
- o worst cases become persistently recumbent.
- o acute laminitis may be hard to identify if the hindlimbs alone are affected.
- bounding digital pulses and warm feet.
- increased sensitivity to percussion of the dorsal hoof wall and application of hoof testers to the dorsal region of the toe.
- initially, no radiographic changes, except slight thickening of the dorsal hoof wall.
- many horses with endocrinopathic laminitis display a more insidious onset of lameness:
  - o early milder lameness may mimic common causes of bilateral foot soreness.
  - o may persist or progress to a more severe level.
  - o course of the disease is often prolonged and with visible changes to the hoof capsule at the initial presentation (Fig. 3.11).

FIG. 3.10 A pony with forelimb laminitis exhibiting the typical stance, with the forelimbs stretched out forwards and the hindlimbs placed underneath the body.

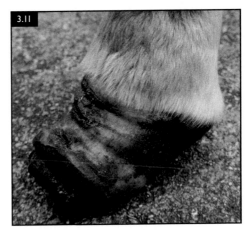

FIG. 3.11 Chronic laminitic foot with long toe, dorsal compression of the hoof wall and hoof-ring formation.

- alternatively, horses with endocrinopathic laminitis may have:
  - prior history of laminitis and present with a more recent acute episode.
  - gradual deterioration of chronic disease.
  - complications such as hoof bruising or abscesses.
- animals displaying moderate to severe lameness in one limb at rest in a stall (e.g. septic arthritis, fracture or cast limb), may present with supporting limb laminitis in the opposite, initially non-lame, limb.
  - first symptoms may be increased weight bearing by the primary lame limb.
- horses without characteristic symptoms or gross morphological changes of the hoof capsule may require regional analgesia to localise the pain to the foot:
  - whether pain is arising from the solar or parietal lamellae:
    - lameness should improve more with either a palmar digital or an abaxial sesamoid nerve block, respectively.

Imaging

- radiography to determine the severity and staging of the disease:
  - sequential radiographs help document disease progression.
- lateromedial and horizontal dorsopalmar/plantar views:
  - radiographic beam centred on the solar margin of the distal phalanx.
    - approximately 1–1.5 cm from the ground surface of the foot.
    - evaluate position of distal phalanx.
  - radiodense marker of known length on the dorsal hoof wall in the median plane that extends from the coronary band distally prior to radiography.
  - capsular rotation:
    - dorsal parietal surface of distal phalanx diverges from dorsal hoof wall.
    - DIP joint flexes so increases angle the solar margin of the distal phalanx makes with the ground.
    - distance from the solar margin in the toe region of the distal phalanx to the ground decreases.
  - distal displacement:

- thickness of the dorsal hoof increases.
- depth of sole decreases.
- extensor process of the distal phalanx moves distally in relation to the coronary band (Fig. 3.12).
  - asymmetric distal displacement:
    - one side (most commonly the medial side) of the distal phalanx is significantly closer to the ground.
    - ipsilateral side of the DIP joint space increases.
    - distance from the distal phalanx to the surface of the adjacent wall increases (Fig. 3.13).
  - chronic rotation, other changes include:
    - remodelling (lipping) of the distal phalanx (Fig. 3.14), especially at the tip.

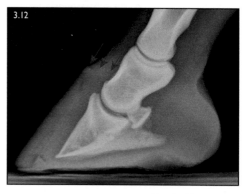

FIG. 3.12 Laminitis. Distal displacement of the distal phalanx demonstrated by decreased depth of sole and a prominent 'sinker line' at the coronary band (arrows).

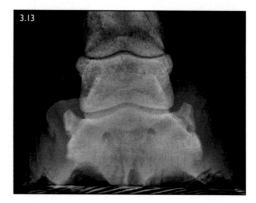

FIG. 3.13 Horizontal dorsopalmar radiograph of a chronic laminitic foot with medial distal displacement of P3.

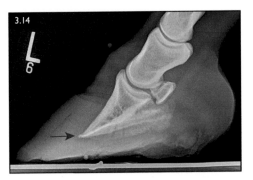

FIG. 3.14 Lateral radiograph of a chronic laminitic foot with subtle bony remodelling of the tip of P3 (arrow). (Photo courtesy Frank Nickels)

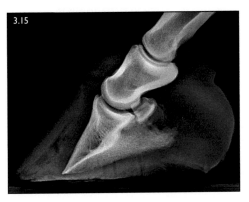

FIG. 3.15 Laminitis. Radiograph after capsular rotation of the distal phalanx. Note the gas line in the dorsal hoof wall.

- ◆ flexural deformity of the DIP joint (also known as phalangeal rotation) (Fig. 3.15).
- ◆ periosteal new bone formation on the dorsal parietal surface of the distal phalanx midway between the extensor process and the dorsal solar margin.
- ◆ 45° dorsoproximal–palmarodistal oblique view is useful to evaluate the dorsal solar margin of the distal phalanx for osteolysis and remodelling.

Differential diagnosis

- acute laminitis is almost unique:
  - ○ must be distinguished from diseases that make a horse move stiffly such as colic, pleuropneumonia, and rhabdomyolysis.

- chronic laminitis must be differentiated from:
  - ○ diseases causing bilateral chronic forelimb lameness:
    - ◆ pedal osteitis
    - ◆ navicular disease   ◆ OA.
  - ○ diseases that cause similar hoof wall defects and displacement of the distal phalanx (e.g. white line disease).

Diagnosis

- acute laminitis – diagnosis is usually straightforward, based on:
  - ○ history   ○ characteristic stilted gait
  - ○ increased digital pulses.
  - ○ heat in the feet
  - ○ withdrawal response to hoof testers.
  - ○ reluctance to lift a limb.
- chronic laminitis – diagnosis can be based on:
  - ○ characteristic concavity of the dorsal hoof wall.
  - ○ laminitic rings: convergence of growth rings dorsally.
  - ○ dropped sole   ○ widened white line.
  - ○ presence of a dorsal coronary band depression.
  - ○ poor horn quality and bruising distal to the solar margin of the distal phalanx, often seen as an arc just in front of the apex of the frog.

Management

- varies with the stage of the disease and can therefore be divided into prophylactic measures, treatment of acute laminitis, and treatment of chronic laminitis.

Prophylactic measures

- removal of the precipitating cause (e.g. taking the horse off lush pasture).
- treatment of primary diseases associated with endotoxaemia:
  - ○ limit systemic absorption and the effects of toxins:
    - ◆ mineral oil and flunixin meglumine (ingestion of excessive amounts of grain).
- endotoxaemia cases should receive appropriate medical therapy including antibiotics, anti-inflammatory drugs, fluids, polymyxin B, and anti-endotoxin hyperimmune serum.

- shoe removal and provision of sole support.
- standing a horse in a slurry of ice and water, during the developmental and early acute stages of the disease:
  - reduce the severity of lamellar inflammation and pathology.
- typically continued until the high-risk period has passed after 2–5 days.

Treatment of acute laminitis

- combination of medical therapy (mainstay) and supportive care:
  - systemic treatment with phenylbutazone (2.2–4.4 mg/kg i/v or p/o q12h):
    - most effective pain control and anti-inflammatory agent early in the disease.
  - Flunixin meglumine – treatment of the primary disease (0.25–1.0 mg/kg i/v q12h).
    - not as effective for analgesia.
  - acepromazine (0.01–0.02 mg/kg i/v or i/m q6–8h), dimethylsulfoxide (0.1–0.2 g/kg i/v or via a nasogastric tube q8–12h), pentoxifylline, isoxsuprine, aspirin, and heparin have all been recommended.
- cryotherapy (standing in ice slurry) in the acute stage of the disease for at least 72 hours.
- supportive therapy to prevent displacement of the distal phalanx:
  - strict stall rest   ○ removing the shoes
  - bedding the stall with peat or sand.
  - packing the ground surface of the feet between the walls, typically with Styrofoam or silicone putty.
  - move the breakover back and elevate the heels by applying a commercial cuff and wedge pad combination (Modified Redden Ultimates, Nanric Inc.).
- stable bandages to control limb oedema.

Treatment of chronic laminitis

- combination of medical therapy and supportive care (mainstay) in the form of corrective shoeing.
- main objectives of corrective shoeing in horses with chronic rotation are:
  - positioning the shoe in relation to the distal phalanx:

- use radiographic control, with specific parameters in mind (Figs. 3.16–3.18).
  - move point of breakover in a palmar direction to decrease the length of the lever arm as the foot breaks over.
  - reduce weight bearing by the most severely affected lamellae:
    - increasing surface contact in palmar aspect of the foot (frog and heels).
  - elevation of the heels decreases tension in:
    - DDFT.
    - dorsal lamellae.
  - preservation of sole thickness.
  - currently, most widely used shoeing techniques for chronic rotation:
    - heart bar shoes.
    - egg bar shoes in conjunction with silicone putty sole support.
    - Aluminum Four Point Rail Shoe/ Rocker shoe (Nanric Inc.) (Fig. 3.19).
    - Equine Digit Support System (Equine Digit Support System, Inc.),
    - wooden shoe (Steward Clog) (Fig. 3.20).
- deep digital flexor tenotomy is occasionally required in:
  - cases showing progressive rotation of the distal phalanx.
  - horses with rotation and thin soles that are displaying persistent pain.
  - cases with secondary flexural deformities of the DIP joint.
- partial hoof wall resections and hoof wall grooving to correct the concavity that occurs in the dorsal hoof wall as the foot grows out.
- drainage of abscesses preferably through the distal wall rather than the sole.
- treatment of horses with distal displacement of the distal phalanx is more problematic:
  - standing horse on a soft deformable surface is best way to improve its comfort.
  - wooden shoes can be helpful but may be insufficient to stop displacement.
  - phalangeal cast with a rounded bottom, for horses that fail to respond.
- horses that sink asymmetrically:

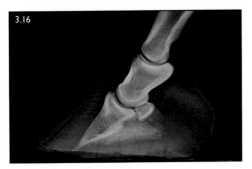

FIG. 3.16 Lateral radiograph showing rotation of the distal phalanx and perforation of the sole.

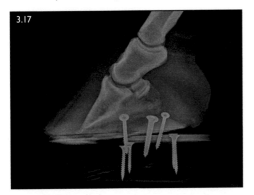

FIG. 3.17 Lateral radiograph of case in 3.16 demonstrating placement of the wooden shoe in relation to the position of the distal phalanx.

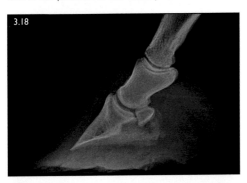

FIG. 3.18 Lateral radiograph of case in 3.16 demonstrating sole growth and significant realignment of the hoof wall with the distal phalanx following a deep digital flexor tenotomy and 5 months' treatment with a wooden shoe.

- o enhance breakover and move weight bearing towards the less affected side of the foot.
  - ◆ extend shoe out further on the unaffected side.
  - ◆ most readily accomplished with a wooden shoe.

FIG. 3.19 A Four Point Rail Shoe formed here by two components of the Equine Digital Support System: an aluminium (or aluminum) Natural Balance Shoe combined with plastic wedges called rails.

FIG. 3.20 A wooden clog shoe.

Prognosis

- varied based on:
  - o severity of the initial disease/symptoms.
  - o type of displacement.
  - o relative stability of the distal phalanx within the hoof capsule at the time of diagnosis.
- severe acute laminitis should always be guarded to poor as the course of the disease is highly unpredictable.
  - o likelihood of survival and return to function is best correlated against the severity of the initial clinical signs.
- chronic laminitis.
  - o best for horses with no rotation.
  - o more guarded with rotation.
  - o worst for horses with symmetrical sinking.
  - o survival is probably best correlated with:
    - ◆ severity of the lameness.
    - ◆ degree of flexion of the DIP joint (phalangeal rotation).
    - ◆ thickness of the sole.

White line disease

Definition/overview

- keratolytic syndrome of the deeper layers of the stratum medium of the hoof wall.

Aetiology/pathophysiology

- exact infectious cause is unknown:
  - both fungi and anaerobic bacteria have been implicated.
- keratolytic process forms cavities within the hoof wall containing air and moist degenerating horn, not associated with inflammation of the underlying tissues.
- loss of lamellar connections in the affected hoof wall may cause the distal phalanx to lose adequate support, resulting in displacement.
- subsequent pressure on the sole may cause bruising and lameness of variable magnitude.

Clinical presentation

- early disease results in a discoloured area or defect in the inner layers of the stratum medium at the white line:
  - often asymptomatic and incidental finding by the farrier.
- lameness may be associated with displacement of the distal phalanx.
  - flattened sole and characteristic discolouration associated with bruising.
- unilateral usually, but multiple-foot cases do occur.

- warm foot (feet).
- application of hoof testers to the sole elicits a withdrawal response.
- occasional bulging of the hoof wall superficial to cavitation in the wall (Fig. 3.21).

Differential diagnosis

- chronic laminitis.
  - characteristic history; deeper defects in the wall; no chalky or moist/waxy horn in defect.
- abscess.
  - more acute history of lameness; no chalky or moist/waxy horn in defect.

Diagnosis

- size and appearance of the defect in the white line not correlated with the degree of hoof wall cavitation.
- horn on the surface and at the margins of the defect is often chalky or moist and waxy.
- exploration of the wall defect with a flexible probe.
- radiography (Fig. 3.22) with signs of:
  - radiolucent areas within the hoof wall.
  - sequence of tangential horizontal views to determine proximal extent of the cavitation.
  - displacement of the distal phalanx is confirmed by:
    - increased thickness of the wall.
    - divergence of the parietal surface of the distal phalanx from the wall.
    - decreased thickness of the sole.

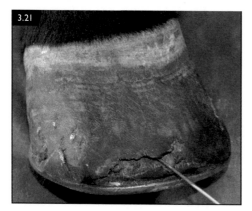

FIG. 3.21 Horse with white line disease at the toe of the hoof with a probe inserted into the defect underneath the hoof wall. (Photo courtesy Dr Stephen O'Grady)

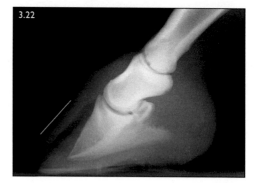

FIG. 3.22 Lateral radiograph with a large radiolucent defect in the inner margin of the dorsal hoof wall indicative of white line disease, and secondary rotation of the distal phalanx.

3.23

FIG. 3.23 White line disease hoof post resection of all overlying hoof wall.

Management

- remove all undermined hoof wall to expose the underlying surface to air to dry (Fig. 3.23).
- stabilisation of the hoof wall if large defects cause the wall on either side of the defect to spread.
- numerous topical medications are used but no evidence of additional benefit to debridement.
- bandaging may be required if it is difficult to keep the debrided surface clean:
  - dry or with a topical astringent such as 2% tincture of iodine or
  - antiseptic such as povidone iodine if germinal epithelium is near the surface of the defect.
- synthetic composite patch (possibly incorporating metronidazole) if return to competition is required.
  - **surface must have been free from evidence of disease for 2 weeks.**
- shoe as for chronic laminitis if displacement of the distal phalanx has occurred.

Prognosis

- complete elimination of the disease may require persistence but should be successful.

- return to full work should be expected in all horses without displacement of the distal phalanx, and in most horses with displacement.

Hoof cracks

Definition/overview

- horizontal or vertical fissures within the hoof capsule.

Aetiology/pathophysiology

- poor mediolateral and/or dorsopalmar hoof balance for vertical cracks.
- sheared heels/quarters for quarter cracks.
- trauma to the coronary band.
- poor quality hoof horn.
- inadequate/infrequent hoof trimming.
- more common in the presence of primary disease of the hoof wall such as laminitis.
- classification by location, depth, and completeness:
  - partial vs. full-thickness cracks.
  - blind hoof cracks start deeper in the hoof capsule.
  - complete (bearing surface to coronary band) vs. incomplete hoof cracks.
  - toe cracks vs. quarter cracks.
- pain and lameness due to:
  - infection or instability of cracks with pinching of the underlying tissues.
    - infection is more common in quarter cracks than toe cracks.
- horizontal cracks are less common than vertical.
  - spread around a variable amount of the circumference of the hoof wall.
  - often sequel of abscesses that drain at the coronary band.
  - rarely caused by selenium toxicity:
    - more severe and more lameness.
    - extend around a greater part of the circumference.

Clinical presentation

- usually present on visual examination.
- some cracks are not significant:
  - regional/intrasynovial analgesia to help differentiate significance.
  - lameness is usually associated with instability or infection.
- hoof testers are useful to assess instability and pain:

  o   increase visibility of purulent or bloody discharges.
  o   may reveal occult cracks especially in the bars.

Differential diagnosis

- none.

Diagnosis

- hoof cracks are usually visibly obvious (Fig. 3.24).
  o   careful inspection for small cracks at the coronary band or cracks in the bars.
- occult hoof cracks at the coronary band may be tentatively identified based on pain on palpation.
- radiographic changes in the distal phalanx possible with chronic cases (Fig. 3.25).

Management

- **not all hoof cracks need treatment.**
- treatment depends on the nature of the crack and the exercise expectations of the horse.
- corrective trimming to restore optimal balance +/- shoeing.
- variable periods of rest from work.
- shoeing to unload the area of the crack (heart bar shoes or silicone sole packing), decreasing movement within the crack and allowing it time to grow out (Figs. 3.26, 3.27).
- crack stabilisation for those horses that must perform soon:
  o   debridement of the crack to remove all foreign or infected material.
  o   stabilisation to prevent compression and expansion of the crack:
    ♦   synthetic polymer patches that incorporate sutures such as Kevlar or steel, with or without sheet metal screws (Figs. 3.28, 3.29).

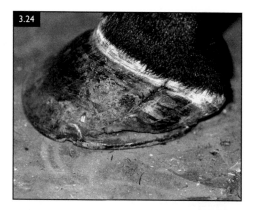

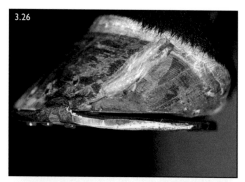

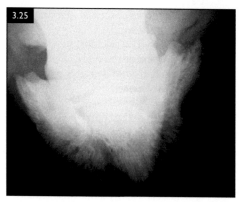

FIGS. 3.24–3.27 Hoof crack. (3.24) Quarter crack. (3.25) Radiograph of the foot showing chronic remodelling of the abaxial margin of the distal phalanx subsequent to a chronic quarter hoof crack. (3.26, 3.27) Quarter crack following debridement, floating of the heel, and shoeing with a heart bar shoe.

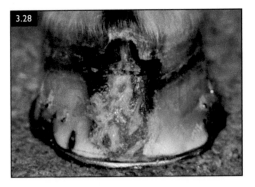

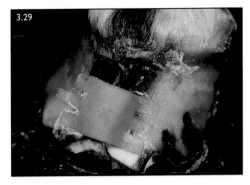

FIGS. 3.28, 3.29 Hoof crack. (3.28) Filling of a toe crack with a thermoplastic compound after the crack has been debrided and thoroughly cleaned. (3.29) Stabilisation of the toe crack in 3.28 by application of a reinforced plastic patch, which is glued to the dorsal wall.

- ○ applying a patch over a drain in infected cracks to allow antiseptic flushing.
  - ○ shoeing for stabilisation.
- cracks from coronary band injury:
  - ○ indefinite management may be necessary.
  - ○ reconstructive coronary band surgical repair to restore continuity of horn growth.
- cracks associated with laminitis may respond to therapeutic shoeing for the primary disease.

Prognosis

- varies depending on location, depth, foot balance and the presence of other foot problems.
- good for elimination of cracks associated with poor conformation, that can be corrected by trimming and shoeing.
- guarded for cracks associated with prior trauma, concurrent disease, or non-correctable foot imbalance or conformation.

Heel bulb lacerations

Definition/overview

- usually form an inverted U-shaped flap (apex of the U proximally).
- distal extent of the wound margins is usually either just proximal to the coronary band or extending for 1–2 cm into the coronary band.

Aetiology/pathophysiology

- acute trauma.
- irregular margins and varying damage to deeper underlying tissues:
  - ○ palmar digital vessels and nerves
  - ○ DIP joint. ○ DDFT and sheath
  - ○ navicular bursa. ○ ungual cartilage.
- prone to excessive granulation tissue formation and slow epithelialisation if untreated:
  - ○ movement between the opposite margins of the wound.
  - ○ prolonged healing time, increased scarring, and possible effect on function.

Clinical presentation

- fresh laceration soon after acute trauma:
  - ○ haemorrhage may be substantial.
- older wound with excessive granulation tissue.
- lameness more likely in:
  - ○ horses with acute wounds or
  - ○ wounds that are associated with injury and infection of deeper structures.

Differential diagnosis

- none.

Diagnosis

- visual inspection.
- careful examination of wound:
  - ○ determine its extent; the involvement of deeper structures; and the severity of contamination (Fig. 3.30).

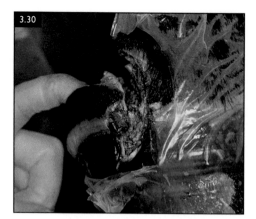

FIG. 3.30 Heel bulb laceration. A severe acute laceration of the heel bulb that has injured the lateral cartilages, coronary band, hoof and heel soft tissues.

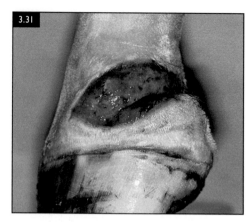

FIG. 3.31 Heel bulb laceration. Chronic heel bulb wound with a healthy granulating surface.

- - assess the wound margins for viability and epithelialisation.
  - assess for the presence of granulation tissue formation (Fig. 3.31).
- biopsy of excessive granulation to exclude the development of a sarcoid or habronemiasis.
- diagnostic lavage of synovial structures to assess synovial penetration.
- diagnostic imaging to assess for fractures or synovial penetration (contrast).

Management

- acute injuries with haemorrhage:
  - ligate affected vessel(s) or apply a pressure bandage before further treatment.
  - debridement.
  - synovial lavage of deeper structures if contaminated or infected.
  - surgical closure, followed by immobilisation with a phalangeal/digital cast for three weeks.
- acute wounds with substantial contamination:
  - debridement.
  - open wound management with topical antimicrobials, wound lavage and bandaging, until healthy granulation tissue is present across the wound.
  - delayed secondary suture closure with removal of granulation tissue.
  - immobilisation in a phalangeal cast.

- chronic heel bulb lacerations (Figs. 3.32–3.35):
  - open wound management until the surface is healthy.
  - may require considerable excision of granulation tissue.
  - delayed complete secondary closure or partial closure with partial healing by secondary intention if wound cannot be closed completely.
  - immobilisation in phalangeal cast:
    - ♦ more than 3 weeks if healing by secondary intention.
- heel bulb lacerations with traumatic coronary band defect:
  - coronary band suture (see hoof wall avulsions below).
- all cases require tetanus prophylaxis, systemic NSAIDs, analgesics, and perioperative antibiotics.

Prognosis

- return to athletic activity is good with variable residual scarring.
- wounds with involvement of deeper structures vary dependent on:
  - which structures are involved and the extent of the damage.
  - presence of infection.

Hoof wall avulsions

Definition/overview

- segment of the hoof wall becomes separated from the underlying tissues.

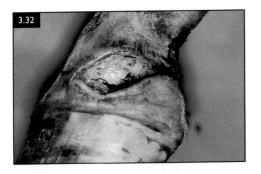

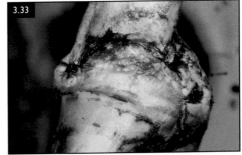

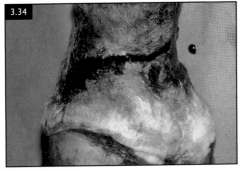

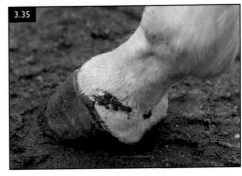

FIGS. 3.32–3.35 Heel bulb laceration. (3.32) Laceration after excision of excess fibrogranulomatous tissue. (3.33) The margins of the laceration are sutured together immediately prior to application of a phalangeal cast. (3.34) The laceration 3 weeks after closure and immediately following cast removal. (3.35) The residual scar after an additional 3–4 weeks.

Aetiology/pathophysiology

- trauma to the wall that causes a segment to fracture and become elevated:
  - complete avulsion when the segment is completely detached.
  - partial avulsion when the segment is still attached along one border.
  - usually in a distal to proximal direction.
- extent of avulsions:
  - confined to the hoof capsule or involving the coronary band and the skin of the pastern.
  - may include any of the underlying structures.
- pattern of healing depends on the depth of the trauma:
  - full-thickness avulsions:
    - extend deep to the basement membrane of the epidermis.
    - heal as full-thickness wounds.
    - epithelialisation from the adjacent margin.
      - nature of new hoof wall varies.

  - surgically created avulsion (hoof wall resection):
    - usually leave sufficient epidermal structures (tips of the epidermal lamellae are broken off and left on the surface).
    - heal as partial thickness wounds.
    - epithelialisation from the epidermal lamellar tips.

Clinical presentation

- acute injuries with varying degrees of lameness and haemorrhage.
- chronic injuries with a granulating surface, plus or minus complications associated with deeper structure involvement.
- healed injuries may leave residual altered function of the hoof and lameness.

Differential diagnosis

- none.

Diagnosis

- assessment of coronary tissues and adjacent skin for viability. (Fig. 3.36).
- careful exploration of deeper structures for involvement:
  - injury to ligamentous or tendinous structures: lameness due to instability.
  - communication with synovial structures: arthrocentesis and sterile lavage.
  - fractures or pedal osteitis (chronic cases): radiography.

Management

- complete avulsions:
  - healing by secondary intention.
  - cleaning and debridement.
  - topical antimicrobials and dressings under bandage.
  - systemic antibiotics until the wound surfaces are granulating.
  - analgesia given as needed, and tetanus prophylaxis provided.
  - foot cast or therapeutic shoe if foot unstable due to amount of wall lost.
- partial hoof wall avulsions in which the coronary tissues are present and viable:
  - resection of hoof wall 1–2 cm distal to the coronary band.
  - reconstruction of the coronary band (Fig. 3.37).
    - drilling holes in the hoof wall either side and suturing with wire.
    - regular suture material can be used if the outer layers of the hoof capsule have been removed first.
  - cast application over the reconstruction for additional stability (Fig. 3.38).

- partial avulsions that are deemed unviable are converted to complete avulsions.
- affected joints and tendons are treated accordingly.
- therapeutic shoeing to support the injured site of the foot following either complete or partial healing of an avulsion.

Prognosis

- depends on the degree of injury to the coronary band and underlying structures:
  - coronary band intact or successfully repaired: good functional prognosis.
  - loss of coronary band tissue: mostly permanent hoof wall defect that requires support with therapeutic shoeing.
  - injury to the deeper synovial or tendinous structures: poorer prognosis.

Keratoma

Definition/overview

- epithelial tumours of the hoof.

Aetiology/pathophysiology

- aetiology is unknown but trauma to the hoof has been associated with keratomas in some horses.
- expansile mass in the wall or the sole usually deep to the hoof capsule (Fig. 3.39).
  - originates from the germinal layers of the epithelium of the hoof.
- frequently associated with recurrent secondary infection of the white line.

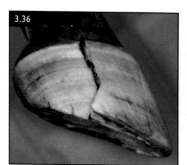

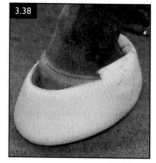

FIGS. 3.36–3.38 Hoof wall avulsion. (3.36) Hoof wall avulsion that is still attached dorsally. (3.37) Resection of the distal portion of the avulsed hoof wall and apposition of the proximal margin of the avulsed wall with a wire suture. (3.38) Stabilisation of the hoof wall with a rim cast.

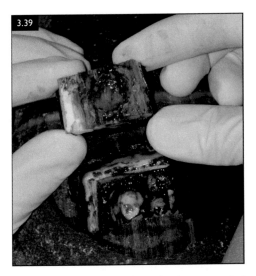

FIG. 3.39 Intraoperative view of a large rounded keratoma immediately underneath the dorsal hoof wall (window resection). Photo courtesy Dr John Peroni).

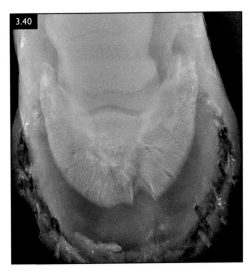

FIG. 3.40 Dorsoproximal–palmarodistal oblique radiographic view of the foot of a horse with a keratoma at the toe of P3. Note the defect in the hoof wall immediately dorsal to the bony lesion.

Clinical presentation

- lameness accompanied by distortion of the hoof capsule.
- recurrent foot abscesses.
- occasional incidental imaging finding without lameness.

Differential diagnosis

- other rare hoof masses and tumours.
- fibrous hoof wall scar from healed white line disease.
- chronic hoof wounds and abscesses.

Diagnosis

- lameness associated with a distorted hoof capsule and infection.
- relatively well-circumscribed round or oval mass of horn on the ground surface of the foot.
  - either in the white line of the distal wall or in the sole.
- well-demarcated circular or oval area of pressure lysis in the solar margin/parietal surface of the distal phalanx on dorsoproximal–palmarodistal oblique radiograph (Fig. 3.40).
- CT or MR imaging to detect occult hoof wall masses and plan a surgical approach.
- definitive diagnosis requires a biopsy and histopathological evaluation.

Management

- no clinical signs of lameness or infection: leave untreated but monitor.
- clinical signs: surgical excision is recommended:
  - through the sole or a partial hoof wall resection.
  - standing horse with sedation, perineural analgesia and a tourniquet.
  - general anaesthesia in lateral recumbency.
  - routine postoperative wound care to allow healing by secondary intention.
  - perioperative antibiotics if the dermis has been disrupted as evidenced by haemorrhage into the resected area.

Prognosis

- good for surgical excision with approximately 80% return to previous athletic performance.

Foot conformation and hoof imbalance as a cause of lameness

Definition/overview

- Balance describes the relationship between the hoof, ground, and the rest of the limb:

- lateromedial and dorsopalmar components.
- dynamic and static components.
- determined by:
  - genetics and development.
  - hoof growth ◆ wear on the foot
  - trimming and shoeing.
  - nature of the horse's work.
- Conformation describes the shape of the limb (and rest of the body) proximal to the coronary band:
  - dictated by the shape and size of the individual structures of the limb.
    - way they relate to each other.
  - determined by genetics and development.
  - established by the time of skeletal maturity.
  - limb conformation should not be altered by altering the distribution of stresses within the hoof in adults as it is likely to cause injury rather than benefit.
- poor conformation and imbalance:
  - changes in the shape of the digit from the normal, near-symmetrical form and in the way the hoof relates to the ground.
  - can cause changes of stresses and injury to structures in the limb.
    - can result in lameness.

Aetiology/pathophysiology

- normal pattern of hoof growth is uniform around the circumference of the limb, the coronary band is symmetrical about the sagittal plane of the limb, and the walls are straight.
- abnormal distribution of stresses on a 'normal' hoof capsule leads to:
  - movement of the coronary band, changes in hoof growth, and distortion of the capsule.
- abnormal conformation and poor trimming or shoeing will cause the redistribution of stresses.
- hoof growth is inversely related to load:
  - greater the load on any given area of the hoof, the slower hoof growth and vice versa.
  - foot attempts to re-establish normal load distribution:
    - leads to changes in shape and relationship to pastern causing an imbalance.

- imbalance is caused by:
  - poor foot trimming and shoe placement/selection in shod horses.
  - inadequate foot care or poor conformation in barefooted horses.
- imbalance may cause lameness by distortion of the hoof capsule and injury to other structures in the limb due to stress redistribution.
- types of imbalance are often discussed separately but in reality they are interrelated:

1. **Dorsopalmar imbalance:**
   - broken back foot–pastern axis is associated with:
     - navicular disease.
     - distal DDFT strain.
     - heel bruising and haemorrhage in the dorsal white line.
     - pedal osteitis of the palmar processes.
   - broken-forward foot–pastern axis (Fig. 3.41) is associated with:
     - dorsal sole haemorrhage and pedal osteitis.
2. **Mediolateral imbalance** (Fig. 3.42) is associated with:
   - sheared heels.
   - quarter or heel cracks.
   - Sidebone.
   - fracture of the palmar processes.
   - asymmetrical bruising and pedal osteitis.
3. **Contracted heels:**
   - abnormal proximity of the heel bulbs and heel buttresses.
   - associated with a narrow frog, and frequently high or underrun heels.
   - secondary to:
     - heel pain.
     - reduced weight bearing by the heels.
     - subsequent reduced expansion of the heels during normal cycle of the stride.
4. **Underrun heels:**
   - heels more acutely angled to the ground and longer than normal.
   - usually occurs in horses with excessive length of toe.
   - weight-bearing surface of the heels displaced dorsal to base of the frog.
   - foot–pastern axis usually broken back.

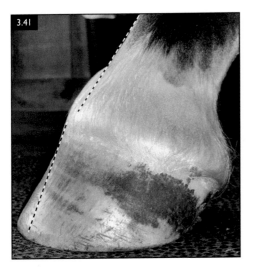

FIG. 3.41 A broken-forward foot–pastern axis with lines drawn on the photograph to accentuate the broken axis.

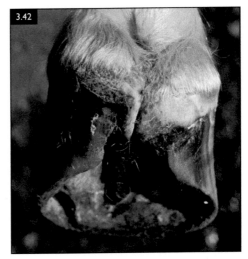

FIG. 3.42 Hoof imbalance. Mediolateral foot imbalance with asymmetry of the heel bulbs and walls of the hoof, sheared and contracted heels, and uneven bearing surface of the wall.

- some breed predispositions, particularly in Thoroughbreds.
- caused by poor trimming leaving excessive length of heel or toe length.
- caused by shoes resulting in excessive weight bearing on the heels, encouraging them to become underrun.

5. **Upright foot with high heels:**
   - usually accompanies either contracted heels or a flexural deformity of the DIP joint.
   - frequently associated with a broken-forward foot–pastern axis.
   - contralateral foot often has low/underrun heels.

6. **Sheared heels:**
   - one heel bulb is displaced proximally compared to the other.
   - proximal displacement of the coronary band frequently begins in the quarter.
   - often associated with contracted heels.
   - secondary to inappropriate trimming prior to shoeing.
   - may cause palmar foot pain and lameness.

Clinical presentation

- hoof imbalance may cause lameness in many horses.
- can be an incidental finding without lameness.

Differential diagnosis

- any chronic lameness originating within the foot.
- healed traumatic hoof wounds.

Diagnosis

- careful visual inspection and measurement (Fig. 3.43) may reveal changes that reflect:
  - movement of the coronary band.
  - distortion of the hoof wall.
  - change in the spacing of the growth rings.

FIG. 3.43 The view from the front of a foot with chronic mediolateral imbalance. Note the sloping coronary band, asymmetrical walls and obvious flaring of the medial wall.

- **examination with the foot both on and off the ground.**
  - loaded examination:
    - ground surface of the foot should be approximately perpendicular to the axis of the limb.
    - observe length and angles of the wall at the heel and toe.
    - approximate symmetry of the medial and lateral walls of the hoof.
      - presence of underrun and flared walls.
  - unloaded examination:
    - from palmar/plantar aspect to better determine the relationship of the foot to the rest of the limb.
    - mediolateral symmetry of the ground surface of the foot.
    - length and width of hoof (usually approximately equal) (Figs. 3.44, 3.45).
- **foot balance radiographs:**
  - radiographs can help determine the relationship between the hoof capsule and the phalanges.
    - even weight bearing on both front feet.
    - metacarpus/metatarsus vertical (stand on blocks of equal height).
  - horizontal dorsopalmar view:
    - interphalangeal joint spaces should be symmetrical and usually parallel to the ground.

- slight tilting of the distal phalanx towards one side of the foot, but with a parallel joint space.
  - most likely to be a normal variation for that horse.
- uneven spacing of the DIP joint is rare in horses with chronic mediolateral imbalance.
  - lateromedial view:
    - solar margin of the distal phalanx should form an angle of 2–10° to the ground (3–5° is more common).
    - subtle misalignment of the foot–pastern is hard to evaluate because it changes as the horse's weight shifts.
    - measure the proportion of the foot in front of the centre of rotation of the DIP joint vs. that palmar to it:
      - normal = 60% of the ground surface dorsal to the centre of rotation.
      - flexural deformity of the DIP joint = 50%.
      - long toe–low heel imbalance = 70% or more.
- **Lameness:**
  - rarely caused by imbalance *per se:*
    - usually caused by a subsequent pathological process in one or more structures other than the hoof capsule.
  - imbalance is a frequent incidental finding in horses without lameness.

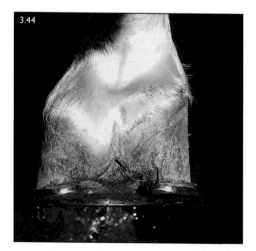

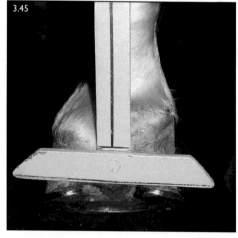

FIGS. 3.44, 3.45 Hoof imbalance. (3.44) Assessing the mediolateral balance of the foot with a shoe in place. Note the asymmetry of the walls and bulbs of the heel. (3.45) Same foot as in 3.44 with a T square in place demonstrating the asymmetry of the bearing surface (long on the lateral side [left side of picture]).

Management

- therapeutic hoof trimming and shoeing for imbalance:
  - benefits can be immediate or take several months.
  - serial radiographs help to monitor changes in the relationship between the hoof capsule and the phalanges.
  - limb conformation cannot be corrected in the adult horse.
- specific guidelines are limited and poorly documented:
  - contracted heels secondary to pain:
    - no effective treatment unless the pain is removed.
  - contracted heels accompanied by a long toe:
    - shortening toe encourages the entire foot, including the heels, to expand.
  - underrun heels with long toe:
    - rasp heels back to base of the frog to encourage the heels to grow wider (Figs. 3.46–3.51)
    - decrease toe length.
  - sheared heels:
    - shorten long heel/quarter so both heels are level with the base of the frog.
    - bar shoe to minimise movement between the heels.

Prognosis

- acute imbalance associated with secondary pathology: good.
- chronic imbalance, especially if associated with degenerative conditions: guarded to poor.

Trimming and shoeing as a cause of lameness

Overview

- poor trimming may cause a foot imbalance:
  - trimming the medial and lateral walls unevenly creates a mediolateral imbalance.
  - inappropriate trimming of the heel and/or toe may create a broken foot–pastern axis:

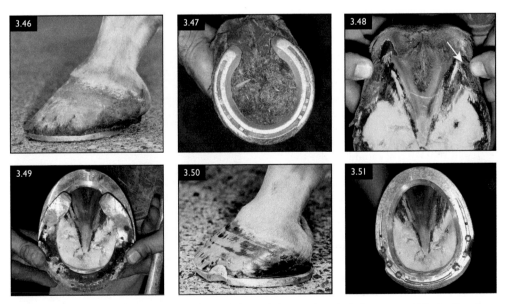

FIGS. 3.46–3.51 Hoof imbalance. (3.46) Lateral view of dorsopalmar imbalance associated with a long toe, low heels, and a broken back foot–pastern axis (3.47). Solar view indicating position of the shoe in relation to the ground surface of the foot. (3.48) The left heel (arrow) has been trimmed back to the base of the frog. Note the difference in length between the trimmed and untrimmed heels. (3.49) The old shoe is superimposed on the new shoe to show the difference in position of breakover and coverage of the heels. (3.50) Lateral view of the foot after shoeing; note the improvement in the foot –pastern angle. (3.51) Solar view showing the position of the new shoe.

- ◆ leaving the toe too long creates a dorsopalmar imbalance.
  - – broken back foot–pastern axis.
  - ◆ leaving the heel too long may create a broken-forward axis.
  - ○ trimming the sole too thin causes undue pressure on the underlying sensitive structures:
    - ◆ predisposes to bruising.
- attaching steel shoes with nails to the foot affects the foot's biomechanics:
  - ○ decreases expansion of the foot.
  - ○ increases maximum deceleration of the foot.
  - ○ increases frequency of vibrations as the foot contacts the ground.
  - ○ increases the maximum ground reaction force.
  - ○ attaching the shoe and the shoe selected can cause lameness to develop.

Attachment of the shoe

- incorrect nail placement can damage the sensitive tissues immediately underlying the hoof capsule, causing immediate or delayed lameness (nail prick or nail bind):
  - ○ localisation of the pain to a specific nail with hoof testers.
  - ○ visible incorrect placement after shoe removal.
- nail which impinges on the sensitive laminae (nail bind) may after several days:
  - ○ cause pain secondary to pressure on the laminae localised by hoof testers.
  - ○ cause microfracture of the hoof wall internally, with subsequent introduction of infection (Fig. 3.52).
- removal of the nail should resolve the problem:
  - ○ nail hole is then infused with an antiseptic.

Shoe selection

- poor shoe choice may predispose to lameness.
- shoe that is too large, too wide, or too long: more likely to be pulled off.
- shoe too small or short:
  - ○ increased pressure at the heels or angle of the sole.
  - ○ predisposes to bruising, hoof cracks, corns and underrun heels.
- shoe with a web that is too wide: pressure on a dropped sole.
- shoe with a web that is too narrow: poor protection of adjacent sole.
- shoe too heavy: fatigue, which may predispose to injury.

Shoe placement

- normal placement is symmetrical about the axis of the frog.
- asymmetrical placement causes greater stress on the lateral or medial wall (Fig. 3.53).
- unintentional rotated shoe: one heel more covered than the other:
  - ○ least covered heel predisposed to bruising, hoof cracks, corns and underrun heels.

Inappropriate use of traction

- traction devices such as toe grabs or calks are widely used to increase a horse's speed or prevent a horse from slipping (Fig. 3.54).
  - ○ concentrate stress in the adjacent wall.
  - ○ unilateral devices induce an imbalance in the limb if the horse is standing on a firm surface.

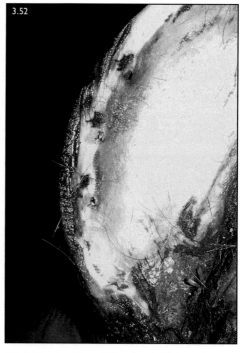

FIG. 3.52 A foot in which several of the nails have been placed too close to the sensitive laminae (nail bind).

○ place undue torque on the limb during rotation and predispose to phalangeal fractures.
• use removable traction devices, use them only when deemed necessary, and use the lowest, broadest devices possible.

FIG. 3.54 Racing plate with toe grab and heel stickers.

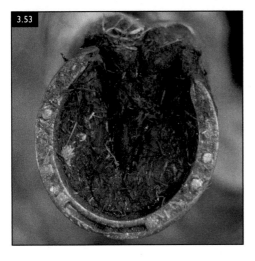

FIG. 3.53 This shoe has been placed asymmetrically, leading to foot imbalance and an abnormal landing position.

DISEASES OF BONES AND JOINTS

Fractures of the distal phalanx

Definition/overview

• classification of distal phalanx fractures (Figs. 3.55–3.57):
  ○ fractures of the body of the distal phalanx: articular or non-articular.
  ○ fractures of the wings (palmar/plantar processes): articular or non-articular.
  ○ fractures of the extensor process.
  ○ fractures of the solar margin.

Aetiology/pathophysiology

• trauma or exercise on hard or uneven ground.
• common racing injury in North America, reflecting counterclockwise racing/training:
  ○ lateral aspect of the left front foot.
  ○ medial aspect of the right front foot.
• pathological fractures following:
  ○ penetration injuries with septic osteitis.
  ○ osteolysis from large keratomas or cysts.
• extensor process fractures can be difficult to distinguish from:
  ○ separate centre of ossification and osteochondrosis fragment of the extensor process.
  ○ both may be incidental radiological findings.
• solar margin fractures:
  ○ abnormal stress in distal phalanx of horses with chronic laminitis.
  ○ in foals, originating and ending at the solar margins:
    ♦ non-articular palmar or plantar process fractures.
• non-displaced incomplete or complete stress fractures of a palmar process at the base of an ossified ungual process (sidebone).
• distal phalanx fractures heal slowly and, in some cases, only by fibrous union.

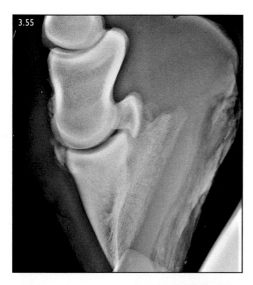

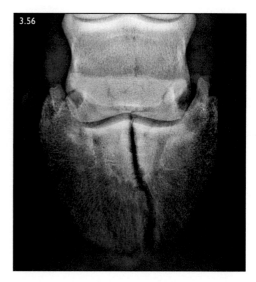

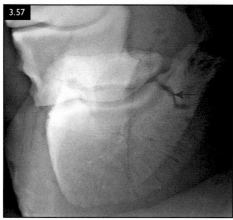

FIGS. 3.55–3.57 Distal phalanx fractures. (3.55) Fracture of the extensor process. (3.56) Mid-sagittal articular fracture. (3.57) Non-articular fracture of the palmar process (arrow).

Clinical presentation

- severe acute lameness.
- long-term fractures may cause chronic lameness of moderate severity.
- articular fractures tend to cause more severe clinical signs:
  - may develop OA and chronic lameness in the longer term.

Differential diagnosis

- any acute foot lameness such as:
  - foot abscesses   ○ puncture wounds
  - navicular bone fracture
  - acute strain or sprain of tendons or ligaments
- separate centre of ossification of the extensor process.

Diagnosis

- acute fractures characterised by:
  - heat in the foot
  - increased digital pulses
  - hoof tester pain
- chronic fracture characteristics:
  - distortion of the coronary band and deformation of the dorsal hoof wall (buttress foot) in chronic extensor process fractures.
  - OA of the DIP joint in chronic intra-articular fractures.
    - ♦ positive distal flexion and distended dorsal pouch of joint.
  - chronic lameness in some types of body/wing fractures healed by fibrous union.

- Radiography:
  - standard projections (dorsopalmar, 45°
    dorsoproximal–palmarodistal oblique,
    lateromedial, and 60° dorsoproximal–
    palmarodistal oblique).
    - additional 30–45° medial or lateral
      oblique views may be required.
  - acute non-displaced fractures may be
    difficult to identify radiographically:
    - repeat radiographs after 7–10 days
      may help identify these fractures.
    - early osseous resorption of the
      fracture margins.
  - OA in chronic fractures.
- regional and intra-articular analgesia
  techniques may be helpful in localising
  pain to specific parts of the foot.
- Scintigraphy, CT or MRI:
  - demonstrate occult fractures.
  - significance of chronic fractures that
    have healed by fibrous union.
  - significance of solar margin fractures.
  - MRI is essential to diagnose non-
    displaced palmar process stress
    fractures (Fig. 3.58).

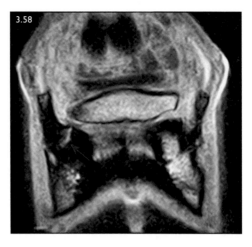

FIG. 3.58 T1-weighted MR image of an incomplete
palmar process stress fracture (arrow).

Management

- acute fractures of the body or the wings:
  - limit movement of the hoof capsule
    to reinforce the natural splinting it
    provides:
    - rim cast, that incorporates the heel
      bulbs, around the perimeter of the
      foot.
    - bar shoe with a continuous rim that
      extends 1–1.5 cm proximally.
    - bar shoes with quarter clips are
      frequently used but are not as
      effective.
  - over a prolonged period may lead to
    marked hoof contraction.
    - may be used during return to
      exercise to decrease recurrence.
- acute articular fractures in the median
  plane of the body in adult horses:
  - lag screw fixation.
  - shorter healing time (3–6 months).
  - complications possible including:
    - infection.
    - screw irritation.
    - step-formation at the articular
      surface.
- extensor process fractures:
  - small fractures:

- can be incidental without lameness.
- can be removed arthroscopically.
  - larger fractures may be reduced by
    internal fixation with a lag screw.
    - fragments up to 20% of distal
      phalanx have been removed
      successfully.
- Solar margin fractures:
  - not associated with infection,
    treatment with rest and NSAIDs.
  - in foals less than 6 months of age, stall
    rest or small round pen turn out allows
    spontaneous healing.
  - fractures associated with infection are
    treated in the same manner as septic
    osteitis.

Prognosis

- extensor process fragments:
  - good after arthroscopic removal of
    small fragments.
  - fair for removal of large fragments.
  - fair to guarded after internal fixation
    of large fracture fragments.
- solar margin and non-articular fractures
  of the wing of the distal phalanx:
  - good.
- articular fractures of the wing/body of
  the distal phalanx:
  - overall, 50% return to soundness after
    6–12 months if treated with rim shoes.
    - good in horses less than 3 years old
      with rim shoes.

- fair to guarded for older non-racing horses with articular fractures treated with rim shoes.
  - ♦ guarded to poor in racehorses.
  - o fair for older horses treated with a lag screw.
- secondary OA is a poor outcome factor and requires treatment:
  - o intra-articular corticosteroids.

Pedal osteitis

Definition/overview

- anatomy of the distal phalanx is variable between horses:
  - o particularly size and number of radiating vascular channels perforating the solar margin.
  - o wide variability makes it difficult to distinguish normal anatomical variation from perceived enlargement of these vascular channels as a sign of suspected generalised inflammation of the distal phalanx.
- two forms of pedal osteitis are described:
  - o **primary form** is a radiographic description of a pattern of generalised resorption of bone around the solar margin of the distal phalanx with assumed enlargement of the radiating vascular channels of the bone.
  - o **secondary form** is any reaction of the bone (bone loss, bone proliferation, bone fragmentation, or bone sequestration) in response to a primary insult.

Aetiology/pathophysiology

- Generalised (primary) pedal osteitis:
  - o certain inflammatory processes or injuries that induce inflammation:
    - ♦ e.g. laminitis and hoof wall avulsions.
  - o recurrent bruising in horses with flat soles and underrun heels.
- Localised pedal osteitis can occur secondary to:
  - o blunt external trauma
  - o penetrating injuries with or without infection.
  - o solar bruising   o abscesses
  - o lamellar inflammation.
  - o keratomas   o surgical implants.

Clinical presentation

- most commonly mild to moderate lameness in both forelimbs.
  - o lameness worse on hard ground.
  - o abolished by a palmar digital nerve block.
  - o both intra-articular and intra-bursal anaesthesia may desensitise the dorsal half of the sole.
- increased sensitivity to the application of hoof testers.
- common incidental radiographic finding in horses with foot lameness.

Differential diagnosis

- other causes of mild to moderate foot lameness:
  - o podotrochlear syndrome   o laminitis
  - o bruising   o imbalance.

Diagnosis

- careful assessment of shape of the sole and response to hoof tester application.
- **Radiography:**
  - o best assessed on lateromedial and 45° dorsoproximal/palmarodistal oblique radiographic projections (Fig. 3.59).
  - o variable demineralisation with irregular contour of the solar margin.
  - o vascular channels fanning out towards the periphery of the bone.
  - o lipping of the dorsal solar margin.
  - o new bone production on the parietal surface of the distal phalanx.
  - o palmar process changes:
    - ♦ scalloping of the solar margin.
    - ♦ remodelling of the process itself.
  - o signs of localised osteolysis, sclerosis, bone fragmentation, bone sequestration and new bone formation of the distal phalanx following treatment of:
    - ♦ penetrating injuries, subsolar bruising, abscesses, laminitis.
    - ♦ keratomas and implant surgery.
- **Scintigraphy:** increased radiopharmaceutical uptake in the solar margin.
- **MRI:** abnormal osseous fluid and an irregular bony surface, especially in one or both palmar processes (Fig. 3.60).

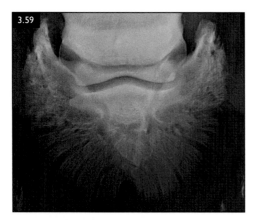

FIG. 3.59 Pedal osteitis. Radiograph demonstrating demineralisation and irregularity of the solar margin of the distal phalanx and widening of the vascular channels.

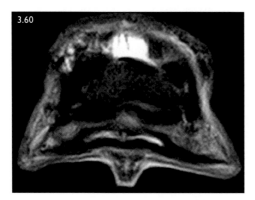

FIG. 3.60 Dorsal fat-suppressed MR image of the foot. There is marked generalised hyperintense signal in the spongiosa of the distal phalanx indicative of bone oedema associated with severe pedal osteitis.

Management
- treatment of the primary cause of secondary osteitis.
- symptomatic treatment in the absence of obvious primary disease:
  - pain control with non-steroidal analgesics.
  - avoid working on hard surfaces.
  - corrective trimming and shoeing:
    - seated-out, wide-webbed shoes and/or pads diminish concussion on the sole.

- indefinite use may be necessary in flat-footed horses with chronic foot soreness and chronic remodelling of the distal phalanx to minimise concussion.

Prognosis
- related to the primary condition.
- chronic lameness without primary cause and poor foot conformation/balance, then guarded.

Septic osteitis of the distal phalanx

Definition/overview
- infection of the distal phalanx.

Aetiology/pathophysiology
- puncture wound to the sole.
- extension of an abscess.
- laminitis as a sequela to infection of the lamellae and sole.
- sequestration possible as the result of infection of a fragment of dead bone.
- pathological fractures possible if a large part of the distal phalanx is affected.
- rarely, infection can spread to adjacent structures (joint, bursa, DDFT).

Clinical presentation
- history of trauma to the foot or other foot lameness (e.g. abscess) is common.
- lameness varies in severity as a function of the opportunity for drainage.
  - severe if poor and mild to moderate if good.
- draining wound may be the primary presenting symptom.

Differential diagnosis
- severe foot lameness:
  - abscess; fractured distal phalanx or navicular bone; other deep digital sepsis.
- drainage from the foot: abscess; other deep digital sepsis.

Diagnosis
- wound and/or drainage from the foot.
- positive to hoof testers.

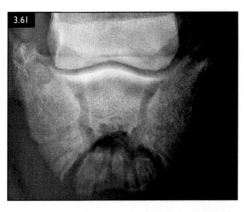

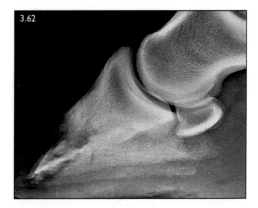

FIGS. 3.61, 3.62 Septic pedal osteitis. Dorsoproximal–palmarodistal oblique (3.61) and lateromedial (3.62) radiographs indicating the presence of septic pedal osteitis and a sequestrum at the dorsal solar margin of the distal phalanx.

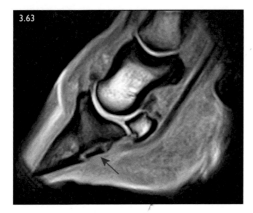

FIG. 3.63 T1-weighted MR image of a sequestrum of the distal phalanx adjacent to the insertion of the DDFT (arrow) following a puncture wound to the frog 2 weeks previously. There is marked and generalised osseous fluid signal in the distal phalanx.

- radiographic signs: (Figs. 3.61, 3.62):
  - o irregularly marginated areas of lysis on the solar margin of the distal phalanx.
  - o occasionally, radiolucent cavities within the substance of the distal phalanx.
  - o sequestrum forms as an osseous radiodensity surrounded by an area of lysis.
- CT and MRI
  - o good for visualising fragments that may not be observed on plain radiographs.
  - o delineate early sequestrum formation.

- o detect the presence of bone oedema in septic osteitis (MRI).
  - o identify tracts in the soft tissues of the foot (MRI) (Fig. 3.63).
- surgical evaluation of a persistent drainage tract in the foot.

Management

- open surgical drainage and debridement under local or general anaesthesia.
- postoperative bandaging with a topical antimicrobial dressing.
- broad-spectrum antibiotics until all exposed surfaces are covered with granulation tissue.
- NSAIDs and tetanus prophylaxis.
- wound healing by secondary intention.

Prognosis

- good.
- negative outcome parameters:
  - o extensive loss of bone.
  - o extension of the infection to the navicular bursa, DIP joint or DDFT.
  - o pathological fracture of the distal phalanx.

Bone contusions/ bruising of the distal and middle phalanges

Definition/overview

- MRI diagnosis characterised by the presence of osseous fluid signal in bone (oedema or haemorrhage or both).

mostly in the region of the:
  - palmar processes of the distal phalanx.
  - dorsodistal aspect of the middle phalanx adjacent to the dorsal half of the DIP joint.

Aetiology/pathophysiology

- impact trauma, either of a monotonic or chronic repetitive nature.
- in the distal phalanx, kicking of a wall or a stall door.

Clinical presentation

- variable lameness from insidious low-grade to acute and severe.
- depending on the site of pathology removal by regional analgesia of the foot.
- dorsodistal bone contusion of the middle phalanx can sometimes be observed in the absence of lameness.

Differential diagnosis

- all other causes of foot lameness without radiographic abnormalities.

Diagnosis

- intraosseous signal hyperintensity on fat-suppressed MR images (Fig. 3.64).
- often associated increase in radiopharmaceutical uptake on scintigraphic images.

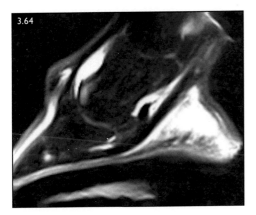

FIG. 3.64 Fat-suppressed MR image of a bone contusion of the middle phalanx, characterised by the presence of an area of high signal intensity near the dorsodistal margin of the DIP joint (arrow).

Management

- stall rest.
- administration of NSAIDs.
- potentially administration of bisphosphonates to limit bone resorption and bone inflammation/pain.

Prognosis

- excellent if no extension to the DIP joint.
- resolution of bone fluid on MR images may lag resolution of lameness:
  - up to 6 months to soundness.

Osseous cyst-like lesions of the distal phalanx

Definition/overview

- may affect horses from a variety of breeds and all ages.
- location varies from the:
  - extensor process to palmar border of the articular surface of the distal phalanx.
  - most in the central weight-bearing portion of the distal phalanx.

Aetiology/pathophysiology

- traumatic or developmental in origin.
- most osseous cyst-like lesions communicate with the DIP joint.

Clinical presentation

- variable lameness from intermittent, insidious low-grade to acute, severe and unilateral.
  - tends to improve with rest but recurs with exercise.
  - improved with palmar digital nerve block or intra-articular anaesthesia.
- occasionally an incidental radiographic finding during pre-purchase examination.

Differential diagnosis

- all other causes of foot lameness.

Diagnosis

- **Radiography** – standard views:
  - centrally located cysts best seen on dorsopalmar or 45° dorsoproximal–palmarodistal oblique views (Fig. 3.65).

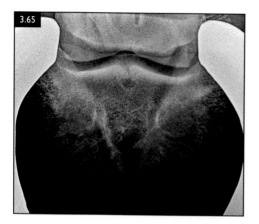

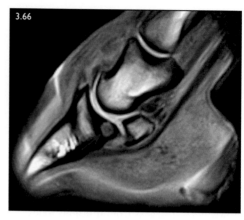

FIG. 3.65 Subchondral cyst-like lesion. Radiograph demonstrating a circular area of demineralisation of the subchondral bone of the distal phalanx immediately distal to the extensor process. A narrow communication with the DIP joint can be observed.

FIG. 3.66 T1-weighted MR image of an osseous cyst-like lesion in the distal phalanx (arrow) that was not visible on radiographs.

- extensor process cysts best seen on lateromedial projections.
- small subchondral cyst-like lesions at the palmar border of the distal phalanx not visible on radiographs.
- CT or MRI for diagnosis of occult cysts (Fig. 3.66).
- Scintigraphy to identify those cysts with increased radionuclide uptake.

Management

- conservative management with:
  - stall rest and anti-inflammatory intra-articular medication.
  - successful in a small percentage of cases.
- injection of the cyst cavity with corticosteroids under arthroscopic or radiographic control.
- surgical debridement of cyst-like lesions if persistent lameness:
  - either arthroscopically or extra-articularly.
- transcortical implantation of osteogenic hydroxyapatite and poly L-lactic acid (Osteotrans).

Prognosis

- good for centrally located cyst-like lesions in young horses that can be debrided arthroscopically.

- guarded for all other lesions that are associated with lameness.

Sidebone

Definition/overview

- ossification of the ungual (collateral) cartilages.
- gradation 1–5 based on the height of ossification relative to the DIP joint.

Aetiology/pathophysiology

- precise aetiology is unknown:
  - concussion in heavy, large horses.
  - more common in older horses.
  - uneven loading of the heels in imbalanced feet.
- progressive ossification:
  - usually occurs in distal to proximal direction from the attachment to the palmar process.
  - centripetal ossification may occur concurrently from a separate proximal centre.
  - persistent non-union between both centres of ossification may appear like a chronic fracture.
- associated with increased incidence of collateral ligament desmitis of the DIP joint.

- moderate to severe ossification predisposes to stress-induced osseous trauma and stress fracture of a palmar process at the base of ossified cartilage (see collateral ligament injury and fractures of the distal phalanx).

Clinical presentation

- mostly an incidental finding not associated with lameness.
- lameness is rare but may be associated with:
  - external trauma with injury of the sidebone.
  - traumatic fracture of the sidebone.
  - stress injury or fracture of the palmar process at the base of the sidebone.

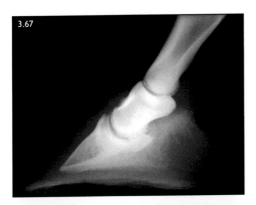

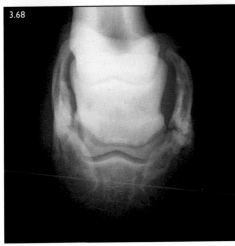

FIGS. 3.67, 3.68 Sidebone. Lateral (3.67) and dorsoproximal–palmarodistal oblique (3.68) radiographs of bilateral ossification in a non-lame horse. Note in 3.68, a radiolucent line going through the base of the sidebone is an area of non-fusion. In a lame horse it could be a fracture.

Differential diagnosis

- none.

Diagnosis

- loss of flexibility of ungual cartilages on palpation proximal to coronary band.
- local heat, swelling, and pain on palpation after external trauma.
- resolution of lameness following a uniaxial palmar digital nerve block.
- **Radiography:**
  - signs of ossification on standing dorsopalmar and lateromedial projections (Figs. 3.67, 3.68).
  - fractures of the distal phalanx or the sidebone may only be visible on 45° oblique views.
  - separate centres of ossification must be distinguished from fractures.
- **Scintigraphy:**
  - increased radionuclide uptake is a feature of normal ossification.
  - injury results in much higher increase in radionuclide uptake in comparison with unaffected sidebones.
- **MRI** (Fig. 3.58) necessary to identify:
  - osseous fluid (bone oedema) associated with trauma of the distal phalanx.
  - stress damage (sclerosis) or stress fracture at the base of the ossified cartilage.
  - collateral ligament injury adjacent to an ossified ungual cartilage.
  - fracture of an unossified ungual cartilage, with displacement of the cartilage from its attachment to the ipsilateral palmar process of the distal phalanx.

Management

- no treatment required for asymptomatic horses.
- if ossified cartilage can be confirmed as the cause of lameness:
  - NSAIDs.
  - long period of rest.
- always correct any imbalance of the horse's feet:
  - stress injury, stress fracture or traumatic fracture.
    - bar shoe with several clips or a rim shoe to immobilise the hoof capsule for 3–6 months.

- o persistent lameness:
  - ♦ light work with judicious use of NSAIDs.
  - ♦ uniaxial palmar digital neurectomy may be helpful.

Prognosis

- good for incidental findings: acute injury; stress injuries and stress fractures.
- guarded for chronic injuries and fractures with persistent lameness.

Quittor

Definition/overview

- septic necrosis of the ungual cartilage.

Aetiology/pathophysiology

- direct external trauma to the ungual cartilage.
- ascending infection from within the foot.
- persistent because the ungual cartilage has a poor blood supply.
- marked unilateral inflammatory response.

Clinical presentation

- unilateral swelling overlying the ungual cartilage proximal to the coronary band.
- one or more discharging skin sinuses within the swelling.
- varying degree of lameness.
- chronic fibrosis of the heel and unilateral deformity of the hoof wall.

Differential diagnosis

- abscesses that drain at the coronary band (gravel); subcutaneous abscess; puncture wound.

Diagnosis

- history of trauma proximal to the coronary band.
- lameness.
- unilateral location of swelling; heat and pain on palpation: one or more draining tracts (Fig. 3.69).
- insertion of a probe into a sinus that reaches the ungual cartilage to differentiate from a gravel or subcutaneous abscess.
- dorsopalmar and lateromedial radiographs with a probe *in situ* to

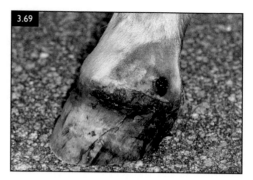

FIG. 3.69 Quittor. Lateral swelling and a draining sinus immediately proximal to the coronet, indicative of quittor.

confirm the origin of the sinus in the ungual cartilage or in the palmar process.
- MRI in complicated cases to determine the structures involved and guide surgical planning.

Management

- conservative treatment with long-term broad-spectrum antibiotics:
  - o may lead to a cure but recurrence of drainage is common.
- surgical excision of the infected tissue:
  - o complicated surgery as ungual cartilage is approximately half inside the hoof capsule and half proximal to the coronary band.
  - o postoperative bandage, systemic antibiotics, analgesia and tetanus prophylaxis.
  - o hoof wall defect may be filled with a composite once dry and keratinised.
- maggot therapy:
  - o promising alternative to surgical resection.
  - o multiple treatments may be necessary.

Prognosis

- fair to guarded because recurrence is possible.
- return to normal athletic activity can be expected.

Navicular disease

Definition/overview

- lameness associated with pain arising from the navicular bone, the collateral

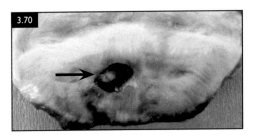

FIG. 3.70 Navicular disease. Erosion on the flexor cortex of the distal sesamoid (arrow)

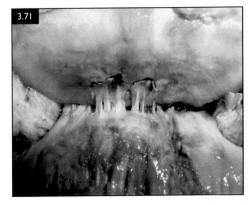

FIG. 3.71 Adhesion between the flexor surface of the distal sesamoid and the DDFT.

sesamoidean and distal impar ligaments, and the navicular bursa.
- o concurrent degenerative change of the DDFT can be present.
- five basic forms of navicular disease based on the pathology and MRI appearance of lesions:

1. **degenerative disease of the palmar surface of the navicular bone:**
   - degeneration and loss of fibrocartilage and cortical bone from the flexor surface.
   - associated with secondary remodelling changes in the spongiosa (including bone oedema) and navicular bursitis. (Figs 3.70–3.72).
2. **osseous fragmentation of the distal border of the navicular bone:**
   - remodelling changes of the adjacent cortical bone and spongiosa at the distal border of the navicular bone.
   - possible degenerative change in the distal impar ligament. (Fig. 3.73).
3. **primary inflammation, contusion or necrosis of the medulla:**
   - with oedema or haemorrhage in the spongiosa of the navicular bone.
   - possibly a result of trauma rather than of degeneration. (Fig. 3.74).
4. **primary desmitis or entheseopathy of the supporting ligaments of the navicular bone:**
   - possible remodelling changes of the proximal and distal borders of the navicular bone and navicular bursitis.
5. **primary navicular bursitis with marked effusion, proliferative synovitis or both.**

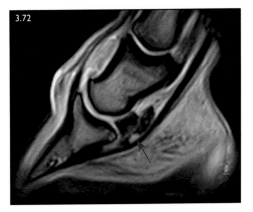

FIG. 3.72 T2*-weighted MR image of an erosion of the flexor cortex of the navicular bone.

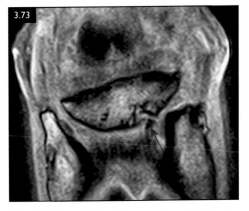

FIG. 3.73 T1-weighted image of a distal border fragment and a corresponding concave defect at the lateral angle of the distal margin of the navicular bone.

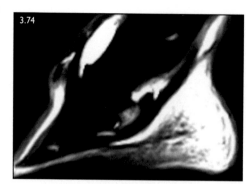

FIG. 3.74 Fat-suppressed MR image of generalised bone oedema in the spongiosa of the navicular bone.

Aetiology/pathophysiology

- variety of pathological conditions of the podotrochlear apparatus suggest the aetiology is likely to be multifactorial:
  - compressive forces and stress on the navicular bone exerted by the DDFT influenced by extraneous factors such as:
    - horse's foot anatomy
    - conformation/balance.
    - equestrian discipline
    - work routine.
    - trimming and shoeing.
  - developmental abnormality during skeletal growth with separate ossification centres along the distal border of the navicular bone:
    - instability developing consequent to repetitive loading.
  - acute or chronic repetitive strain injury in supporting ligaments of the navicular bone:
    - consequence of acute mechanical overload or repetitive loading.
  - acute monotonic or chronic repetitive compressive trauma resulting in primary medullary oedema, haemorrhage or necrosis.
  - weak heritability influence has been identified for navicular disease which may explain some breed predispositions.

Clinical presentation

- all ages affected but mostly early middle age (7–12 years old).

- any breed but Warmbloods and Quarter Horses most frequently affected.
- associated with a variety of foot conformations/imbalances.
- history:
  - loss of performance, stiffness, stumbling, shortening of the stride, and mild shifting bilateral forelimb lameness, especially on circles or on hard ground.
  - mostly insidious onset, bilateral lameness.
  - may become apparent after a change in ownership, change in shoeing routine, workload or turnout routine, or an enforced period of rest for an unrelated injury.
  - lameness may be most noticeable when the horse first leaves the stall or at the beginning of exercise.
  - horses with pain at rest may be seen to point one foot or to heap up the bedding to stand with the heels elevated.
- response to application of hoof testers:
  - unremarkable except when applied across the middle third of the frog in some small-footed Quarter Horses.
  - may be positive when navicular disease is accompanied by poor foot balance.
- lameness characteristics:
  - short-strided trot, with a shortened cranial phase and toe-first foot placement.
  - mostly bilateral with one dominant limb resulting in a visible head nod.
  - worse on hard ground and circles (lame limb on the inside of the circle).
  - may switch between limbs when a horse is exercised in different directions on a circle or following nerve blocks.
  - distal limb flexion tests may be positive in the flexed limb.
    - occasionally horse trots off worse in the stance limb (increased loading).
  - hyperextension of DIP joint with a wooden wedge or board (toe elevation test).
    - may increase the severity of lameness (not specific for navicular disease).

Chapter 3 The Foot **89**

3

Differential diagnosis

- any cause of chronic bilateral foot lameness:
  - bruising   ○ pedal osteitis
  - collateral ligament desmitis.
  - DIP joint pain.
  - tendinopathy of the DDFT.
  - collapsed, underrun and sheared heels.

Diagnosis

- **diagnostic analgesia** should result in significant improvement of lameness after:
  - palmar digital nerve block, intra-articular analgesia of the DIP joint and analgesia of the navicular bursa.
    - latter is the most specific for localisation to the podotrochlear apparatus.
    - many horses with deep flexor tendinopathy share the same blocking pattern.
  - approximately 20% of horses with navicular disease will be much improved or sound following intrabursal analgesia after they have first failed to improve to intra-articular analgesia of the DIP joint.
- **Radiography:**
  - main diagnostic tool in practice to demonstrate morphological changes in the navicular bone (Figs. 3.75–3.78).
  - minimum of three high quality radiographic projections required:
    - 60° dorsoproximal–palmarodistal oblique view.
    - lateromedial view.
    - palmaroproximal/palmarodistal oblique (flexor) view.
  - only allows the recognition of signs of advanced bone disease:
    - defects in the flexor cortex
    - medullary trabecular disruption.
    - medullary cyst-like lesions
    - medullary sclerosis.
    - poor flexor corticomedullary demarcation.
    - proximal/distal extension of the flexor border of the bone (entheseophytes).
    - distal border fragments.
    - bipartite or tripartite navicular bone.

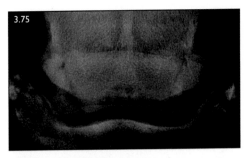

FIGS. 3.75, 3.76 Dorsoproximal–palmarodistal (3.75) and palmaroproximal–palmarodistal oblique (3.76) radiographs demonstrating an increase in number and size of the distal nutrient foramina/synovial invaginations.

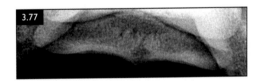

FIG. 3.77 Palmaroproximal–palmarodistal oblique radiograph showing loss of the corticomedullary junction, increased radiodensity of the medullary cavity, and erosion of the palmar cortex of the distal sesamoid centred on the sagittal ridge.

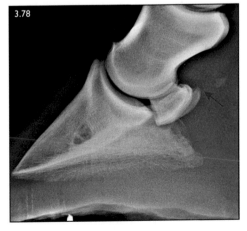

FIG. 3.78 Close-up lateral radiograph demonstrating a large entheseophyte in the collateral sesamoidean ligament of the distal sesamoid (arrow).

- size, shape and number of the synovial invaginations of the distal border.
  - unreliable parameters for the diagnosis of navicular disease.
  - wide variation in the appearance of the distal border between horses.
  - absence of radiographic abnormalities does not rule out presence of disease.
- **Nuclear scintigraphy:**
  - more sensitive than radiography for diagnosing increased metabolic turnover.
  - increase in radionuclide uptake in navicular bone not necessarily the source of pain.
- **CT and MRI:**
  - more sensitive in detecting structural bony lesions including distal border fragments.
  - MRI is the preferred technique for evaluation of palmar foot pain:
    - increased bone fluid signal in the spongiosa on fat-suppressed (STIR) images is the most common finding in horses with navicular disease.
    - only technique that can identify the five basic forms of navicular disease and distinguish them from deep digital flexor tendinopathy.

Management

- specific treatment protocol depends on ability to determine exact cause of pain with MRI.
- general management/treatment strategies:
  - horses with no or few radiological abnormalities are managed medically:
    - supportive trimming and shoeing.
    - Isoxsuprine hydrochloride.
    - bisphosphonates.
    - NSAIDs.
    - return to regular exercise as soon as possible.
  - horses that fail to respond to medical management:
    - intra-articular triamcinolone and hyaluronic acid.
    - intrabursal triamcinolone and hyaluronic acid.
  - horses with radiological abnormalities or refractory to corticosteroid medication and medical management:

- palmar digital neurectomy (**not if deep digital flexor tendinopathy**).
- navicular suspensory (collateral sesamoidean) desmotomy (rarely used).
- specific treatment strategies:
  - navicular bone contusion or primary inflammation (bone oedema):
    - medical management as above, in combination with:
      - shoeing with heel elevation via a wedge or graduated branches.
      - 2–3 months of stall rest with controlled in-hand exercise.
  - degenerative disease of the palmar surface (fibrocartilage and flexor cortex):
    - supportive trimming and shoeing.
    - intrabursal triamcinolone and hyaluronic acid.
  - distal border fragmentation:
    - shoeing with heel elevation with a wedge or graduated branches.
    - intrabursal triamcinolone and hyaluronic acid.
    - 1–2 months stall rest with controlled walking.
  - desmitis of the supporting ligaments of the navicular bone:
    - medical management as above, in combination with:
      - shoeing with heel elevation with a wedge or graduated branches.
  - primary navicular bursitis:
    - shoeing with heel elevation with a wedge or graduated branches.
    - intrabursal triamcinolone and hyaluronic acid.
    - 1–2 months stall rest with controlled walking.
- specific treatments:
  - trimming directed at improving dorsopalmar and mediolateral hoof balance.
  - no universal shoeing solution exists for navicular disease:
    - important to try and meet the needs of the individual horse.
    - potentially successful shoeing modifications include:
      - rolling of the toe
      - elevation of the heels:
        - $6°$ elevation reduces compressive force by 24%.

◇ long-term heel elevation may overload the heels.
- extending the ground surface contact with an egg bar shoe:
  ◇ acts like a 6° wedge on soft arena surface.
○ NSAIDs as needed are helpful to maintain horses in regular daily exercise.
○ vasodilator isoxsuprine hydrochloride causes peripheral vasodilation:
  ♦ evidence suggests not absorbed and does not improve blood flow to foot.
  ♦ may be effective in milder cases.
○ Calcium dobesilate:
  ♦ synthetic venoactive drug to reduce venous congestion.
  ♦ further research is needed.
○ Bisphosphonates (tiludronate and clodronate):
  ♦ widely used to treat navicular disease.
  ♦ relative lack of data on their efficacy.
○ injection of corticosteroids:
  ♦ with or without sodium hyaluronate.
  ♦ DIP joint or navicular bursa.
  ♦ lameness improved for 2–6 months.
  ♦ intrabursal injection twice as effective as intra-articular injection.
    - repeat intrabursal injections progressively less successful.
○ desmotomy of the collateral sesamoidean ligaments:
  ♦ fallen out of favour because of the unpredictable results.
  ♦ high rate of recurrence of lameness around 6–12 months after surgery.
○ palmar digital neurectomy:
  ♦ usually effective for 1–2 years in horses with navicular disease.
  ♦ **avoid in horses with deep digital flexor tendinopathy:**
    - fast recurrence of lameness.
    - may lead to catastrophic tendon rupture.

Prognosis
- depends on MRI diagnosis:
  ○ fair for primary navicular bone inflammation, navicular ligament desmopathy, primary navicular bursitis with effusion.
  ○ guarded for distal border fragments and primary navicular bursitis with synovial proliferation and adhesions.
  ○ poor for degenerative disease of the palmar surface of the navicular bone.
- fair if therapeutic shoeing and medical management are sufficient to return a horse to exercise.
- guarded to poor if intrasynovial corticosteroid medication needs to be repeated.
- poor if concurrent deep digital flexor tendinopathy.

Fracture of the distal sesamoid or navicular bone

Definition/overview
- most common in parasagittal plane, approximately 1–2 cm medial or lateral to the sagittal ridge:
  ○ usually simple or rarely comminuted (displaced).
- distinguish from congenital, non-fused centres of ossification of the navicular bone:
  ○ bipartite or tripartite.
  ○ may occur unilaterally, bilaterally or occasionally in three or four limbs.

Aetiology/pathophysiology
- trauma such as kicking at a wall or landing hard on an uneven surface.
- slow to heal and usually with a fibrous union if untreated.
- OA of the DIP joint usually develops as a sequela.

Clinical presentation
- acute-onset, moderate to severe lameness.
- chronic fractures evolve to mild to moderate lameness.
- instability between separate centres of ossification can cause:
  ○ mild to moderate, intermittent lameness.

Differential diagnosis
- all causes of acute severe lameness or chronic moderate foot lameness.
- differentiate from bipartite navicular bone.

Diagnosis

- history of acute-onset, moderate to severe lameness.
- withdrawal response to hoof testers placed across the heels or on the frog.
- absence of swelling or other physical findings.
- palmar digital nerve block is often needed to localise the pain.
- **Radiography** is essential for definitive diagnosis (Figs. 3.79, 3.80):
  - 60° dorsoproximal–palmarodistal oblique view
    - superimposition of a paracuneal (collateral) sulcus of the frog on the navicular bone can mimic a fracture.
    - repacking the sulcus and repeating the radiograph should remove the artefact.
  - Palmaroproximal–palmarodistal oblique view.
- **CT and MRI** are useful to:
  - determine the true fracture configuration and the presence of comminution.
  - MRI to identify concurrent damage to the dorsal surface of the DDFT.

Management

- rest and immobilising the hoof:
  - rim cast, that incorporates the heel bulbs, around the perimeter of the foot.
  - alternatively, a bar shoe with a continuous rim that extends 1–1.5 cm proximally.
  - **bar shoes, with quarter clips, are frequently used but are not as effective.**
  - fibrous union and subsequent OA are common.
- elevation of the heels by 12° for 2 months, with gradual reduction by 3° at a time over a 4-month period.
- internal lag screw fixation:
  - fractures less than 4 weeks old.
  - healing by osseous or fibrous union.
  - protective against the onset of OA of the DIP joint.
  - difficult technique which requires special equipment and expertise.
  - complications include:
    - infection; rotational instability; splitting of the fragment.
    - creation of a permanent step defect.
- palmar digital neurectomy recommended for:
    - horses with persistent lameness associated with a fibrous union.
    - persistent lameness due to OA of the DIP joint.

Prognosis

- poor for return to exercise with conservative management.
- guarded to fair for internal fixation of acute fractures.

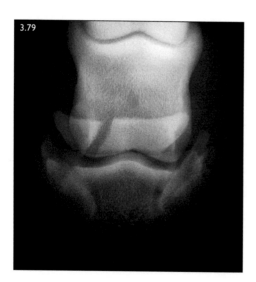

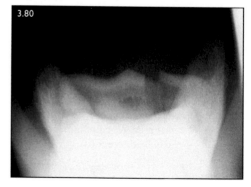

FIGS. 3.79, 3.80 Navicular bone fracture. Dorsoproximal/palmarodistal (3.79) and palmaroproximal–palmarodistal oblique (3.80) radiographs showing a fracture through the body of the navicular bone.

Synovitis, Capsulitis and Osteoarthritis of the distal interphalangeal joint (DIP joint)

Definition/overview

- synovitis, traumatic arthritis and OA occur in the DIP joint.
  - relatively low incidence on MR images of horses with foot pain.
- inability to localise pain exclusively to the DIP joint with local analgesia prevents a conclusive diagnosis.

Aetiology/pathophysiology

- secondary to acute trauma, joint sprain, intra-articular fractures, collateral desmitis, infection, OCD/subchondral bone cysts.
- secondary to repetitive, low-grade trauma associated with constant high athletic performance.
  - exacerbated by poor foot conformation.
- secondary to repeated joint medication with corticosteroids.

Clinical presentation

- acute or insidious in onset, unilateral or bilateral lameness.
- almost exclusively in the forelimbs.

Differential diagnosis

- acute lameness of the foot:
  - collateral desmitis; deep digital flexor tendinitis; fracture; bruising; abscess.
- chronic lameness of the foot:
  - chronic bruising (pedal osteitis); navicular disease; chronic laminitis; hoof imbalance.

Diagnosis

- synovial distension of the dorsal pouch of the DIP joint may be palpated:
  - dorsodistal aspect of the pastern, just above the coronary band in some horses (Fig. 3.81).
- lameness is usually worse on hard ground and when circled with affected limb on the inside.
- pain on flexion of the digit is variable and depends on the stage of the disease.

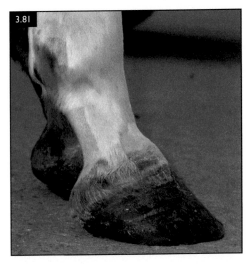

FIG. 3.81 This horse has a distended DIP joint visible as a swelling just above the dorsal coronary band.

- **regional analgesia** can be unhelpful in localising pain to the DIP joint:
  - palmar digital nerve block desensitises the entire DIP joint in most horses.
    - plus, the entire foot (except dorsal aspect of the coronary band and sensitive lamellae).
  - intra-articular analgesia of the DIP joint desensitises the DIP joint as well as:
    - navicular bone and bursa.
    - DDFT.
    - distal phalanx and the dorsal portion of the sole.
  - intrasynovial analgesia of the navicular bursa does not alleviate pain arising from the DIP joint.
    - lameness that is abolished by intra-articular analgesia of the DIP joint but not by intrabursal analgesia suggests primary DIP joint pain.
- **Radiography:**
  - no radiographic changes may be present in synovitis, capsulitis or early OA.
  - radiographic changes of OA (Figs. 3.82–3.85):
    - periarticular osteophytes and joint capsule entheseophytes:
      - dorsodistal and palmarodistal aspects of P2; extensor process P3; dorsoproximal margin of the navicular bone.

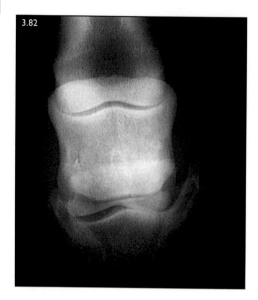

FIG. 3.82 Dorsopalmar radiograph showing loss of joint space and subsequent rotation in the frontal/dorsal plane of the distal phalanx in relation to the middle phalanx.

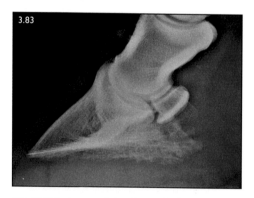

FIG. 3.83 Lateromedial radiograph demonstrating decreased width of the DIP joint joint and exostoses (osteophytes and entheseophytes) on the dorsal surface of the middle phalanx.

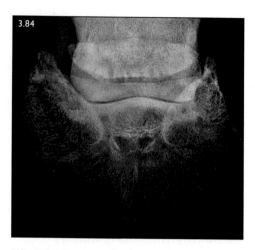

FIG. 3.84 Dorsoproximal–palmarodistal oblique radiograph of same horse as Figure 3.83 demonstrating severely decreased width of the DIP joint due to near complete loss of articular cartilage.

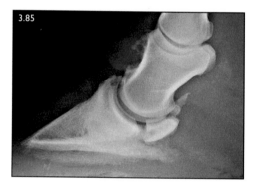

FIG. 3.85 Lateromedial radiograph of a horse with chronic OA of the DIP joint. The new bone formation on the dorsal aspect of the middle phalanx consists of capsular entheseophytes and marginal osteophytes.

- care with interpretation at the level of the extensor process as it is highly variable between horses.
  ♦ loss of joint space.
  ♦ subchondral bone sclerosis and lysis.
  ○ chronic synovitis may result in enlargement of the synovial invaginations of the distal border of the navicular bone.

- **MRI** (Figs. 3.86, 3.87) is most useful to identify:
  ○ focal or generalised cartilage loss and joint space narrowing.
  ○ focal or generalised subchondral bone change.
  ○ soft tissue changes that are not visible radiographically.

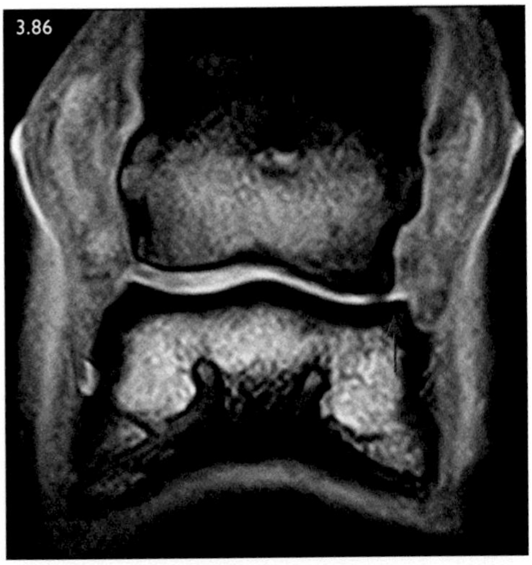

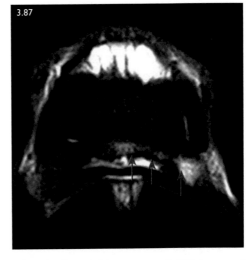

FIG. 3.86 Dorsal T1-weighted MR image of the DIP joint. There is narrowing of the medial half of the joint space with loss of the normal 3-layer appearance of the articular cartilage and synovial fluid (see lateral half of the joint space). There are focal hypointense areas of cartilage signal loss, where cartilage is replaced by pooling of synovial fluid.

FIG. 3.87 Dorsal fat-suppressed MR image of the same DIP joint. There is hyperintense osseous fluid signal indicative of subchondral bone oedema in the medial half of the distal phalanx adjacent to the DIP joint.

Management

- shoeing for horses with confirmed DIP joint pain:
  - correct any foot imbalance.
  - use shoes that ease breakover and pads that diminish the concussion.
- oral NSAIDs where lameness is mild and competition rules permit.
- intra-articular medication for DIP joint pain:
  - mild synovitis/capsulitis, without degenerative changes of cartilage or subchondral bone:
    - hyaluronic acid
    - polysulphated glycosaminoglycan.
    - autologous conditioned serum (ACS/interleukin 1-receptor-antagonist (IRAP)).
      - injected once every 10–14 days for 3–4 injections.
      - horse kept on stall rest with controlled in-hand walking only.
  - moderate to severe synovitis/capsulitis or mild OA:
    - intra-articular triamcinolone.
      - **not methylprednisolone as it is detrimental to cartilage.**

- corticosteroid controversy:
  - repeated injection and continued exercise may lead to accelerated cartilage degeneration.
  - low dosages better tolerated (6 mg or less of triamcinolone).
    - ◇ less effective.
  - total-body dose of triamcinolone should not exceed 18 mg.
    - ◇ possible link with iatrogenic laminitis, especially in large, overweight Sport horses.
- non-responsive cases to above recently been treated successfully with:
  - Stanozolol.
  - Polyacrylamide gels.

Prognosis

- good for primary synovitis/capsulitis without radiographic abnormalities,
- poor for horses with:
  - radiographic signs of OA.
  - MRI signs of cartilage loss.
  - non-responsive to joint medication.

Collateral desmitis of the distal interphalangeal (DIP) joint

Definition/overview

- second most important soft tissue injury responsible for foot lameness.
- entheseopathy refers to injury at the bone/ligament interface.
- desmitis or desmopathy affects the body of the ligament.
- injury ranges from:
  - mild to moderate fibre damage (sprain).
  - severe fibre damage with elongation or separation resulting in joint instability (partial or complete rupture).

Aetiology/pathophysiology

- acute traumatic strain injury.
- chronic repetitive strain injury more common.
- predisposed by pre-existing degenerative changes within the ligament and at its insertion.
- medial collateral ligament most affected.

Clinical presentation

- rarely any localising signs:
  - discrete, palpable swelling at the level of the origin of a collateral ligament.
  - immediately proximal to the dorsomedial or dorsolateral aspect of the coronary band (Fig. 3.88).
  - distension of the dorsal pouch of the DIP joint is not a characteristic finding.
- unilateral or bilateral mild to moderate lameness of variable severity:
  - worse on a circle, especially on hard ground.
  - when the injured ligament lies on the outside (increased strain).
- acute-onset, severe lameness with a history of trauma is atypical.
- digital flexion may be painful.
- elevation of the opposite hoof wall with a wedge, under the foot, may cause increased strain, be painful and exacerbate lameness.

Differential diagnosis

- any cause of moderate lameness arising within the foot: navicular disease;

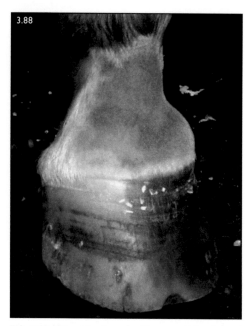

FIG. 3.88 View of a horse with a medial collateral ligament injury of a forelimb showing clearly a unilateral swelling above the coronary band which was confirmed by ultrasonography and MRI.

bruising; DIP joint pain; pedal osteitis; sheared heels.

Diagnosis

- response to diagnostic analgesia is variable:
  - 25% of horses with desmopathy improve following intra-articular anaesthesia of the DIP joint.
  - 66% improve following a palmar digital nerve block.
  - 100% improve following an abaxial sesamoid nerve block.
  - analgesia of the navicular bursa does not cause any improvement in lameness.
- **Radiography** is rarely useful but may occasionally identify:
  - osteolysis or focal osseous cyst-like lesion at the insertion of a collateral ligament in the collateral fossa of the distal phalanx on a dorsoproximal–palmarodistal oblique view (Fig. 3.89).
  - proliferative new bone (entheseopathy) at the origin of a collateral ligament on the middle phalanx on 45° oblique views (Fig. 3.90).

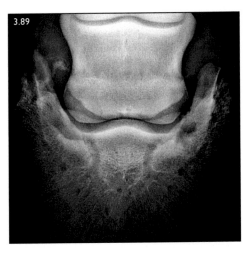

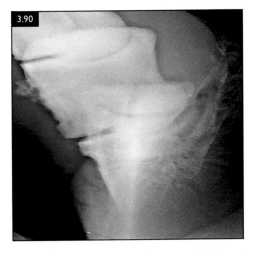

FIG. 3.89 Dorsoproximal–palmarodistal oblique view of a foot with collateral entheseopathy at the insertion of a collateral ligament (right side of the image). There is focal osteolysis bordered by an irregular margin of sclerosis at the site of insertion of the collateral ligament to the distal phalanx.

FIG. 3.90 Dorsolateral–palmaromedial flexed oblique view of a foot with collateral entheseopathy at the origin of a collateral ligament (left of the image). There is entheseous new bone formation at the site of the proximal attachment of the collateral ligament to the middle phalanx.

- **Ultrasonography** is difficult and limited to the proximal 25% of the collateral ligament.
- **Scintigraphy** in some cases:
  - increased radionuclide at the insertion site of the ligament on the distal phalanx.
  - low sensitivity but high specificity.
- **MRI** abnormalities include:
  - increased cross-sectional area, irregular contour and increased signal intensity of the ligament.
  - osseous abnormalities of the distal or middle phalanx in 40% of cases:
    - cortical irregularities or defects.
    - osseous cyst-like lesions.
    - abnormal mineralisation or bone oedema associated with the origin or insertion of the diseased collateral ligament (Figs. 3.91, 3.92).
  - different imaging artefacts causing signal increase in the collateral ligament.

Management

- rest is the most important treatment:
  - minimum of 4–6 months of stall rest.

  - during this time the foot must be kept level and well balanced.
  - graduated programme of walking after 2 months of rest.
    - circles must be avoided (no horse walkers).
- asymmetrical shoes:
  - double width branch at the side of the injury to prevent the injured side of the foot from sinking excessively into the soft arena surfaces and straining the ligament.
  - narrow branch opposite to the injury.
  - roll (bevelling) of the shoe at the toe and the branch opposite to the injury.
- pastern–foot limb cast:
  - when the injury is severe with discontinuity of the ligament.
  - 4 to 6 weeks of immobilisation.
- Extracorporeal shock-wave therapy:
  - three applications are spaced 2 weeks apart over a period of 4 weeks.
  - proximal to the coronary band at the level of the affected collateral ligament.
- intra-articular medication of the DIP joint:
  - Autologous conditioned serum (IRAP) to suppress concurrent synovitis and capsulitis.
- intralesional injection:

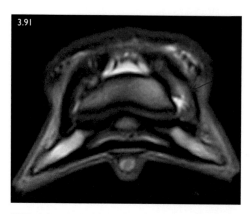 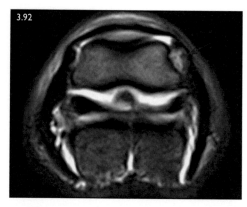

FIGS. 3.91, 3.92 Dorsal (3.91) and transverse (3.92) T2-weighted MR images showing signal increase characteristic of collateral desmitis of the DIP joint (arrows).

- o imaging guidance required (radiography, ultrasonography, CT or MRI).
- o mesenchymal stem cells or PRP.

Prognosis

- guarded to fair for return to athletic function.
  - o approximately 50% of horses recover successfully.
  - o not adversely affected by the presence of osseous abnormalities.
- joint instability, OA or other concurrent injuries in the foot are indicators of a poor outcome.

Septic arthritis of the distal interphalangeal (DIP) joint

Definition/overview

- infection of the DIP joint.

Aetiology/pathophysiology

- traumatic penetrating injuries or pastern lacerations:
  - o may also infect the digital flexor tendon sheath and navicular bursa.
  - o solar puncture wounds that enter the DIP joint may also enter the bursa.
  - o neurectomised horses may develop septic arthritis from expansion of an unnoticed subsolar abscess.
  - o usually, a mixed bacterial population with a predominance of gram-negative Enterobacteriaceae.

- iatrogenic infection following a joint injection, mostly with *Staphylococcus aureus* contamination.
- haematogenous spread of infection from the umbilicus, respiratory or digestive system in foals.

Clinical presentation

- severe lameness.
- swelling proximal to the coronary band, particularly dorsally.
- presence of a wound or solar puncture.

Differential diagnosis

- all causes of severe foot lameness with soft tissue swelling; abscess; fracture of the distal phalanx or navicular bone; severe strain or sprain; other deep digital sepsis.

Diagnosis

- clinical findings of severe lameness with diffuse swelling proximal to the coronary band dorsally:
  - o marked pain on flexion of the digit.
  - o history of a recent joint injection.
  - o presence of a wound:
    - ♦ dorsal to middle third of the frog.
    - ♦ proximal to the coronary band adjacent to the joint capsule.
- confirmation of sepsis is achieved by arthrocentesis:
  - o elevated WBC count ($>10\times10^9$/l) and protein (>60 g/l).
  - o identification of bacteria on joint fluid smear – culture and sensitivity.

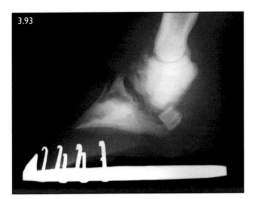

FIG. 3.93 Septic DIP joint. Lateral radiograph of the DIP joint demonstrating extensive loss of the articular surfaces, exostoses on the dorsal and palmar surfaces of the middle phalanx, a large sequestrum of the extensor process of the distal phalanx, and subluxation of the joint.

- o demonstration of communication with an external wound:
  - ◆ distension of the joint with sterile saline from a site remote to the wound.
  - ◆ leakage from the wound.
- **Radiography:**
  - o not usually helpful in the diagnosis of acute joint sepsis.
  - o signs of septic arthritis later in the course of the disease:
    - ◆ loss of joint space; subchondral lysis; periarticular new bone (Fig. 3.93).
- **MRI** may allow earlier and more accurate diagnosis of septic arthritis in horses.

Management

- arthroscopic joint lavage:
  - o both the dorsal and palmar/plantar pouches of the joint.
  - o removal of intra-articular pannus with grasping forceps.
- antimicrobial therapy based on sensitivity results if available:
  - o systemic broad-spectrum antibiotics.
  - o regional intravenous perfusion.
  - o intrasynovial antibiotics via ingress drain in the dorsal pouch of the DIP joint.
    - ◆ facilitates continuous or repeated intra-articular administration.
  - o continued for at least 2 weeks after closure of any wound communication

with the joint, the lameness has resolved, or after the DIP joint synovial fluid analysis is normal.
- NSAIDs.
- if chronic debilitating OA ensues, surgical arthrodesis of the DIP joint can be attempted to obtain pasture (breeding) soundness.

Prognosis

- fair to good if infection controlled before significant degeneration of the articular surface.
- poor if significant degeneration of the articular surface and OA.
- hopeless if failure to eliminate infection.

Septic navicular bursitis ('Streetnail injury')

Definition/overview

- septic synovitis of the navicular bursa.

Aetiology/pathophysiology

- puncture wound in an area centred on the middle third of the frog and its collateral sulci.
  - o injury of at least 15 mm depth and orientated in the direction of the navicular bone.
- elastic nature of the frog usually seals over any puncture wound and prevents natural drainage.
- nail traverses the tendon on its way into the bursa.
  - o concurrent septic tendonitis of the DDFT.
- nail often impacts on the flexor cortex cartilage and bone.
  - o concurrent damage and sepsis of the navicular bone and overlying cartilage.
- life-threatening injury with 35% of horses with a puncture wound to the frog being euthanased.
- DIP joint and digital flexor tendon sheath may become infected by the original injury or by secondary spread of the infection.
- hoof and heel bulb lacerations/avulsions can occasionally be deep enough to involve the bursa.

- iatrogenic infection following a bursal injection (mostly *Staphylococcus aureus* contamination).

Clinical presentation

- history of a known solar nail puncture or bursal injection.
- severe lameness in which the horse will not put its heel down.
- swelling between the heel bulbs and ungual cartilages in subacute and chronic cases.
- entire distal limb may be diffusely swollen up to and beyond the level of the fetlock.

Differential diagnosis

- all causes of acute, severe foot lameness:
  - abscess; fracture of the distal phalanx or navicular bone; severe strain or sprain; other deep digital sepsis.

Diagnosis

- severely lame horse that is not putting its heel to the ground.
  - history and/or visible evidence of a puncture wound to the frog or adjacent sulci.
  - recent navicular bursa injection.
- swelling between the heel bulbs and ungual cartilages.
- pain on flexion of the digit.
- marked withdrawal response following application of hoof testers over the frog.
- thorough exploration of the solar surface of the foot after a palmar digital nerve block:
  - identify possible nail entry site.
- confirmation of sepsis by a palmar centesis of the bursa between the heel bulbs:
  - elevated WBC count ($>10 \times 10^9$/l) and protein (>60 g/l).
  - identification of bacteria on a synovial fluid smear.
- confirmation of communication of the bursal cavity with the external wound:
  - remote pressure injection of the bursa with sterile saline followed by fluid exiting from the wound.
  - injection of the bursa with radiographic contrast and lateromedial radiograph.
- **Radiography:**

  - minimum of three high-quality radiographic projections required to identify bone damage:
    - 60° dorsoproximal–palmarodistal oblique view.
    - lateromedial view.
    - Palmaroproximal–palmarodistal oblique (flexor) view.
  - lateromedial view with either a solid probe (Fig. 3.94) or radiographic contrast medium inserted into the wound.
    - **careful not to force contaminants deeper into the tract.**
  - preferable to inject radiocontrast in the bursa from a remote site and observe communication with the tract on a lateromedial radiograph (Fig. 3.95).
  - later in the disease, there may be radiographic evidence of lysis of the flexor cortex of the navicular bone (Figs. 3.97, 3.98).
- ultrasonography is rarely useful.
- **MRI** is the imaging technique of choice because it can determine:
  - full extent of the injury, the need for surgical intervention and the prognosis much earlier than radiography (Fig. 3.96).

Management

- **emergency that requires referral to a specialist facility for further assessment.**
- systemic antibiotics, NSAIDs and tetanus prophylaxis.
- navicular bursoscopy under GA is the preferred surgical technique:
  - enlargement of entry wound and exposure of the tract.
  - visualise the bursa, entry wound and damage to the navicular bone.
  - collection of samples for bacterial culture.
  - debridement, copious lavage and removal of fibrin and contaminants.
  - postoperative care:
    - waterproof bandage changed in a sterile fashion every 2–3 days.
    - elevation of heels by 6–12° to decrease stress on the injured DDFT.
    - systemic antibiotics.
    - intravenous regional perfusion with antibiotics.

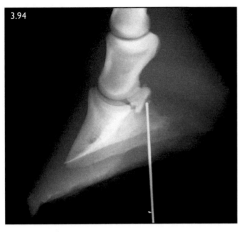

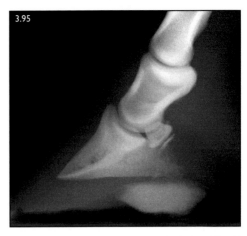

FIGS. 3.94, 3.95 Septic navicular bursa. Radiographs showing the use of a solid probe (3.94) and liquid contrast medium (3.95) to confirm communication between a wound in the ground surface of the foot and the navicular bursa.

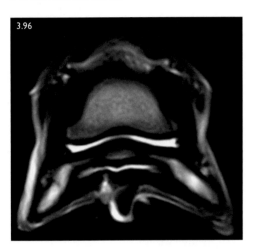

FIG. 3.96 Dorsal T1-weighted MR image showing a hyperintense tract extending from the intense keratinised region of the frog, through the solar corium and the DDFT towards the navicular bursa.

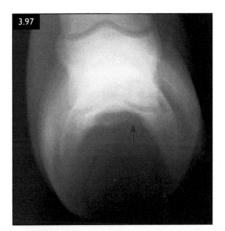

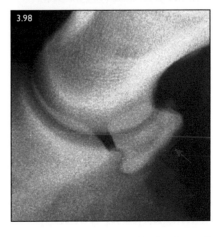

    ◦ many horses improve initially but
      suffer a relapse 4–5 days after
      bursoscopy:
        ♦ recurrence of severe lameness.
        ♦ drainage from the penetrating
          tract.
• navicular bursotomy ('streetnail'
  procedure) can alternatively be used:
    ◦ surgical fenestration of the frog, digital
      cushion and DDFT.
    ◦ direct lavage, limited debridement and
      ventral drainage (Fig. 3.99).

FIGS. 3.97, 3.98 Palmaroproximal–palmarodistal oblique (3.97) and close-up lateral (3.98) radiographs demonstrating erosion of the palmar cortex of the navicular bone (arrows).

FIG. 3.99 Solar view of the foot a few days after a 'streetnail' procedure has been performed.

  ○ similar postoperative care but with a patten shoe.
  ○ used where bursoscopy is not possible or available.
    ♦ good outcomes are possible, but the aftercare is more complicated.
  ○ potential complications:
    ♦ adhesions between the DDFT wound and navicular bone.
    ♦ prolonged aftercare of a larger solar wound required.
    ♦ opportunity for ascending retrograde infection through solar wound.

Prognosis

- depends on time interval between injury and surgical treatment.
- historically, guarded prognosis for survival (50%) and poor for return to athletic function (30%) for septic navicular bursitis:
  ○ improved with immediate referral and bursoscopic lavage with survival rates of 80% and return to function rates 60%.
- recent report of good prognosis also for bursotomy.
- excellent for puncture wounds of the foot that do not involve a synovial structure.

Deep digital flexor tendinopathy in the foot

Definition/overview

- most common soft tissue injury causing lameness in the horse's foot:

  ○ 30–50% of all foot lameness without radiographic abnormalities.
- four basic lesion types in the navicular bursa portion of the DDFT:
  ○ core lesions
  ○ dorsal abrasions and fibrillations.
  ○ sagittal plane splits and tears
  ○ insertional lesions (including entheseopathy).

Aetiology/pathophysiology

- acute traumatic fibre tearing at the end of the stance phase.
- repetitive overload stress.
- possible predisposition due to age-related degenerative changes.
- possible risk factors:
  ○ hoof conformation; angle of the distal phalanx; palmar digital neurectomy.
  ○ horse's athletic discipline (jumping over fences).

Clinical presentation

- lameness varies from mild, insidious bilateral lameness to acute unilateral severe lameness:
  ○ more obvious lameness cases may be seen pointing the affected foot at rest.
  ○ worse trotting in a circle on hard ground, especially affected limb on inside.
- no visible or palpable abnormalities.
- rarely:
  ○ mild distension of the digital flexor tendon sheath.
  ○ focal, painful, firm soft tissue swelling in the palmar hollow of the pastern.
- flexion and extension tests produce variable results that are not specific:
  ○ toe elevation with a wooden board makes some horses worse.
  ○ elevation of the heel with a wooden wedge of 15–20° makes some horses worse.

Differential diagnosis

- all other causes of foot lameness without radiographic abnormalities:
  ○ either acute and severe (fracture, severe collateral desmitis, solar bruise or abscess).
  ○ mild and insidious (navicular disease, pedal osteitis, DIP joint pain, collateral desmitis).

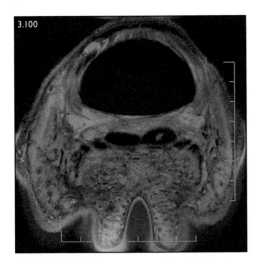

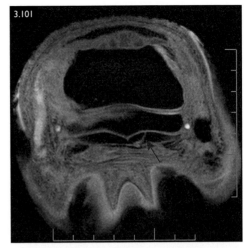

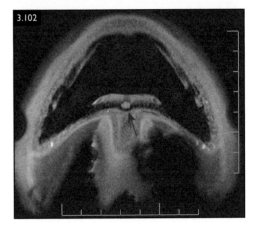

FIGS. 3.100–3.102 T1-weighted fat-suppressed MR images of different lesion types of the DDFT in the foot (arrows): (3.100) core lesion; (3.101) parasagittal split; (3.102) insertional tear.

Diagnosis

- diagnostic analgesia:
  - palmar digital nerve block causes significant improvement in 66% of horses.
  - intra-articular analgesia of the DIP joint: significant improvement in 66% of horses.
  - analgesia of the navicular bursa: significant improvement in 66% of horses.
  - abaxial sesamoid nerve block: resolution of lameness in all horses.
  - intrathecal analgesia of the digital flexor tendon sheath: significant improvement in <50% of horses.
- foot radiographs are generally normal.
  - rare exception of dystrophic tendon mineralisation in a few horses.
- **Ultrasonography:**
  - limited to the palmar aspect of the pastern and the proximal recess of the navicular bursa.
  - underestimates the extent and the true prevalence of injury.
  - poor sensitivity and specificity.

- **MRI** is the imaging technique of choice (Figs. 3.100–3.103):
  - focal signal increase on both T1- and T2-weighted sequences.
  - enlargement of the affected lobe.
  - herniation of torn fibres and granuloma formation into the navicular bursa.
  - concurrent navicular bursitis with adhesion formation between:
    - dorsal surface of the tendon and collateral sesamoidean and impar ligaments.
  - monitor healing by following fibroplasia.
- **CT:** also visible on soft tissue windowing of images as:
  - focal hypoattenuation
  - enlargement of a tendon lobe.
  - contrast enhancement in acute injuries.

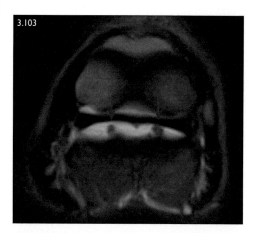

FIG. 3.103 T2-weighted MR image showing dorsal surface granulomas (arrows) on each lobe of the DDFT.

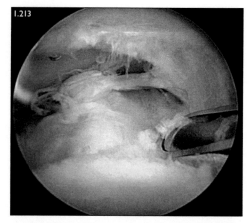

FIG. 3.104 Navicular bursoscopy showing a surface tear in the DDFT with a protruding disorganised mass of torn tendon fibres.

Management

- rest (6 months or more) is the most important treatment:
  - stall confinement with a low-grade maintenance daily exercise programme of:
    - 10–15 minutes walking, after the first or second month of strict stall rest.
    - depending on the severity of the injury.
  - walking exercise can be increased gradually in duration but not intensity over the ensuing 4 months.
- shoeing:
  - egg bar shoes with rolled toes or back-to-front shoes.
    - limits hyperextension of the DIP joint.
  - heel elevation useful in horses with severe lameness to improve lameness quickly.
    - **heel elevation can, however, exacerbate lameness in some horses.**
- additional therapies:
  - intrabursal injection of triamcinolone and hyaluronic acid.
  - Extracorporeal shock-wave therapy.
  - intralesional injection of biological and regenerative products (PRP, stem cells).
  - inferior check ligament desmotomy.
  - bursoscopic debridement of dorsal surface tendon tears and granulomas (Fig. 3.104).

Prognosis

- overall, with conservative treatment:
  - 25% of horses returned to previous levels of athletic activity within 18 months.
  - 60% of horses suffered persistent lameness.
- conservative treatment of dorsal border tears:
  - 27 % of horses with persistent lameness.
- conservative treatment of parasagittal splits and longitudinal tears:
  - 50% of horses with persistent lameness.
- surgical treatment of intrabursal dorsal splits, tears and granulomas:
  - 37% of horses with persistent lameness.
  - 42% of horses able to return to previous levels of activity.
- conservative treatment of core lesions:
  - 60% of horses with persistent lameness.
- conservative treatment of core lesions, with a cross-sectional area of more than 10% or a total length of more than 30 mm:
  - unable to regain previous level of performance.

The Forelimb

PASTERN

Fractures of the proximal and middle phalanges

Definition/overview

Fractures of the proximal phalanx (P1) can be:

- sagittal (incomplete or complete) (Figs. 4.1–4.3).
- dorsal frontal.
- comminuted (Fig. 4.4).
- complete transverse (Fig. 4.5).
- osteochondral fragmentation of dorsal or palmar/plantar aspect of fetlock joint.
- palmar/plantar eminence fractures within the fetlock joint.

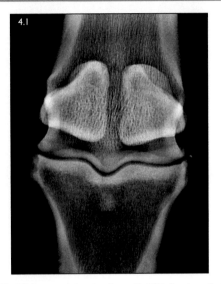

FIG. 4.1 Incomplete short sagittal P1 fracture just lateral to the sagittal groove.

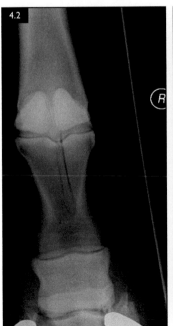

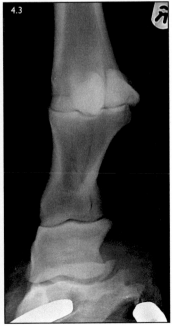

FIGS. 4.2, 4.3 A complete biarticular and minimally displaced P1 fracture in a dorsopalmar radiograph (4.2) and a dorso-lateral–palmaromedial oblique view (4.3).

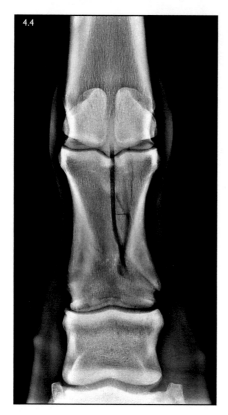

FIG. 4.4 A complete, comminuted, sagittal, uniarticular and displaced P1 fracture in a dorsopalmar radiograph.

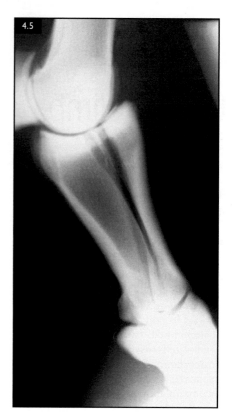

FIG. 4.5 Preoperative radiograph of an unusual complete displaced transverse fracture of P1.

Fractures of the middle phalanx (P2) include:
- osteochondral fragmentation of dorsal or palmar/plantar aspect of the pastern joint.
- uniaxial fractures of palmar/plantar eminence (Fig. 4.6).
- biaxial eminence fractures.
- comminuted fractures (Figs. 4.7, 4.8).
- less commonly, oblique or transverse fractures.

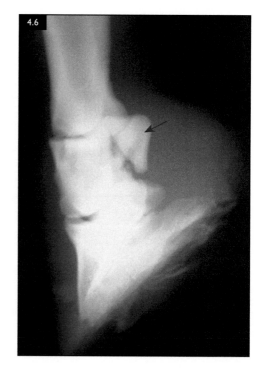

FIG. 4.6 Dorsolateral–palmaromedial oblique radiograph of the right forelimb showing a P2 fracture involving the PIP joint. Note the separated palmarolateral eminence of P2 (arrow).

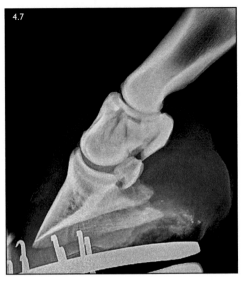 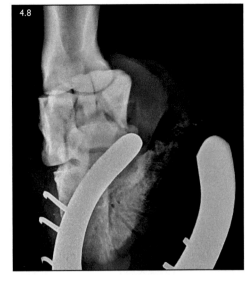

FIGS. 4.7, 4.8 A comminuted biarticular P2 fracture shown in lateromedial (4.7) and oblique (4.8) radiographs of the distal limb.

Aetiology/pathophysiology

- P1 fractures are one of the most common fractures in racehorses.
  - other breeds and types can be affected at work and pasture.
- often an acute manifestation of chronic ongoing changes.
- classification:
  - sagittal: start at proximal articular surface of P1 and extend variably distally.
    - short incomplete: <30 mm.
    - long incomplete: >30 mm.
    - complete:
      - exit at lateral cortex.
      - biarticular – extending to proximal interphalangeal (PIP) joint.
  - frontal: start at proximal articular surface of P1 and extend variably distally.
    - incomplete or complete.
  - multiplanar or comminuted (≥3 pieces).
  - P2 fractures usually occur due to acute trauma and can be:
    - uniaxial/ biaxial of palmar/plantar eminence due to avulsion forces.
    - comminuted ('ice bag' fractures) due to high energy torquing and compressive forces.

- may be associated with subluxation or severe soft tissue trauma.
- osteochondral fragments may be related to avulsion within associated soft tissues following joint hyperextension.
- axial fragments may be in the straight sesamoidean ligament and/or axial palmar/plantar ligament of the PIP joint.
- abaxial fragments may be within the insertion of the superficial digital flexor tendon (SDFT) or abaxial palmar/plantar ligament of the PIP joint.

Clinical presentation

- depending on configuration:
  - acute-onset severe lameness (complete).
  - mild intermittent lameness (incomplete).
- incomplete P1 fractures usually moderate to marked lameness with effusion/pain of the metacarpo/tarsophalangeal joint.
  - pain on direct palpation over the fracture site may be present.
- complete/comminuted P1and P2 fractures usually present as:
  - severe lameness referable to proximal phalanx with instability.

- P2 fractures, especially uniaxial fractures, may be less unstable on palpation:
  - due to their position mainly within the hoof.
- osteochondral fragments present as incidental findings or low degree of lameness:
  - diagnostic analgesia of pastern joint may be necessary to prove significance.

Differential diagnosis

- distal metacarpal/metatarsal fracture
- fractures of the proximal sesamoid bone.
- subluxation (pastern/fetlock)
- synovial sepsis
- severe soft tissue injury.
- subsolar abscessation.

Diagnosis

- **Radiography** – standard radiographic projections – always four views.
  - document configuration of fracture(s).
- incomplete short fractures:
  - periosteal new bone on the proximodorsal aspect of P1.
  - subchondral sclerosis and cyst-like lesions.
- **CT** may be helpful if reconstruction is considered.

- **MRI** has been used to describe prodromal fracture-like changes in P1.
- **Nuclear scintigraphy** has documented focal increased radiopharmaceutical uptake in the dorsoproximal aspect of P1.

Management

- short, incomplete P1 fractures:
  - can be managed conservatively.
  - surgical repair gives improved healing rates:
    - ◆ repair in standing, sedated horse reduces risk associated with GA.
- complete sagittal P1 fractures usually amenable to internal fixation (Figs. 4.9, 4.10).
- comminuted P1 fractures without an intact strut are serious injuries:
  - guarded prognosis for survival and are often euthanased.
  - salvage of horse through use of transfixation casts or external skeletal devices has been described with/without internal fixation:
    - ◆ high complication rates.
- comminuted P1 fractures with intact strut can be repaired with internal fixation often through open reduction.
- uniaxial eminence fractures can be repaired by internal fixation or treated conservatively if the PIP joint is stable.

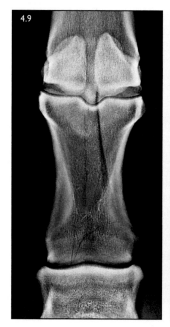 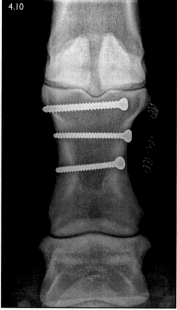

FIGS. 4.9, 4.10 A complete, biarticular P1 fracture with mild displacement prior to repair (4.9) and following the placement of three lag screws placed through stab incisions (4.10).

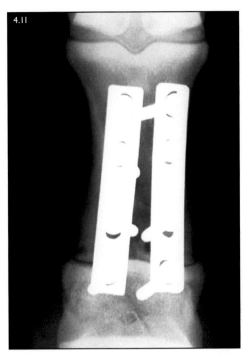

FIG. 4.11 Dorsopalmar radiograph of two narrow DCPs used for a PIP joint arthrodesis following a biaxial P2 fracture.

- biaxial fractures with instability of the PIP joint may require pastern arthrodesis (Fig 4.11).
- comminuted P2 fractures can be managed by transfixation cast and/or pastern arthrodesis depending on degree of comminution, or euthanasia.
- osteochondral fragmentation of the pastern joint can be treated conservatively or surgically depending on the horse's use and degree of lameness:
  - palmar/plantar fragments can be removed arthroscopically.
  - greater soft tissue dissection is required with abaxial compared to axial fragments.

Prognosis

- short, incomplete fractures: favourable with surgery and fair without, with risk of fracture propagation:
  - success rates of up to 61–88% of horses returning to racing.

- osteoarthritis (OA) within the fetlock joint at time of fracture decreases the prognosis.
- survival for comminuted P1 fractures is guarded:
  - those surviving often remain lame.
- fracture healing and arthrodesis of the pastern following management of P2 fractures with joint instability is associated with a fair outcome.

Pastern subluxation

Definition/overview

- subluxation of the PIP joint.

Aetiology/pathophysiology

- usually, due to acute trauma, particularly in relation to collateral ligament avulsion.
- reported in the hindlimb of juvenile horses, without evidence of trauma.
- subluxation in a dorsal plane:
  - usually due to severe disruption of the suspensory apparatus.
- subluxation in a palmar/plantar plane:
  - due to disruption to the distal sesamoidean ligaments and/or insertion of the SDFT.
- avulsion of the collateral ligaments can result in medial or lateral instability.

Clinical presentation

- acute lameness.
- severe soft tissue swelling +/- incongruity of the pastern region (Fig. 4.12).
- avulsion fragments should alert to the possibility of a subluxation having occurred.

Differential diagnosis

- P1/P2 fractures
- severe soft tissue injury
- synovial sepsis.

Diagnosis

- **Radiography** of pastern (four views required):
  - stress-radiographs may be helpful (Fig. 4.13).
- **Ultrasonography** of the pastern to assess the integrity of the soft tissues:
  - e.g. distal sesamoidean ligaments.

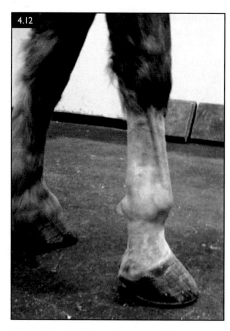
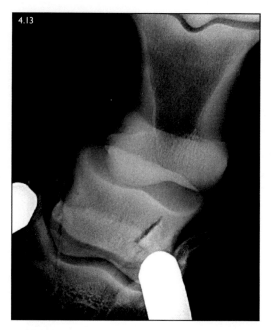

FIGS. 4.12, 4.13 View of a horse with an acute-onset subluxation of the pastern joint (4.12) and a stressed view radiograph (4.13) confirming the PIP joint subluxation.

Management

- traumatic subluxation can be initially managed through external coaptation.
- ongoing instability and/or the presence of fractures:
  - surgical management through arthrodesis of the pastern joint.
- mild cases of dorsal subluxation in young horses are usually managed conservatively.

Prognosis

- following pastern arthrodesis fair–good outcome particularly in the hindlimb.

Osteoarthritis

Definition/overview

- common condition of the pastern joint.

Aetiology/pathophysiology

- progressive condition leading to:
  - loss of articular cartilage, subchondral thickening and osteophyte production.
- advanced cases can lead to cystic formation and joint collapse:
  - eventual ankylosis occurs rarely.
- insidious in onset or a sequel to other joint insults (e.g. fracture, sepsis, subluxation).
- poor limb conformation can be associated with uneven loading across the joint and contribute to its development.

Clinical presentation

- low-grade lameness/performance issues:
  - occasionally associated with an acute flare-up.
  - progression to persistent lameness is common.
- bony thickening, particularly dorsally, over the pastern articulation ('ringbone').
- joint effusion rarely detected but may be positive to distal limb flexion (Fig. 4.14).

Differential diagnosis

- foot or fetlock lameness.
- pastern subluxation and soft tissue injury in pastern region.
- synovial sepsis.

4

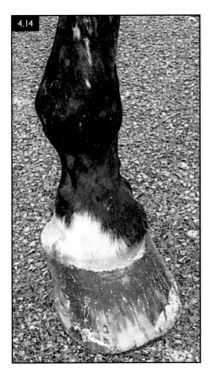

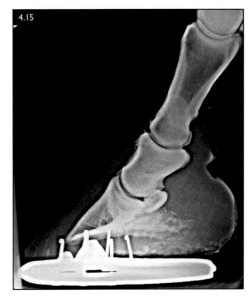

FIG. 4.15 Lateromedial radiograph of the pastern region of a horse with OA of the PIP joint. Note the spikey new bone dorsally on P1 and P2.

FIG. 4.14 A view of the right forelimb of a horse with OA of the PIP joint. Note the swelling around the pastern joint on medial, lateral and dorsal aspects.

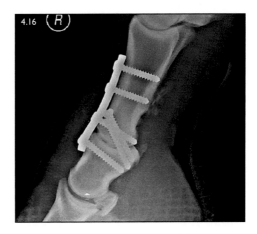

FIG. 4.16 Postoperative lateromedial radiograph of a case of pastern OA treated by pastern arthrodesis.

Diagnosis

- positive intra-articular anaesthesia - can be difficult to enter in chronic cases.
- standard radiographic projections usually show:
  - classic signs of osteoarthritis, particularly dorsally (Fig. 4.15).
  - changes can be moderate/severe:
    - ♦ not necessarily correlated to degree of lameness.

Management

- early stages:
  - rest/light exercise, intra-articular medication and corrective trimming/shoeing.
- later stages, a significant proportion of cases may be helped by:
  - palliative treatment using systemic NSAIDs.
  - joint supplementation.

- end-stage:
  - chemical (70% ethanol) ankylosis, laser facilitated and surgical arthrodesis. (Fig. 4.16).

Prognosis

- chemical ankylosis 60% horses returning to previous or higher levels of work after 8-12 months.

- surgical arthrodesis horses have a good prognosis for return to function:
  - hindlimb success rates of 85% of horses performing their intended use.
  - lower in the forelimb.

Osteochondrosis

Definition/overview

- uncommon in the pastern joint.
- manifestations include osseous cyst-like lesions (OCLL) and osteochondral fragmentation.

Aetiology/pathophysiology

- palmar/plantar osteochondral fragmentation is considered a traumatic incident (see fractures of the proximal and middle phalanges).
- osteochondral fragments on the dorsoproximal margin of P2 have been described but are of unknown clinical significance.
- OCLLs occur in distal P1 or, less commonly, proximal P2 (Fig. 4.17).

Clinical presentation

- low-grade insidious lameness or occasionally as an acute lameness.
- clinical signs relating the lameness to the pastern may be scant.

Differential diagnosis

- pastern OA
- synovial sepsis
- pastern subluxation
- foot/fetlock lameness.

Diagnosis

- localisation by diagnostic analgesia techniques.

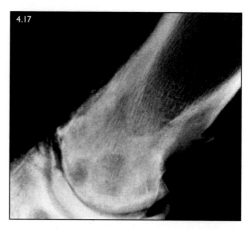

FIG. 4.17 A case of distal proximal phalangeal OCLL with secondary severe OA of the PIP joint.

- radiography or MRI of the pastern region.
- secondary joint disease (OA) may be associated with chronic lesions.

Management

- conservatively with:
  - intra-articular medication and/or palliative treatment.
- surgical debridement, bone grafts or other bone-replacements have been used to resolve the cyst in some cases.
- PIP arthrodesis has also been used.

Prognosis

- generally guarded, and concurrent joint disease significantly reduces the prognosis.

FETLOCK

Sesamoid bone fractures

Definition/overview

- Fractures of the proximal sesamoid bone (PSB) include (Figs. 4.18–4.21):
  - apical (<30% of the height of the bone)
  - mid-body
  - axial
  - basilar
  - abaxial
  - comminuted forms.
  - uniaxial or biaxial.

Aetiology/pathophysiology

- single traumatic incident.
- pre-existing stress adaptive remodelling due to exercise and suspensory ligament tension.

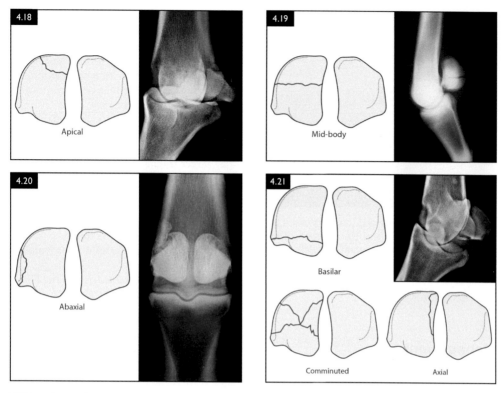

FIGS. 4.18–4.21 Classification of fractures of the PSB with representative radiographs: apical (4.18); mid-body (4.19); abaxial (4.20); basilar (4.21). Comminuted and axial forms only shown in diagram form.

Clinical presentation

- acute-onset lameness with swelling and focal pain on palpation.
- effusion of the fetlock joint.
- biaxial PSB fracture usually severely lame:
  - when weight bearing may show evidence of severe fetlock drop.

Differential diagnosis

- P1 or distal third metacarpal/metatarsal fracture.
- fetlock subluxation.
- severe fetlock joint injury.
- suspensory branch injury.
- palmar annular ligament (PAL) injury and/or digital flexor tendon sheath pathology.
- synovial sepsis.

Diagnosis

- clinical examination although in some cases lameness can resolve rapidly.

- **Radiography** to identify:
  - nature of the fracture.
  - presence of other damage (e.g. small metacarpal/metatarsal bone fracture).
  - multiple views and angles may be required for full assessment:
    - ◆ including proximal-to-distal angulation to separate the affected bone from overlying bone/joint margins.
  - axial fractures may occur with condylar fractures and are easily overlooked.
- **Nuclear scintigraphy** is sensitive for:
  - complete fractures, and bone with evidence of stress remodelling.
- **Ultrasonography** is important to evaluate the involvement of:
  - suspensory ligament branch.
  - intersesamoidean and distal sesamoidean ligaments (Fig. 4.22).

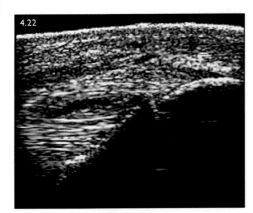

FIG. 4.22 Longitudinal ultrasound image of the lateral proximal sesamoid and the insertion of the associated suspensory ligament branch. Note the mid-body fracture of the sesamoid bone and the associated hypoechoic tear in the suspensory ligament branch.

Management

- uniaxial PSB fractures, particularly in foals or young horses may heal with conservative management and result in elongated bones.
- apical fractures can be removed arthroscopically.

- mid-body fractures can be repaired by:
  - circumferential wiring or screw fixation placed in lag-fashion.
    - latter associated with improved return rates to athletic function.
  - concurrent cancellous bone grafting may assist bone healing.
- basilar fractures are difficult to manage:
  - smaller fragments can be burred/removed arthroscopically.
  - larger fragments: moderate dissection and associated damage to ligaments.
  - internal fixation may not be possible due to the thinness of the bone and risk of splitting the fragment.
- abaxial fragments can be treated conservatively if non-articular or removed arthroscopically.
- fractures on the axial margin of the PSB (Figs. 4.23, 4.24).
  - difficult to manage and associated with:
    - condylar fractures or avulsion fractures of the intersesamoidean ligament.
    - may be possible to debride arthroscopically if visible.
- biaxial PSB fracture cases are often euthanased or salvaged through fetlock arthrodesis (Fig. 4.25).

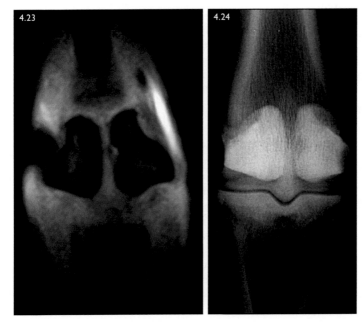

FIGS. 4.23, 4.24 MRI image (4.23) and dorsopalmar radiograph of the fetlock region (4.24) revealing a intersesamoidean ligament avulsion and associated bony changes in the axial PSB.

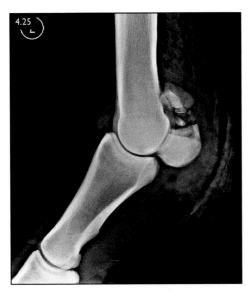

FIG. 4.25 A biaxial PSB fracture and suspensory apparatus breakdown in a Thoroughbred racehorse.

Prognosis

- apical fractures following removal – good, particularly in the hindlimb.
- mid-body fractures better prognosis following lag screw fixation.
- basilar fractures are associated with guarded prognosis for return to athletic function:
  - fractures that do not extend to palmar/plantar surface better prognosis.
- axial fractures with a condylar fracture is associated with a reduced outcome.
- abaxial PSB fracture following removal – good.
- biaxial PSB fractures with a severe breakdown injury – poor prognosis.

Sesamoiditis

Definition/overview

- common finding in young racehorses.
- inflammation of the soft tissues around the palmar aspect of the fetlock and associated bony changes in the PSBs.

Aetiology/pathophysiology

- repeated stresses on palmar aspect of the fetlock and PSBs can lead to tearing of the soft tissue attachments.

- cycle of incomplete healing and further damage can result in:
  - entheseopathy and bony modelling of the sesamoid bones.
- may be associated with chronic suspensory ligament injury.
- acute single overload injury may result in acute inflammation of this region.

Clinical presentation

- unilateral low-grade chronic lameness:
  - thickening/heat over the palmarolateral/palmaromedial fetlock region.
- acute tearing may show more overt lameness and pain localised to this region.
- **always examine the suspensory ligament structures for concurrent injury.**

Differential diagnosis

- PSB fracture.
- PAL injury and avulsion.
- SL branch injury.

Diagnosis

- regional perineural analgesia may be required to localise the lameness.
- **Radiography** with four standard views:
  - assess palmar aspect of PSB on oblique views.
  - changes in bone density with radiating lytic lines around vascular channels.
  - entheseopathy resulting in bony enlargement and change in shape of PSB:
    - usually lengthening (Fig. 4.26).
  - compare to the contralateral limb to determine:
    - degree of change.
    - assess the joint for concurrent radiographic changes (e.g. OA).
- **Ultrasonography** of the region and suspensory apparatus is advised (Fig. 4.27).

Management

- reducing inflammation and preventing recurrence:
  - local cold therapy o rest o NSAIDs.
  - controlled exercise programme.
- Extracorporeal shock-wave therapy may be helpful in refractory cases.
- concurrent suspensory ligament injury will also require appropriate management.

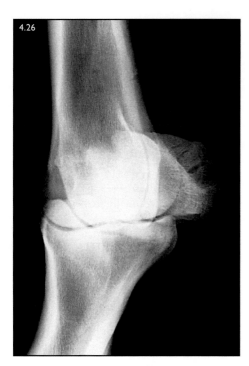

FIG. 4.26 Oblique radiographic view of the right foreleg fetlock joint of a horse with sesamoiditis revealing radiolucent lines radiating across the sesamoid bone and bony enlargement resulting in change in shape.

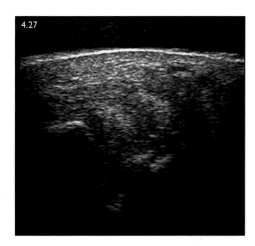

FIG. 4.27 Transverse ultrasound view of the medial distal suspensory ligament branch insertion onto the sesamoid bone in a case of severe sesamoiditis. Note the damaged ligament and the irregular bone interface.

Prognosis

- young racehorses with moderate sesamoiditis (enlarged vascular canals >2 mm in width) are associated with reduced performance.
- concurrent soft tissue injury (e.g. suspensory ligament injury) is a negative finding.

Osteochondral fragmentation and palmar/plantar eminence fractures of proximal P1

Definition/overview

- osteochondral fragmentation of the proximal phalanx occurs either dorsally or palmarly/plantarly.
- palmar/plantar eminence fractures describe acute traumatic avulsion fractures of the palmar or more commonly, plantar, eminence of proximal P1.

Aetiology/pathophysiology

- osteochondral fragmentation is thought to have a traumatic origin:
  - ○ fragments may be clinically silent and found on survey radiographs.
  - ○ may contribute to lameness, particularly at high speed.
- acute fracture of the palmar/plantar eminence of P1 is usually traumatic in origin.

Clinical presentation

- joint effusion and/or lameness related to the fetlock joint.
- diagnostic analgesia is important to determine clinical significance of the finding of an osteochondral fragment.
- acute fractures of the palmar/plantar eminence of proximal P1 usually present with acute lameness and pain/swelling of the palmar/plantar fetlock.

Differential diagnosis

- OCD
- PSB fracture.

Diagnosis

- standard radiographic evaluation of the fetlock:
  - ○ different angulations may be required to demonstrate fragment/configuration.

- osteochondral fragments of dorsoproximal P1 are demonstrated as:
  - discrete osseous bodies on the medial or lateral proximal rim of P1.
  - usually uniaxial but occasionally biaxially (Figs. 4.28, 4.29).
  - radiographs of the contralateral limb are recommended.
- palmaro/plantaroproximal osteochondral fragments of P1 can be described as:
  - axial or abaxial.

- may be articular/non-articular and often seen in the hindlimbs (Figs. 4.30, 4.31).
- acute fractures present as bony fragments with:
  - sharp marginations.
  - degree of distraction through the insertion of:
    - distal sesamoidean or proximal digital annular ligaments.

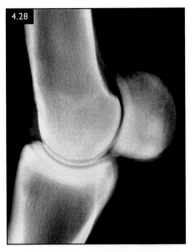 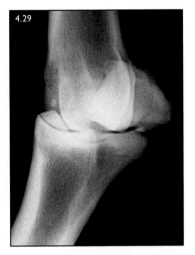

FIGS. 4.28, 4.29 Lateromedial radiograph of a fetlock joint showing a small rounded osteochondral fragment off the proximal aspect of P1 (4.28). A dorsolateral/palmaromedial view confirms it has arisen from the dorsomedial aspect (4.29).

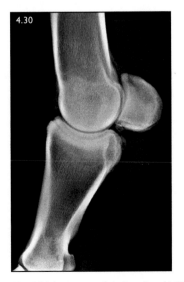 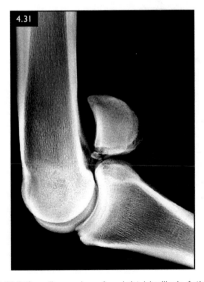

FIGS. 4.30, 4.31 Lateromedial standing (4.30) and flexed (4.31) radiographs of a right hindlimb fetlock joint revealing a small rounded osteochondral fragment just distal to the PSB and proximal to the plantar aspect of P1.

Management

- conservative, if the clinical significance of a fragment is questionable.
- surgical removal of the fragment if causing lameness or future sale considered.
  - arthroscopically, plus assessment of cartilage.
- large palmar/plantar fragments are often chronic and managed conservatively.
- acute fractures of the palmar/plantar eminence of P1 (Figs. 4.32–4.34) can be:
  - stabilised through internal fixation.
  - if small enough, removed.

Prognosis

- good for osteochondral fragments if removed and no comorbidity (e.g. arthritis) is present.

Osteochondrosis

Definition/overview

- OCD of the sagittal ridge of the distal third metacarpus and OCLL (osteochondral lytic lesions) of the condyle of distal third metacarpus.

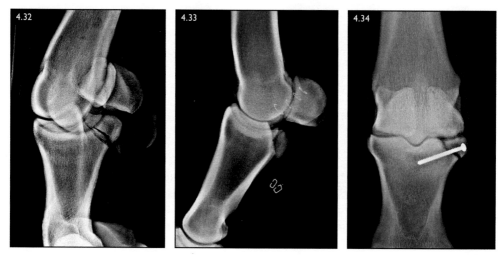

FIGS. 4.32–4.34 Acute traumatic fractures of the plantar eminence of P1 (4.32) can be treated by removal (4.33) or by internal fixation (4.34).

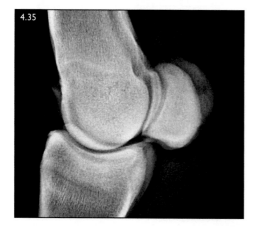
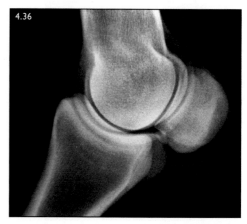

FIGS. 4.35, 4.36 Standing lateromedial radiographs of two fetlock joints with proximal dorsal sagittal ridge OCD lesions. The small OCD fragment is sat within a defect in the proximal ridge (4.35) or visible more clearly as a rounded, apparently separate fragment (4.36).

Aetiology/pathophysiology

- may be seen in all four fetlock joints:
  - flattening or dissecting lesions of the sagittal ridge (Figs. 4.35, 4.36).
- fragmentation in young horses (6–24 months old) including the palmar region (Fig. 4.37).
- OCLLs usually involve the condyle of the distal third metacarpus (Figs. 4.38, 4.39) and present in young horses (1–2 years old).
  - rarely involve proximal P1 near the sagittal groove and may be associated with short, incomplete P1 fractures.

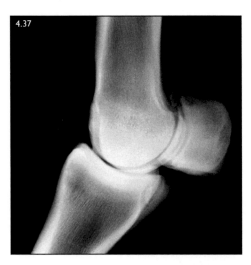

FIG. 4.37 Standing lateromedial radiograph of a palmar fetlock OCD fragment just dorsal to the PSB.

Clinical presentation

- young horses.
- variable lameness and/or fetlock joint effusion(s).
  - dorsal sagittal ridge lesions present with mild lameness and effusion.
  - palmar aspect problems tend to be lamer.
  - OCLL may present with intermittent moderate–severe lameness:
    - exacerbated by distal limb flexion.

Differential diagnosis

- fetlock joint trauma.
- PSB fracture (e.g. apical/abaxial).
- osteochondral fragmentation of proximal P1.
- soft tissue mineralisation (e.g. suspensory branch, distal sesamoidean ligaments).

Diagnosis

- **Radiography** using a full set of views:
  - dorsal sagittal ridge best evaluated by:
    - lateromedial and flexed lateromedial views.
    - dorsoproximal/dorsodistal view skylines ridge in the forelimb.
  - flattening of sagittal ridge, subchondral bone lucency and fragmentation.
  - assessment of the other joints is recommended.

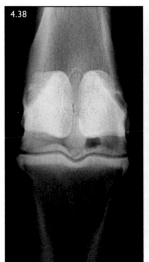
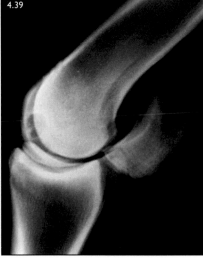

FIGS. 4.38, 4.39 Dorsopalmar (4.38) and flexed lateromedial (4.39) radiographs of a fetlock joint with a well-demarcated osseous cyst-like lesion just medial to the sagittal ridge.

- **Ultrasonography** is also useful in assessment of the sagittal ridge.
- OCLLs can be evaluated through standard radiographic views:
  - location of OCLL on lateromedial and flexed lateromedial views.
    - important if surgical intervention is planned.
  - increased radiopharmaceutical uptake on scintigraphy in active OCLLs.

Management

- dorsal sagittal ridge lesions:
  - managed surgically with arthroscopic removal of fragments and debridement of poorly attached cartilage/subchondral bone.
  - mild flattening of the sagittal ridge may be initially managed conservatively.
- palmar lesions may or may not be accessible surgically and OA is often a sequela.
- OCLLs can be managed conservatively or surgically:
  - surgical treatment involves curettage of the cyst or intracyst injection of corticosteroids but depends on cyst location and accessibility.

Prognosis

- good for OCD of the dorsal sagittal ridge.
- guarded for palmar lesions.
- fair prognosis following debridement for OCLL.
- all cases, the presence of OA reduces the outcome.

Palmar/plantar osteochondral disease (POD)

Definition/overview

- POD is a degenerative condition affecting the distal condyles of the third metacarpal/metatarsal bone almost exclusively seen in young racing Thoroughbreds.

Aetiology/pathophysiology

- possibly caused by repetitive high strains on bone and articular tissues in the distal condyles of the third metacarpus/metatarsus experienced in training and racing.

- post-mortem findings:
  - cartilage wear lines and loss and collapse of the articular surface.
  - underlying subchondral bone pathology.
- POD grades reported to be:
  - higher following use of intra-articular corticosteroids (triamcinolone) in training.
  - lower in horses with longer intervals between races.

Clinical presentation

- suspected in young racehorses with lameness:
  - one or more limbs involving the fetlock region.
  - **some cases have no signs referable to the fetlock.**
  - mild pain and performance-limiting to overt pain localised to the fetlock joint.

Differential diagnosis

- fetlock OA
- osteochondrosis
- osteochondral fragmentation.
- PSB fracture.

Diagnosis

- often better response to perineural rather than intra-articular analgesia:
  - probably due to the underlying pathophysiology of the condition.
- **Radiography** including flexed dorsopalmar/plantar and elevated oblique views.
  - minimal changes.
  - changes in bone density in the palmar condyles.
    - focal radiolucencies ◆ sclerosis.
  - alteration to the outline of the subchondral bone.
  - joint incongruity with evidence of flattening of the condyle (Fig. 4.40).
  - uncommonly radiographic evidence of OA may be present.
- **Nuclear scintigraphy** shows moderate to marked increased radiopharmaceutical uptake in the condylar region of the fetlock.
- **MRI** examination can be used to differentiate from other fetlock pathology (e.g. prodromal condylar fracture).

○ over 50% of racing Thoroughbreds with lameness localised to the fetlock region have signs consistent with POD (often biaxially) (Figs. 4.41, 4.42).

Management

- intra-articular corticosteroids are commonly used in these cases but are known to lead to exacerbated changes and poorer prognosis.

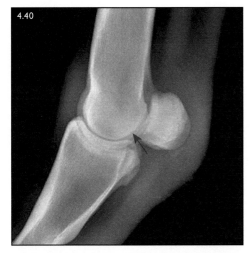

FIG. 4.40 Lateromedial radiograph of the left hind fetlock joint in a 5-year-old Thoroughbred flat racehorse with bilateral hindlimb lameness. Note the flattening of the plantar condyle of the distal third metatarsal bone (arrow), with wedge-shaped subchondral bone sclerosis.

- alterations to training regimes can be associated with positive outcomes.
  ○ limited periods of rest between races may be associated with increased severity of POD.
- common finding at post-mortem and significant contributor to wastage amongst Thoroughbred racehorses.

Prognosis

- variable.
  ○ mild cases often resolve but can be recurrent.
  ○ severity has been linked to reduced race starts/earnings.

Chronic proliferative synovitis (villonodular synovitis)

Definition/overview

- soft tissue mass in the dorsal aspect of the fetlock joint, more common in the forelimb(s).

Aetiology/pathophysiology

- chronic repetitive trauma to the dorsal aspect of the fetlock from hyperextension:
  ○ leads to hyperplasia of the dorsal bilobed synovial pad over the sagittal ridge.

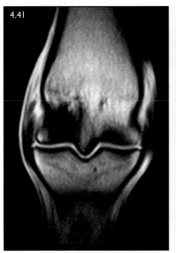

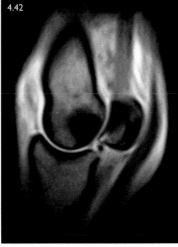

FIGS. 4.41, 4.42 MRI clearly demonstrates the pathology of POD affecting the lateral condyle on the dorsoplantar view (4.41) and the plantar condyle distribution on the lateromedial view (4.42).

- osteoclastic resorption leads to supracondylar lysis of the underlying bone.
- chronic dorsoproximal P1 fragmentation may also result in synovial pad hyperplasia.

Clinical presentation

- palpable thickening over the dorsal aspect of the affected fetlock joint with or without joint effusion.
- reduction in range of motion of the joint in some cases.
- lameness usually worsens following exercise and distal limb flexion.

Differential diagnosis

- fetlock OA and joint trauma.
- synovial sepsis.
- injury or infection of the common digital extensor bursa.

Diagnosis

- **Radiography** may show a crescent-shaped radiolucency on the dorsal aspect of the distal third metacarpus due to cortical lysis:
  ○ dystrophic mineralisation may also be present.
  ○ contrast arthrography may provide further detail.
  ○ evidence of OA, especially palmar aspect of the joint (Fig. 4.43).
- **Ultrasonography** is sensitive in demonstrating:
  ○ thickening of the synovial pad (>10 mm).
  ○ associated joint changes at the dorsal aspect.

Management

- medical therapy and changes to the exercise regime may help in some cases.
- intra-articular hyaluronan and corticosteroids, with rest initially, followed by alteration in training pattern, are often helpful.
- surgical excision by sharp debridement or laser dissection performed arthroscopically:
  ○ removal of associated fragmentation performed concurrently (Fig. 4.44).

Prognosis

- variable but generally guarded to poor and depends on the presence of OA.

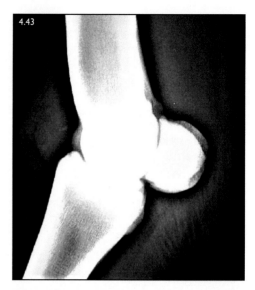

FIG. 4.43 Lateromedial radiograph of a metacarpophalangeal joint with chronic proliferative synovitis showing a crescent-shaped radiolucency on the dorsal aspect of the sagittal ridge due to cortical lysis. Note also the dorsal fetlock soft tissue enlargement with mild ossification, and signs of OA of the fetlock joint (remodelling of the proximal aspect of the PSB).

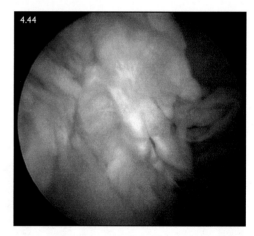

FIG. 4.44 Arthroscopic view of the dorsal pouch of the fetlock joint of a 2-year-old Thoroughbred racehorse. Note the enlarged dorsal plica with synovial proliferation and mineralisation (bright white substance).

- resolution can be temporarily achieved, but recurrence is common unless an underlying cause can be found.

Fetlock joint osteoarthritis

Definition/overview

- degenerative joint disease involving the metacarpo/tarsophalangeal joint.

Aetiology/pathophysiology

- landmarks of this disease are:
  - articular cartilage loss.
  - subchondral bone sclerosis.
  - periarticular osteophyte and entheseophyte formation.
  - subchondral cystic formation.
  - joint collapse.
- horses of any age.
- sequel to, or associated with intra-articular insults such as:
  - sprain/luxation, villonodular synovitis, intra-articular fracture, synovial sepsis and osteochondrosis.

Clinical presentation

- present with variable and fluctuating lameness:
  - exacerbated by distal limb flexion.
- synovial effusion.
- periarticular fibrosis may restrict range of motion and cause pain in advanced cases.

Differential diagnosis

- Villonodular synovitis   • joint sprain.
- trauma or fracture.

Diagnosis

- clinical examination, particularly response to joint flexion.
- intra-articular or perineural analgesia usually assist in the localisation of lameness.
- **Radiographic** signs of OA in the fetlock include (Figs. 4.45, 4.46):
  - periarticular osteophyte formation.
    - ♦ proximodorsal aspect of P1.
    - ♦ dorsoproximal/dorsodistal margins of PSBs.
  - remodelling of proximal aspects of the dorsal and palmar sagittal ridges.
  - subchondral bone sclerosis.

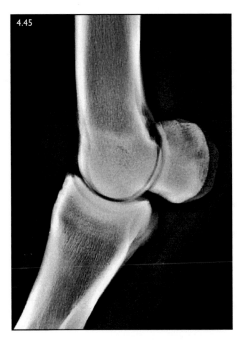
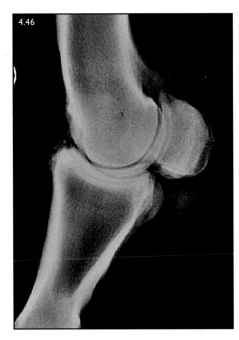

FIGS. 4.45, 4.46 Examples of lateromedial radiographs of fetlock joints with mild (4.45) and severe (4.46) changes of OA. In the mild case there is an example of subtle periarticular osteophyte formation on dorsoproximal P1 and remodelling of the proximal dorsal part of the sagittal ridge. In the severe case, these are accompanied by dorsoproximal PSB osteophytes, subchondral bone sclerosis, and reduction in the joint space.

- o irregular basilar fragments and joint space reduction.
- o **radiographic changes always lag behind clinical disease.**
- **Nuclear scintigraphy** may show increased radiopharmaceutical uptake in the fetlock region with OA.

Management

- mild cases often respond to intra-articular medication (hyaluranon/corticosteroids).
- intra-articular stem cells or other biological agents (such as autologous conditioned serum) as well as hydrogels may provide temporary improvement of signs.
- disease process is often progressive:
  - o later stage cases, treatment is often palliative.
- surgical fetlock arthrodesis:
  - o severe cases (salvage of the horse at the expense of rideability).

Prognosis

- long-term for horses with fetlock OA is poor.

Fetlock joint subluxation (luxation)

Definition/overview

- subluxation may be open or closed.

Aetiology/pathophysiology

- usually occurs due to trauma (e.g. foot trapped in a cattle grid or due to fall).
- severe traumatic disruption:
  - o medial or lateral collateral ligaments.
  - o tearing of the joint capsule and other supporting structures.
  - o leads to incongruity of the articular margins.
- avulsion fractures may also occur.
- osteoarthrosis is common in the longer term.

Clinical presentation

- acutely severe lameness:
  - o obvious anatomical anomalies.

- o in closed subluxations, the articular surfaces may be congruent.
- marked soft tissue swelling.
- manipulation often reveals joint instability.
- open subluxations show overt derangements and often severe contamination.

Differential diagnosis

- severe joint sprain/trauma
- intra-articular fracture   • synovial sepsis
- suspensory apparatus breakdown
- pastern subluxation

Diagnosis

- clinical examination including severe disruption to the joint.
- **Radiography** (Fig. 4.47) will reveal:
  - o presence of the luxation.
  - o stress views may be required for a definitive diagnosis.
  - o check for concurrent injuries such as avulsion fractures.
- **Ultrasonography** is indicated to assess:
  - o damage to the collateral ligaments (Fig. 4.48).
  - o other soft tissues associated with the fetlock joint.

Management

- closed luxations can be managed by casting for 6–8 weeks (Fig. 4.49).
- open luxation requires:
  - o early aggressive lavage and debridement.
  - o followed by external coaptation until any infection is resolved.
  - o casting for at least 6–8 weeks.
- ongoing instability requires fetlock arthrodesis (Fig. 4.50).
- open, contaminated cases, particularly with concurrent injuries (e.g. avulsion fracture):
  - o euthanasia may be necessary.

Prognosis

- favourable for closed fetlock subluxation.
- guarded to poor for open, contaminated luxation.

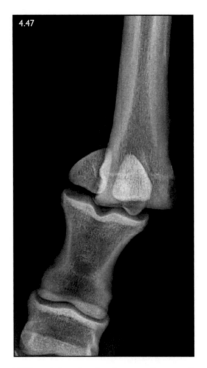

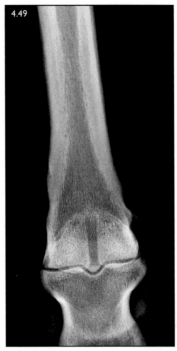

FIGS. 4.47 and **4.49** Dorsopalmar radiographs of a closed fetlock subluxation pre (4.47) and 5 months post reduction (4.49).

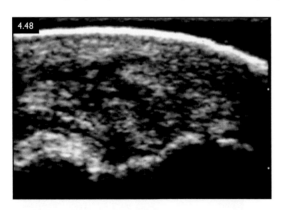

FIG. 4.48 Longitudinal ultrasound scan of the medial collateral ligament of the fetlock joint which has been severely damaged. Note the grossly disrupted heterogenous pattern of the ligament overlying the joint. Proximal is to the right with the medial distal MC3 and proximal P1 bone outlines visible at the bottom of the image.

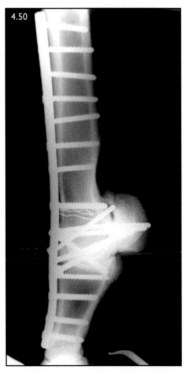

FIG. 4.50 Intraoperative lateromedial radiograph of a horse undergoing a fetlock arthrodesis.

METACARPALS/METATARSALS

Fractures of the third metacarpal/metatarsal bone

Definition/overview

- fractures of the third metacarpal/metatarsal bone include:
  - condylar (lateral and medial).
  - diaphyseal in different configurations.
  - transverse   ○ distal physeal
  - proximal articular fractures.

Aetiology/pathophysiology

- all (except condylar fractures) are a result of single acute overload injury or external trauma (e.g. kick injury).
- condylar fractures are the result of bone failure:
  - repetitive strain cycles altering the biomechanical properties of the bone.
  - particularly at or close to the parasagittal groove of the articular surface:
    - where fracture often originates.
  - lateral condylar fractures (Fig. 4.51):
    - commonly exit through the lateral cortex 1–3 cm above the physeal scar:
      - complete/incomplete.
      - displaced/non-displaced.
    - may be associated with additional injuries such as:
      - axial fractures of the PSB.
      - fragmentation at articular surface of distal palmar condyle.
  - medial condylar fractures:
    - tend to extend into the diaphysis of the bone in a spiral or less commonly Y-shaped pattern (Fig. 4.52).
  - occasionally biaxial condylar fractures may occur.
- diaphyseal fractures may be:
  - complete/incomplete through direct trauma
  - open/closed and simple/transverse/comminuted (Fig. 4.53).
- transverse fractures of the distal third part of the metacarpal bone are:
  - non-articular fractures occurring through or close to the metaphysis.

- usually involving the palmar (or dorsal) cortices.
  - cause is unknown:
    - possibly due to a single overload/bending of the bone at exercise.
- distal diaphyseal fractures generally occur in foals following:
  - trauma from the mare.
  - often Salter–Harris type II.
- proximal articular fractures:
  - involve the articulation between the third metacarpus/metatarsus and carpometacarpal/ tarsometatarsal joint.
  - following an acute single overload injury (Fig. 4.54).

Clinical presentation

- condylar fracture cases are usually:
  - lame with fetlock effusion.
  - and/or pain localising to the distal metacarpus/metatarsus or fetlock region.
  - possible previous history of lameness involving the fetlock prior to fracture.

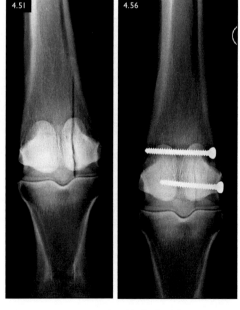

FIGS. 4.51 and **4.56** (4.51) Complete, minimally displaced lateral condylar fracture pre internal fixation. (4.56) Post lag screw internal fixation.

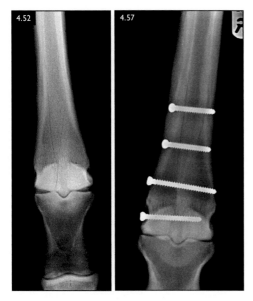

FIGS. 4.52 and **4.57** (4.52) Medial condylar fracture pre internal fixation. (4.57) Post lag screw internal fixation.

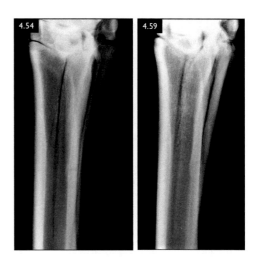

FIGS. 4.54, 4.59 Incomplete, non-displaced, articular proximal fracture (4.54) and post conservative management (4.59).

FIG. 4.55 This Thoroughbred racehorse was involved in a road traffic accident and has sustained an open comminuted fracture of the distal third metatarsus. It was euthanased immediately.

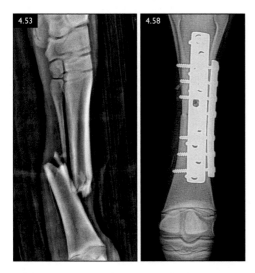

FIGS. 4.53, 4.58 Complete displaced mid-diaphyseal fracture pre internal fixation (4.53). Post double-plate internal fixation (4.58)

- complete diaphyseal fracture cases show:
  ○ severe lameness.
  ○ bone fragments may be displaced, overridden and with a wound (Fig. 4.55).

- incomplete diaphyseal, transverse or proximal articular fracture cases will be:
  ○ acutely lame, but the lameness may resolve quickly with rest.
  ○ clinical examination may reveal focal soft tissue swelling and pain.
    ♦ occasionally there are few localising signs present.
- foals with distal diaphyseal fractures usually have:
  ○ overt lameness and instability associated with the fracture.

Differential diagnosis

- P1 fracture   • fetlock subluxation
- severe soft tissue injury synovial sepsis.

Diagnosis

- physical examination, particularly with complete, unstable fractures.
- **Radiographs** are required for a definitive diagnosis:
  - additional views to completely assess and type the fracture:
    - particularly those involving articulations.
- **Standing CT** or **CBCT** (cone-beam computerised tomography)
  - best method to assess the fracture and popular at referral centres.
- occasionally for incomplete fractures:
  - scintigraphy or MRI may be required to diagnose.
- **responses to local analgesia may be confusing in some cases.**
  - **discouraged if a fracture is suspected.**

Management

- initially, stabilisation of the fracture and assessment as to suitability to travel.
- **Open or severely comminuted fractures may require immediate euthanasia.**
- radiographic assessment to determine initial management and prognosis.
- Condylar fractures are usually managed through internal fixation.
  - lateral condylar fractures:
    - two cortical screws placed in lag-fashion are usually sufficient to deal with most fractures (Fig. 4.56).
    - non-displaced fractures can be treated through stab incisions:
      - commonly performed in the standing sedated horse.
    - displaced fractures require reduction before fixation:
      - arthroscopic assessment of joint surfaces.
  - medial condylar fractures:
    - internal fixation either using:
      - screws alone (Fig. 4.57) or screws and a plate.
      - plating techniques are superior as fractures tend to propagate proximally.

- Complete diaphyseal fractures can be managed through double-plating technique (Fig. 4.58).
- Incomplete diaphyseal fractures can be managed conservatively.
  - radiographic monitoring of healing (Fig. 4.59).
- Transverse and proximal articular fractures are usually managed conservatively.
  - displacement can occur during the rehabilitation period.
  - may then require internal fixation or euthanasia.
- Distal physeal fractures in foals can be managed:
  - conservatively in foals <6 weeks with casting for 2–3 weeks followed by bandaging.
  - older foals (>6 weeks) or unstable fractures by internal fixation.

Prognosis

- return to athletic use following a fracture involving the third metacarpal/metatarsal bone is dependent on:
  - type of fracture.
    - articular/non-articular
    - simple/comminuted.
    - open/closed
    - complete/incomplete
    - displaced/non-displaced.
  - presence of contamination.
  - soft tissue and vascular compromise.
  - size, age and temperament of patient.
- 70–80% of racehorses return to some function (less earnings) following management:
  - non-displaced lateral condylar fractures.
  - reduces to around 50% if displaced.

Dorsal metacarpal disease

Definition/overview

- covers the 'sore' or 'bucked shin' complex and dorsal cortical stress fractures.

Aetiology/pathophysiology

- bone is a dynamic tissue and responds to stress by:
  - remodelling leading to deposition in areas of tension.

o thickening and reduction in the inertial properties of the bone.

- if bone adaptation does not keep up with increased stress (increased training level):
  o painful periosteitis occurs ('sore or bucked shins').
- occurs mainly in 2-year-old racehorses with a recently increased exercise level.
- as 3-year-olds, some of these horses can develop dorsal cortical 'stress' fractures.

Clinical presentation

- pain/heat or swelling over the dorsal (often dorsomedial) metacarpus (Fig. 4.60).
- **lameness is uncommon.**
- may present with reduced performance or an unwillingness to train.

Differential diagnosis

- exostosis
- local trauma.

Diagnosis

- history, signalment and physical examination.
- **Radiography** may confirm findings:
  o soft tissue swelling and irregular periosteal new bone.
  o oblique radiolucent lines confirm dorsal cortical stress fracture (Fig. 4.61).
    ♦ usually at a 30–40° angle to the cortex.
    ♦ extend proximally (occasionally distally).
    ♦ through about 60% of the bone.
    ♦ rarely the line continues and curves back to the surface.
      – 'saucer' fracture configuration.
- **Nuclear scintigraphy** shows increased uptake in dorsal metacarpus (Fig. 4.62).

FIG. 4.60 Two-year-old Thoroughbred racehorse in dorsal recumbency under general anaesthesia about to have arthroscopic surgery of its intercarpal joints. Note the extreme bony swellings on the dorsal aspect of both cannon bones caused by bucked shins.

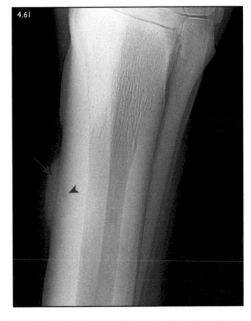

FIG. 4.61 Dorsomedial–palmolateral oblique radiograph of the cannon region of a young Thoroughbred horse showing a dorsolateral third metacarpal stress fracture. Note the obliquely running fracture line in the dorsolateral cortex (arrowhead) and the smooth periosteal new bone (arrow).

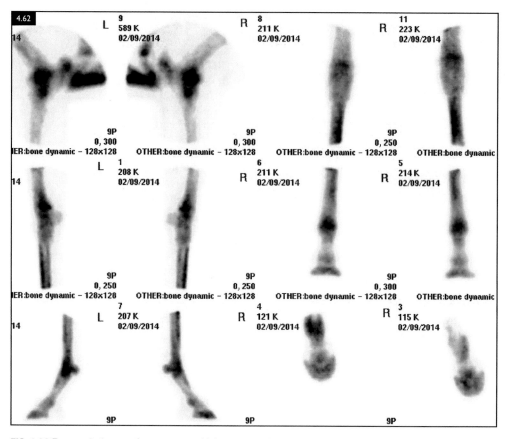

FIG. 4.62 Bone scintigram of an eventer which presented with a bilateral forelimb lameness and sore cannon bones on palpation. Note the marked radiopharmaceutical uptake down the dorsal aspects of both dorsal third metacarpal bones on the lateral and dorsopalmar views.

Management

- mild cases – review and alter training.
- moderate cases – a period of rest, NSAIDs and local treatment (acute phase cold therapy).
- persistent cases, non-responsive to management changes – extracorporeal shock-wave therapy has been used successfully.
- surgical management includes osteostixis combined with screw placement in the dorsal cortex.

Prognosis

- outcome following management changes is generally good.

Palmar cortical stress fractures

Definition/overview

- caused by abnormal bone loading in the palmar aspect of the proximal third metacarpus/metatarsus (usually forelimb).

Aetiology/pathophysiology

- usually occur in the proximal third metacarpus/metatarsus in the region of the attachment of the suspensory ligament (usually medial side).
- abnormalities range from stress reactions to a fulminant cortical fracture.

4

Clinical presentation

- moderate lameness.
- heat/pain, particularly on direct pressure over the medial head of the suspensory ligament.

Differential diagnosis

- avulsion fracture.

Diagnosis

- clinical signs may suggest injury in the proximal metacarpal region.
- diagnostic analgesia is important in localising lameness to this region:
  - unable to discriminate between this and other causes of pain in the area.
- **Radiography** may show a linear radiolucency in the proximal metacarpal region:
  - medial to midline with sclerosis adjacent to the line (Fig. 4.63).

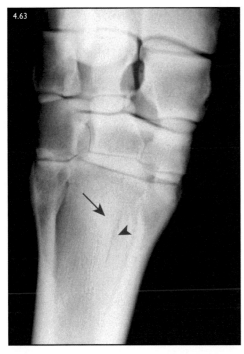

FIG. 4.63 Dorsopalmar radiograph of the carpus and proximal cannon of a horse with a lameness localised to the proximal cannon region by peri-neural analgesia. Note the longitudinal palmar cortical fracture (arrowhead) and surrounding subchondral bone sclerosis (arrow).

- **Ultrasonography** may show:
  - periosteal changes at or near the origin of the suspensory ligament.
  - often with no abnormalities in the ligament.
- **MRI** may provide further evidence for a palmar cortical injury:
  - evaluate suspensory ligament and other structures (e.g. interosseous ligament).

Management

- conservative treatment with rest and NSAIDs.

Prognosis

- generally good.
  - concurrent suspensory ligament injury reduces the outcome.

Fractures of the second/ fourth metacarpal/ metatarsal bones

Definition/overview

- commonly known as 'splint bone' fractures.

Aetiology/pathophysiology

- common in equine practice and many cases are due to external trauma (kick injury).
- most common fracture is of the lateral or fourth metatarsal bone.
- fractures can occur proximally, mid-body or distally.
- **carpometacarpal/ tarsometatarsal joint can be involved with proximal splint bone fractures.**
  - **important both in the management and prognosis of the case.**
- fractures can be open or closed, simple or comminuted and may be associated with a wound/contamination.
- fractures involving the distal part of the splint bone may occur without overt trauma.
  - concurrent suspensory desmitis and reduced fetlock support:
    - ♦ can lead to chronic stress remodelling on the distal aspect of the bone.
    - ♦ may lead to failure of the bone and fracture.

Clinical presentation

- fractures due to external trauma usually present with acute lameness:
  - initially, with swelling centred over the affected region.
- open fractures, a wound may be present (Fig. 4.64).
  - proximal fractures, involvement of the carpometacarpal/ tarsometatarsal joint requires early clarification (Fig. 4.65).

Differential diagnosis

- third metacarpal/metatarsal fracture
- synovial sepsis.
- exostosis of second/fourth metacarpal/metatarsal bone.

Diagnosis

- clinical examination including palpation of fracture fragments.
- **Radiography** using standard projections may confirm the fracture and configuration:
  - additional projections may be required to fully type the fracture and concurrent fractures of the third metacarpus/metatarsus (Figs. 4.66–4.69).

- fracture lines may be obscured by:
  - soft tissue swelling and gas.
  - presence of lines overlying the margins of the bone and the associated third metacarpal/metatarsal bone.
  - repeat radiographs may be necessary.
  - sequestration formation may be evident in chronic fractures.
  - distal aspect of the splint bone fractures may show:
    - evidence of chronicity and callus formation.
    - diagnostic analgesia may be required to determine clinical significance.
- **Ultrasonographic assessment of associated soft tissues is advised.**
- standing MRI or CT/CBCT may be indicated in acute cases to rule out a third metacarpal/tarsal fracture prior to surgery under general anaesthesia.

Management

- open fractures initially:
  - lavage and local debridement.
  - removal of debris and small loose fragments.
  - limb supported appropriately.

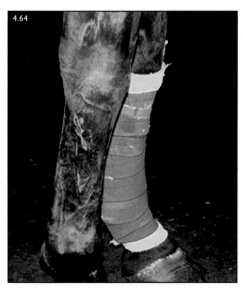

FIG. 4.64 Oblique laceration following a kick to the right forelimb lateral splint leading to an MC 4 comminuted fracture.

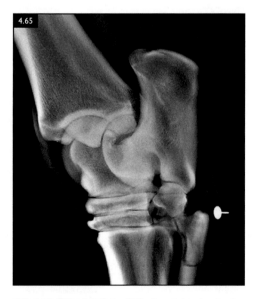

FIG. 4.65 **[Fig 1.278 in 2E]** A proximal articular, minimally displaced MT 4 fracture.

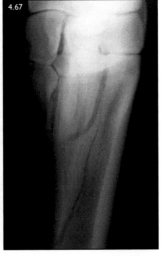

FIGS. 4.66, 4.67 (4.66) Dorsolateral/palmaromedial radiograph of the proximal cannon region of the case shown in Fig 4.64. Note the comminuted fracture of MC 4, with displacement of part of the proximal fragment to the palmar aspect. (4.67) Dorsomedial/palmarolateral radiograph of the same horse. Note the non-displaced fracture of the dorsolateral aspect of the proximal MC 3.

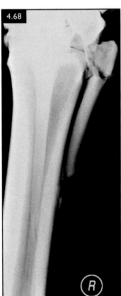

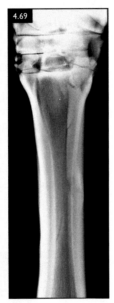

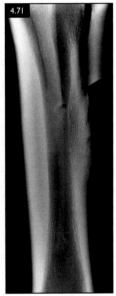

FIGS. 4.68, 4.69 Dorsolateral–plantaromedial oblique radiographic view (4.68) of a conminuted MT 4 fracture with a decreased exposure to highlight the smaller less radiodense bone. The dorsoplantar view (4.69) reveals multiple fracture lines along the length of MT 4.

FIGS. 4.70, 4.71 A distal-third MT 4 fracture before (4.70) and after (4.71) surgical removal.

- sequestrum formation leading to chronic draining tract can be managed through local debridement in many cases.
- refractory open cases or in chronic, non-healing closed fractures of the distal portion of bone:
  - partial ostectomy of the distal portion should be considered.

  - remove affected piece of bone and associated infected soft tissues (Figs. 4.70, 4.71).
- removal of the entire fourth metatarsal bone is possible due to its shared articulation with the fourth tarsal bone:
  - luxation of the tarsometatarsal bone during recovery from general anaesthesia has been recorded in some cases.

- internal fixation of the splint bone is indicated where there is a risk of instability at the articulation proximally.

Prognosis

- conservative management of splint bone fractures results in a good outcome in many cases.
- partial ostectomy (or complete with the fourth metatarsal bone) has a good outcome.

Exostosis of the second/ fourth metacarpal/ metatarsal bones

Definition/overview

- exostosis or 'splints' of the second/fourth metacarpal/metatarsal bones are common in the horse.

Aetiology/pathophysiology

- trauma to the bone leads to haemorrhage and lifting of the periosteum, with subsequent bone formation.
- horses that 'dish' in front may traumatise the second metacarpal bone with the opposing forelimb.
- damage to the interosseous ligament because of instability or trauma can lead to 'splints', particularly between the second and third metacarpal bones.
- young immature animals, especially those with 'bench-knees' (lateral positioning of the third metacarpus in relation to the carpus) are particularly predisposed to 'splints'.

Clinical presentation

- acute cases – trauma will lead to swelling, pain and lameness.
- chronic cases – soft tissue swelling, and oedema subsides leaving a firm, often non-painful fibrous/bony mass on the bone.
- axial impingement on the suspensory ligament can lead to lameness.
- may worsen with exercise and improve with rest.

Differential diagnosis

- fractures of second/fourth metacarpal/ metatarsal bones.
- suspensory ligament injury.

Diagnosis

- clinical examination.
- local anaesthetic infiltration to assess clinical relevance in chronic cases.
- **Radiography** (Figs. 4.72, 4.73) is important to:
  ○ assess size and potential interference with other structures.
  ○ rule out fractures.
- **Ultrasonography** to assess the locality of the exostosis in relation to:
  ○ soft tissue structures such as the suspensory ligament.

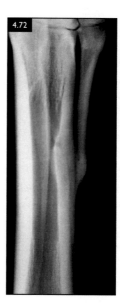 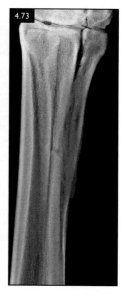

FIG. 4.72 Dorsomedial–palmarolateral oblique radiograph of the left forelimb metacarpus showing smooth new bone formation in the mid-aspect of MC 2 consistent with an exostosis.

FIG. 4.73 Dorsomedial–palmarolateral oblique radiograph of the right fore metacarpus showing extensive new bone formation between the dorsal aspect of MC 2 and the palmar aspect of MC 3 relating to ossification of the interosseous ligament.

Management

- most cases of 'splints' resolve with rest and conservative management.
- acute cases or flare-ups – cold therapy and NSAIDs may be useful.
- local infiltration with corticosteroids:
  - may reduce soft tissue inflammation in the short term.
- Extracorporeal shock-wave therapy has been used in some cases.

- recurrent or severe cases with axial impingement of the suspensory ligament:
  - surgical removal or partial osteotomy may be required.

Prognosis

- good.
- poor limb conformation and recurrence carry a less favourable outcome.

CARPUS

Carpal osteoarthritis

Definition/overview

- degenerative joint disease of one or more carpal joints.

Aetiology/pathophysiology

- due to joint trauma:
  - sprain/subluxation   ○ synovial sepsis
  - intra-articular fracture.
  - soft tissue injury (e.g. medial palmar intercarpal ligament (MPICL) injury).
  - osteochondrosis.
  - osteochondral fragmentation:
    - ◆ can be a cause as well as a result of OA.

- poor distal limb conformation contributes to abnormal loading forces in the carpus.
  - 'back-at-the-knee' or 'bench-knee'.
- certain breeds (e.g. Arabs) have a predisposition to OA of the carpometacarpal joint (Fig. 4.74).

Clinical presentation

- moderate lameness with joint effusion/thickening of the joint capsule (Fig. 4.75).
- positive carpal flexion and may resent passive flexion.
- irregularity and pain along the dorsal margins of the carpal bone may be palpable.

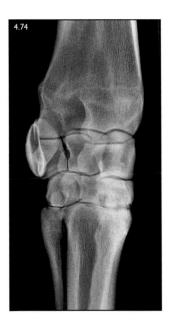

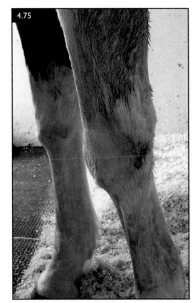

FIG. 4.74 Dorsolateral–palmaromedial oblique radiograph of a horse with OA of the carpo-metacarpal joint. Note the collapse of the joint medially and the surrounding increased subchondral bone sclerosis.

FIG. 4.75 Photo showing swelling over the dorsal aspect of both carpi which were related to palpable new bone formation as well as joint effusion of the carpal joints.

Differential diagnosis

- osteochondral fragmentation
- carpal fracture   • soft tissue injury.
- subluxation.

Diagnosis

- localisation of the lameness through intra-articular analgesia of the carpal joint(s).
- standard radiographic projections will demonstrate radiographic signs of OA (Fig. 4.76).

Management

- mainly palliative through intra-articular medication and/or NSAIDs.
- stem cells as well as other orthobiologics have been used to provide improvement.
- arthroscopy of joints:
  ○ removal of osteochondral fragments may reduce irritation in the joint.
  ○ debridement of loose cartilage flaps, forage and lavage may help short term.
- surgical ankyloses of the carpometacarpal joint has led to good improvement.

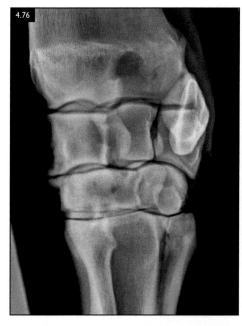

FIG. 4.76 Dorsolateral–palmaromedial oblique radiograph of a horse with chronic OA of the mid-carpal joint. Note the extensive osteophyte remodelling on the distal radiocarpal and proximal third carpal bones.

- partial or pancarpal arthrodesis of the carpus is described for advanced cases.

Prognosis

- established carpal OA is poor:
  ○ conformational abnormalities may develop as a sequela.

Osteochondral fragmentation

Definition/overview

- also known as 'chip fractures'

Aetiology/pathophysiology

- part of the pathophysiology of OA, or a distinct entity.
  ○ particularly in racing breeds or sports horses (training regimens).
- fragmentation typically affects the dorsal articular margin(s) of the:
  ○ distal aspect of the radial carpal bone and proximal third carpal bone.
  ○ can also involve the articular margins of:
    ♦ distal intermediate carpal bone
    ♦ distal lateral (or medial) radius.
    ♦ proximal radial carpal bone
    ♦ proximal intermediate carpal bone.
- subchondral bone of the parent bone may show pre-existing pathology.
- palmar carpal bone fragments are uncommon and usually traumatic.
  ○ reported in horses recovering from general anaesthesia.

Clinical presentation

- acute swelling/lameness involving the carpal joint(s).
- palpation may reveal irregularity/crepitus along dorsal margin of the affected bone.
- lameness exacerbated by carpal flexion, and passive flexion of the carpus resented.

Differential diagnosis

- carpal bone slab or other types of fracture.
- carpal OA   • MPICL injury.

Diagnosis

- response to intra-articular analgesia.
- **Radiographic** evaluation:
  ○ may be present bilaterally.

- radiography of contralateral carpus is recommended.
- conventional views may be supplemented by oblique and special skyline projections (Fig. 4.77, 4.78).

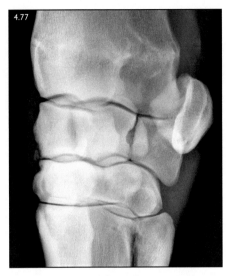

FIG. 4.77 Dorsolateral–palmaromedial oblique view of the left carpus of a horse with a smooth triangular-shaped osteochondral fragment at the dorsomedial margin of the middle carpal joint. There is bony modelling of the dorsodistal aspect of the radial carpal bone and associated soft tissue swelling.

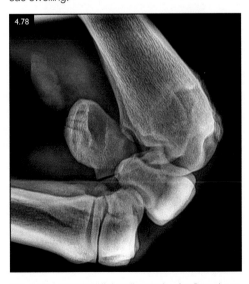

FIG. 4.78 Lateromedial radiograph of a flexed carpus demonstrating a chronic osteochondral chip fracture on the distal medial radius within the antebrachiocarpal joint.

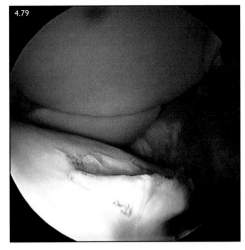

FIG. 4.79 Arthroscopic view of the medial mid-carpal joint clearly showing a chronic distal radiocarpal bone chip fracture prior to removal.

Management

- arthroscopic removal of osteochondral fragments is recommended:
  - evaluation of cartilage and subchondral bone quality.
  - presence of other lesions may affect prognosis (Fig. 4.79).
- removal of palmar fragments can be a challenge.

Prognosis

- following fragment removal, has been linked to cartilage quality at surgery.
  - increasing cartilage loss is associated with reduced return to racing, particularly in the middle carpal joint.

Carpal fractures

Definition/overview

- slab (frontal and sagittal configurations) and comminuted fractures.
- 'Chip' fractures are dealt with in previous section on osteochondral fragmentation.

Aetiology/pathophysiology

- most commonly acute manifestation of a chronic stress-adaptation maladjustment

and repetitive loading, especially third carpal bone (e.g. slab fractures):
- ○ preceded by subchondral bone sclerosis/lucency and loss of overlying articular cartilage.
- ○ occur in the frontal plane through the radial facet of the third carpal bone.
- sagittal plane fractures are less common.
- comminuted carpal bone fractures generally occur from acute trauma/ impact:
  - ○ can involve more than one bone/carpal row, leading to carpal instability.
- fractures of the accessory carpal bone occur through:
  - ○ acute single-impact overload usually at speed.
  - ○ horse landing on the affected limb in flexion.
  - ○ generally, occur in the frontal plane and occasionally comminuted.

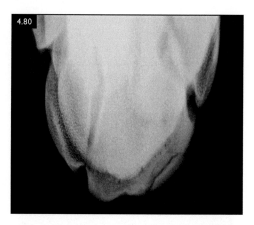

FIG. 4.80 Dorsoproximal–dorsodistal oblique view of the carpus highlighting the distal row of carpal bones. There is a complete, non-displaced fracture of C3.

Clinical presentation

- acute lameness with carpal joint effusion.
  - ○ incomplete fractures, lameness may resolve quickly.
- generally, stand with the carpus in slight flexion to reduce loading through the limb.
- pain on palpation and passive flexion is usually present:
  - ○ less so in incomplete fractures:
    - ♦ may require further investigations.
- comminuted or multiple carpal bone fractures:
  - ○ may have overt limb deformity and carpal instability on manipulation.
- accessory carpal bone fractures:
  - ○ usually stand with carpus semi-flexed.
  - ○ pain/swelling over the palmar aspect of the carpus.
  - ○ may present with carpal sheath effusion if the carpal sheath is involved.

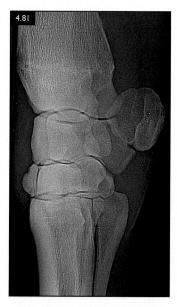

FIG. 4.81 Lateromedial radiograph of a Thoroughbred racehorse with a C3 radial facet slab fracture.

Differential diagnosis

- carpal OA
- osteochondral fragmentation • luxation.

Diagnosis

- **Full radiographic** series, including oblique skyline projections of the carpal rows is required to assess horses with carpal fracture(s) (Figs. 4.80, 4.81).

- incomplete fractures may require diagnostic analgesia to localise lameness.
- **Nuclear scintigraphy** will highlight incomplete fractures as well as subchondral bone disease in the third carpal bone.
- **MRI** may clarify carpal bone injury and associated soft tissue involvement.

Management

- non-surgical management of carpal bone fractures involves rest and controlled exercise.
- surgical management provides a more definitive outcome even with incomplete fractures:
  - small fragments may be removed/burred at the articular margin.
  - complete fractures may also be removed through careful dissection.
- subchondral bone disease without fulminant fracture can be debrided back to healthy margins.
  - slab fracture treatment involves interfragmentary compression under arthroscopic guidance (Figs. 4.82, 4.83).
- accessory carpal bone fractures are usually managed conservatively.
  - heal by fibrous union (Figs. 4.84, 4.85).
    - ◆ internal fixation has been described but is difficult to perform.

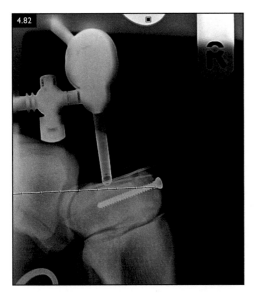

FIG. 4.82 Intraoperative radiograph of the placement of a 3.5 mm lag screw into C3 to repair a radial facet slab fracture.

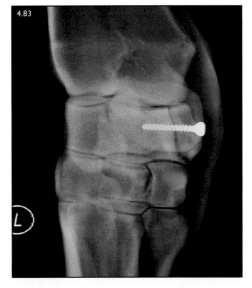

FIG. 4.83 Dorsomedial/palmarolateral radiograph of the carpus following fracture repair of the palmar aspect of the radial carpal bone.

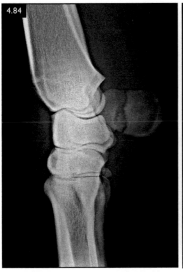
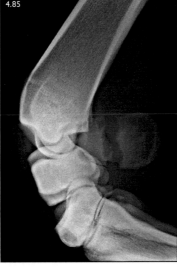

FIG. 4.84, 4.85 Lateromedial (4.84) and flexed lateromedial (4.85) views of the carpus showing a simple, complete non-articular, vertically orientated accessory carpal bone. Note the distraction of the fracture when the limb is flexed.

Prognosis

- slab fractures treated surgically have improved outcomes compared to non-surgical management:
  - return to racing is fair following internal fixation.
  - dependent on presence of cartilage loss and fragmentation at surgery.
  - long-term management of subsequent carpal-related lameness often required.
- non-surgical management of palmar carpal fractures is associated with poor outcome:
  - surgically managed cases – still guarded prognosis.
  - secondary joint disease often requires long-term medical management.
- fractures involving the accessory carpal bone have a fair outcome.
- comminuted or multiple carpal bone fractures (Fig. 4.86) generally have a poor outcome requiring euthanasia or arthrodesis.

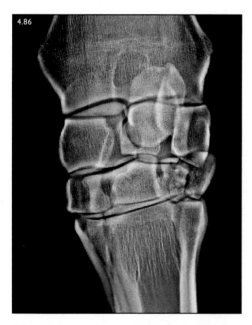

FIG. 4.86 Dorsopalmar radiograph of the carpus showing comminution and displacement laterally of C4.

Intercarpal ligament injury

Definition/overview

- injury to the intercarpal ligaments (including the medial and lateral palmar intercarpal ligaments) has been recognised following the use of arthroscopy.
- palmar intercarpal ligaments provide dorsal stability to the middle carpal joint.

Aetiology/pathophysiology

- tearing of the medial (or less commonly lateral) palmar intercarpal ligament can be due to a traumatic carpal injury:
  - often associated with other carpal pathology:
    - osteochondral fragmentation and cartilage damage.
- herniated ligament fibres act as a source of inflammation and pain within the joint.

Clinical presentation

- effusion and lameness localised to the middle carpal joint.

Differential diagnosis

- osteochondral fragmentation
- carpal bone fracture.
- OA

Diagnosis

- physical examination may point towards involvement of the middle carpal joint.
- local analgesia of the middle carpal joint confirms the location of the lameness.
- **Radiography** is often unrewarding.
- **Ultrasonography** of the intercarpal ligaments has been described and may be enhanced by concurrent joint effusion.
- **MRI** can visualise both ligament and any associated bone pathology.
- arthroscopy of the mid-carpal joint is usually diagnostic (Fig. 4.87).

Management

- debridement of torn fibrils at arthroscopy is recommended.

Prognosis

- depends on the degree of concurrent carpal pathology present.

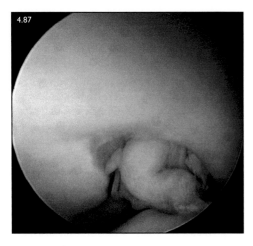

FIG. 4.87 Arthroscopic view of the mid-carpal joint of a Thoroughbred racehorse with chronic osteochondral chip fractures and secondary cartilage pathology of the distal radiocarpal bone. The MPICL is elongated, loose on palpation and shows areas of chronically damaged fibres.

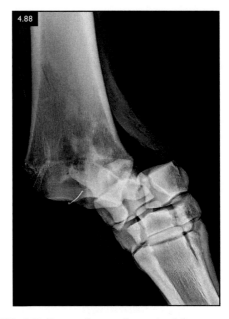

FIG. 4.88 Dorsopalmar radiograph of the carpus showing subluxation of the antebrachiocarpal joint sustained following anaesthetic recovery. This horse had been managed for a chronic wound involving the common digital extensor tendon sheath with a bony sequestrum on the dorsal lateral aspect of the distal radius.

Carpal subluxation

Definition/overview

- subluxation of the carpal joint(s).

Aetiology/pathophysiology

- usually due to a traumatic incident (e.g. high-speed fall or kick).
- may be seen concurrently with multiple carpal bone fractures.
- luxation can occur at any level of the carpus.

Clinical presentation

- severe lameness with overt anatomical deviation of the carpal region.
- palpation and manipulation usually show gross instability.

Differential diagnosis

- proximal metacarpal, carpal or radial fracture
- severe soft tissue injury.
- severe carpal OA.

Diagnosis

- clinical signs.

- **Radiography** confirms the level of the luxation and additional carpal bone fracture(s) (Fig. 4.88).

Management

- initially assessment and stabilisation:
  - subluxations where the bony column is intact and stable.
    - ♦ may respond to a full-limb cast.
  - where instability is present, partial or pancarpal arthrodesis may be considered.
  - carpal collapse and instability due to concurrent carpal bone fractures:
    - ♦ euthanasia should be considered.

Prognosis

- guarded.
  - marked periarticular fibrosis and carpal OA are common sequelae.
    - ♦ loss of function of the carpal joint(s) affected.

Osseous cyst-like lesions (OCLLs)

Definition/overview

- unusual site for osteochondrosis.
- can occur in the small carpal bones:
  - particularly the ulnar carpal bone (Fig. 4.89).
  - also, distal radius (Fig. 4.90), C2, proximal second metacarpal bone.

Aetiology/pathophysiology

- may be an incidental radiographic finding.
  - large cysts may communicate with the joint and be a source of pain.
- cystic changes in the ulnar carpal bone may occur secondarily to avulsion of the lateral palmar intercarpal ligament.

Clinical presentation

- lameness localised to the carpus.

Differential diagnosis

- palmar intercarpal ligament injury.

Diagnosis

- **Radiographically** where a cystic lucency in the affected (usually ulnar) carpal bone is present.
- differentiation from avulsion fractures of the lateral palmar intercarpal ligament is made based on the location and appearance of the lucency.
- evaluation for concurrent joint disease is important for prognostic purposes.

Management

- most are untreated.
- if clinically significant:
  - medical management for joint disease may be helpful.
  - surgical debridement is described:
    - may leave a large cloaca in bone.
    - packing the cyst (bone graft or other substitute) has been undertaken.

Prognosis

- unknown at present and depends on the presence of other carpal joint disease.

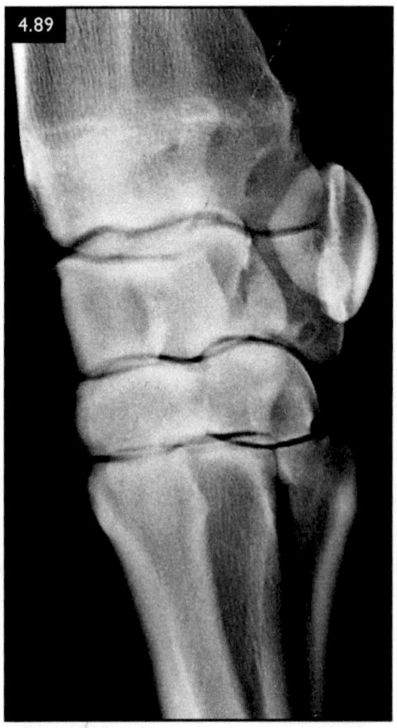

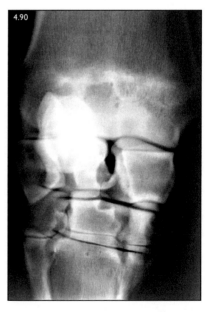

FIG. 4.89 Dorsolateral–palmaromedial oblique radiograph showing a well-circumscribed radiolucency in the distal aspect of the ulnar carpal bone with an area of surrounding sclerosis.

FIG. 4.90 Dorsopalmar radiograph of the carpus of a 9-year-old riding horse with lameness localised to the antebrachiocarpal joint. Note the OCLL in the distal medial radius with a wide entrance into the joint and prominent pericyst bone sclerosis.

Carpal hygroma

Definition/overview

- subcutaneous swelling over the dorsal aspect of the carpus.

Aetiology/pathophysiology

- repetitive trauma to the dorsal aspect of the carpus leads to:
  - chronic inflammation and a fluid-filled subcutaneous structure.
- lameness is usually minimal unless it becomes infected.

Clinical presentation

- fluid-filled, non-painful, fluctuant swelling over the dorsal aspect of the carpus (Fig. 4.91).
- full carpal flexion may be restricted with marked soft tissue swelling.
- if infected, the swelling may be more pronounced, with serous oozing and pain.

Differential diagnosis

- extensor tendon sheath effusion
- carpal joint effusion/herniation.

Diagnosis

- clinical palpation and knowledge of anatomy are required to differentiate carpal hygroma from effusions of the extensor tendon sheaths or carpal joints.
- **Ultrasonography** is useful for examining the hygroma, nearby structures and looking for a foreign body.
- **Radiography** can rule out any bony involvement and contrast medium is used to investigate joint or tendon sheath involvement.

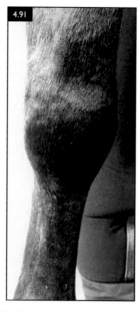

FIG. 4.91 This horse has formed a carpal hygroma over the dorsal aspect due to persistent trauma.

Management

- conservative treatment involves:
  - rest 　 local injections of steroids
  - drainage and bandaging.
  - often unsuccessful.
- surgical treatment requires:
  - *en bloc* resection of the tissue.
  - avoiding penetration of the extensor tendon sheath or joint capsule.
  - sleeve cast or Robert Jones bandage for 7–10 days.

Prognosis

- guarded for complete resolution as recurrence is common.

ANTEBRACHIUM AND ELBOW

Carpal canal syndrome

Definition/overview

- conditions leading to restriction or pain as the carpal sheath passes through the carpal region.

Aetiology/pathophysiology

- causes include:
  - idiopathic or septic tenosynovitis.
  - tendonitis or tearing of the superficial or deep digital flexor tendons/muscle bellies.
  - desmitis of the accessory ligament of the SDFT (AL-SDFT).
  - radial physis exostosis (Fig. 4.92).
  - accessory carpal bone fracture.
  - osteochondroma of the distal radius (Fig. 4.93).

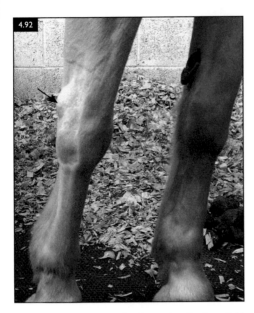

FIG. 4.92 Carpal sheath distension in the right foreleg. Note the golf ball-sized swelling proximal lateral to the carpus (arrow).

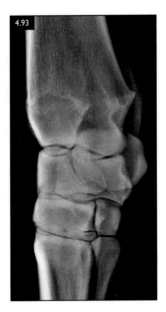

FIG. 4.93 Dorsolateral–palmaromedial oblique view of the carpus showing irregular protruberances on the distal palmarolateral aspect of the radius consistent with a radial physeal exostosis.

- separate area of endochondral ossification from the caudal (usually caudomedial) aspect of the distal radius.
- irritation of the DDFT and accompanying tenosynovitis.
- exostosis or spikes from the radial physis can result in similar clinical signs.

Clinical presentation

- carpal sheath effusion (Fig. 4.94), thickening or pain:
  - effusion can be easily overlooked, particularly on the lateral aspect.
- mild to severe lameness depending on the underlying cause.
- flexion of the carpus usually exacerbates the clinical signs.

Differential diagnosis

- carpal joint pathology
- extensor tendon sheath pathology.

Diagnosis

- clinical examination.
- diagnostic intrasynovial analgesia may be required in some cases.

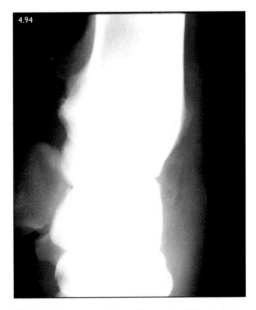

FIG. 4.94 Lateromedial radiograph of the carpus and distal radius deliberately under-exposed to highlight the relatively radiolucent caudal distal radial osteochondroma.

- **Radiography.**
- **Ultrasonography** of the carpal/distal antebrachium (Fig. 4.95).
- **MRI** may be useful in some cases.
- Tenoscopy of the sheath may be required for definitive diagnosis and treatment (Fig. 4.96).

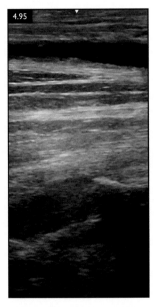

FIG. 4.95 Ultrasonography of the carpal/distal antebrachium confirms the physeal exostosis to be involving the deep digital flexor muscle belly in the carpal sheath.

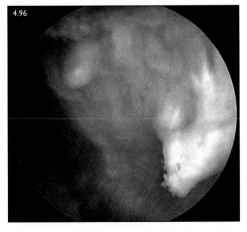

FIG. 4.96 Carpal sheath tenoscopic image of the caudal distal radius showing two enlarged exostoses protruding from the caudal bone surface.

Management

- tenoscopy of the carpal sheath allows:
  - removal of osteochondromas and radial physeal exostoses.
  - management of concurrent damage to the deep digital flexor tendon (DDFT).
  - treatment of accessory carpal bone fracture complications.
- AL-SDFT desmitis and tendonitis of the SDFT or DDFT can be managed conservatively.
- carpal canal syndrome caused by constriction of the carpal flexor retinaculum can be sectioned tenoscopically.
- idiopathic tenosynovitis may respond to rest and controlled exercise with intrathecal hyaluronan and/or corticosteroids.

Prognosis

- following removal of osteochondroma and radial physeal exostosis, often depends on the degree of damage to other structures (e.g. DDFT).
- routine carpal sheath tenoscopy is associated with an increased risk of postoperative sepsis compared to other elective procedures.

Radial fractures

Definition/overview

- in adults, fractures include:
  - incomplete or complete spiral, oblique and comminuted diaphyseal.
- in younger animals:
  - physeal fractures.
  - complete diaphyseal fractures (simple, transverse or spiral) with varying degrees of comminution.

Aetiology/pathophysiology

- usually a result of external trauma (e.g. kick).
  - distomedial aspect of the radius is sparsely covered with soft tissues.
- caudal aspect of the radius is the most common point of failure.
- comminution is unusual, but butterfly fragments can displace leading to catastrophic failure.
- radial sequestra may also develop after trauma.

- foals may have:
  - simple transverse or oblique diaphyseal fractures due to trauma:
    - with or without comminution.
  - physeal fractures can occur at the proximal or distal growth plates:
    - Salter–Harris type I (physeal only) or II (physeal and metaphyseal) configurations.

Clinical presentation

- incomplete radial fractures – usually moderately lame but able to bear weight.
- soft tissue swelling over the fracture site is usually evident:
  - if distal can be mistaken for a carpal injury (Fig. 4.97).
- complete, unstable fractures – severely lame with clear instability and crepitus.
- wound may be present:
  - palpation of bony fragments may indicate an open radial fracture.

Differential diagnosis

- carpal or ulna fracture
- severe soft tissue injury.
- synovial sepsis

Diagnosis

- physical examination.
- diagnosis confirmed through **radiography**:

  - multiple views may be required to characterise the type and length of fracture (Figs. 4.98, 4.99).
  - most fractures propagate proximally and spiral through the bone.
  - occasional involvement of the distal radial articular surface (Figs. 4.100, 4.101).

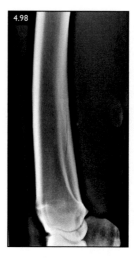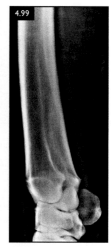

FIG. 4.98, 4.99 Lateromedial (4.98) and craniolateral/caudomedial (4.99) radiographs of the distal antebrachium showing an incomplete radial fracture in the caudomedial aspect of the radius. Note also that the fracture line disappears and reappears at the distal condyle of the radius, showing it to have an articular component.

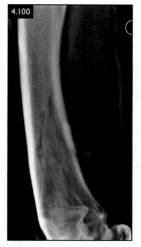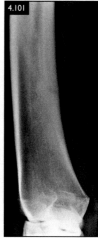

FIGS. 4.100, 4.101 (4.100) Craniolateral–caudomedial oblique radiograph of a horse with an incomplete spiral fracture of the radius 3 weeks following a kick injury to the right fore antebrachium. (4.101) Radiograph 1 month later shows increased bony density related to further fracture healing.

FIG. 4.97 Photo of the cranial antebrachium showing moderate soft tissue swelling over the distomedial radius following a kick injury.

- any discontinuity along the caudal
  cortex or presence of fragmentation
  in a non-displaced fracture should be
  viewed with extreme caution:
  - bone may subsequently fail at this
    point, particularly if a wound
    communicates with the bone.
  - proximal radial fractures may be
    associated with fractures of the ulna
    and/or elbow luxation.

Management

- horses >250 kg with complete, displaced
  fractures, particularly if open, are often
  euthanased (Fig. 4.102).
  - double-plate fixation has been
    described particularly in smaller or
    lighter breeds (<250 kg).
- complete fractures in foals are amenable
  to repair by internal fixation.
- incomplete radial fractures can be
  managed conservatively.
  - caudal and lateral splints to stabilise
    the limb:
    - lateral splint should extend to at
      least the level of the shoulder.
    - prevents lateral abduction of the
      limb (Fig. 4.103).
  - sling to spare weight bearing.

- regular monitoring with radiography
  and wound management is important:
  - cyclical loading/infection can lead
    to bone failure several days later.

Prognosis

- adult horses (>250 kg) with displaced
  complete radial fractures have an

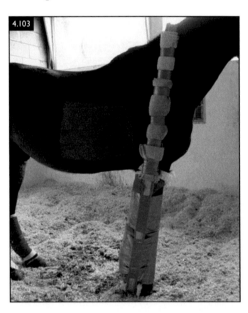

FIG. 4.103 Full-limb Robert Jones bandage on a horse with an incomplete radial fracture. Note that the lateral splint extends to the proximal scapula.

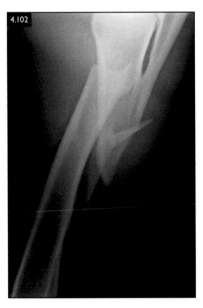

FIG. 4.102 A complete displaced comminuted mid-body radial fracture in a donkey following a kick injury.

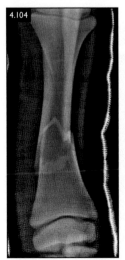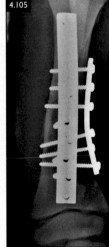

FIGS. 4.104, 4.105 Complete simple transverse radial fracture in a 2-day old foal (Fig. 4.104). Repair was achieved using a double-plate technique (Fig. 4.105).

unfavourable prognosis and are usually euthanased.
- incomplete fractures in adult horses can heal completely barring complications such as infection or displacement.
- complete closed fractures in foals have a favourable outcome following internal fixation provided comminution is not present and infection is avoided (Figs. 4.104, 4.105).

Enostosis-like lesions

Definition/overview
- identified in the radius, as well as other long bones especially in the hindlimb (e.g. femur/tibia).
- single or multifocal areas of increased bone density in long bones.

- incidental findings or a cause of lameness, particularly if involving the femur or humerus.
- main differential is stress fractures.
- increased radiopharmaceutical uptake on scintigraphy in the diaphysis of one or more bones (Fig. 4.106).
- on radiography, oval or indistinct areas of increased radiopacity within the medulla.
  - often associated with nutrient foramen (Figs. 4.107, 4.108).
- clinical relevance is usually determined through ruling out other causes of lameness.
- lesions appear to be self-limiting, and cases are usually managed conservatively.
- recurrence may occur and still unclear as to the long-term effect of these lesions.

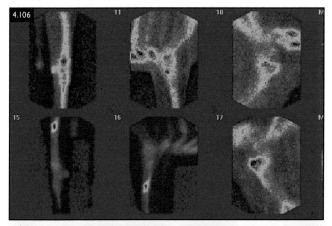

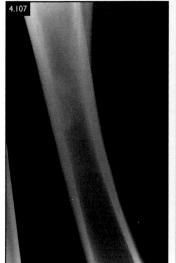

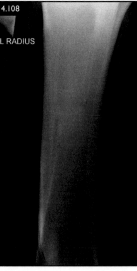

FIGS. 4.106–4.108 This 9-year-old Warmblood dressage horse presented with lameness of the left fore- and left hindlegs. The lameness was not localised in either leg by diagnostic regional or intra-articular analgesia techniques, but a bone scan (4.106) revealed focal increased areas of radiopharmaceutical uptake in the mid-radius and mid-tibia. Lateromedial (4.107) and craniocaudal (4.108) radiographs of the radius revealed a roughly oval-shaped area of increased radiopacity within the medulla – an enostosis-like lesion. The same type of lesion was present in the tibia.

Ulnar fractures

Definition/overview

- fractures of the proximal ulna involving the olecranon.
- 'Monteggia' fractures (ulnar and radius with elbow luxation) have been described but are very uncommon.

Aetiology/pathophysiology

- ulnar fractures are common.
- most cases are due to direct trauma (e.g. kick) or impact at speed.
- five main types of fracture configuration described with or without articular involvement (Fig. 4.109 and Table 4.1).

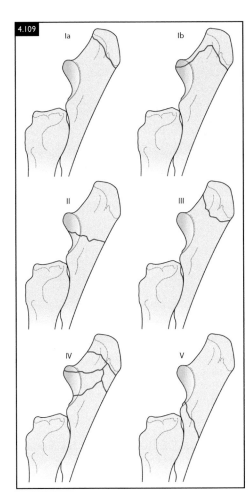

FIG. 4.109 Schematic diagram of proximal ulnar fracture classification. (See **Table 4.1** for description.)

TABLE 4.1 Classification of ulnar fractures

TYPE	DESCRIPTION
Ia	Non-articular fracture involving separation of the proximal ulnar physis, seen in young foals.
Ib	Articular fracture through the proximal ulnar physis and metaphysis, entering the joint near to the anconeal process. Seen in older foals.
II	Simple transverse fracture entering the elbow joint at the mid-point of the trochlear notch.
III	Non-articular fracture across the olecranon, proximal to the anconeal process.
IV	Comminuted, articular fracture.
V	Distal caudal fracture of the olecranon/ulnar shaft, traversing proximally and cranially to enter the distal aspect of the trochlear notch.

Clinical presentation

- usually, severe lameness as cases are unwilling to extend the elbow, and therefore stand with the elbow dropped and carpus flexed (Fig. 4.110).
- palpation and manipulation of the elbow region is resented and there is usually soft tissue swelling, with or without a wound present.

Differential diagnosis

- humeral fracture
- radial nerve paralysis
- elbow luxation.
- synovial sepsis.

Diagnosis

- suspected from clinical examination.
- **Radiography:**
  - required to determine fracture configuration (Fig. 4.111).
  - mediolateral images are the most useful.
  - craniocaudal views are important to check for comminution and unusual fracture configurations.

FIG. 4.110 Ulnar fracture in a mature horse. The horse will often stand with its elbow dropped and the carpus in a semi-flexed position due to pain and loss of the stay apparatus.

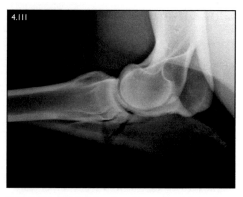

FIG. 4.111 Mediolateral radiograph of the elbow region of a horse showing a complete, comminuted articular (type IV) olecranon fracture.

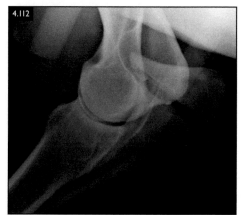

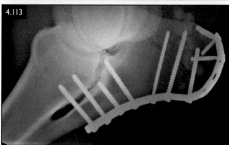

FIGS. 4.112, 4.113 Mediolateral radiographs of a complete, displaced articular proximal (type Ib) olecranon fracture pre (4.112) and post repair (4.113).

Management

- stabilise the stay apparatus of the forelimb by the application of a splint to the caudal aspect of the limb from the olecranon to the ground to lock the carpus in extension.
- non-surgical management is possible in non-displaced fractures of foals.
  - in adults often associated with delayed or non-union and a poor outcome.
- surgical repair and stabilisation techniques include wire, pins and plating:
  - plate fixation is more versatile and stronger:
    - single narrow plate on caudal margin of the olecranon and proximal radius is sufficient in most cases (Figs. 4.112, 4.113).
    - additional lateral plate used in comminuted/unstable fractures in adults.
  - wire and pins used in younger animals where plating may compromise the proximal radial growth plate.
- 'Monteggia' fractures require caudal and lateral plates.

Prognosis

- fixation with wires (with or without pins) or plate fixation has a good outcome in most cases:
  - reduced by the presence of concurrent infection/sepsis.
- unstable fracture repairs will lead to contralateral limb overuse problems.
- large horses with 'Monteggia' fractures have a poor prognosis and are often euthanased.

Elbow osteoarthritis

Definition/overview

- degenerative joint disease of the cubital joint.

Aetiology/pathophysiology

- develops secondary to joint trauma, collateral ligament injury, OCLL, intra-articular ulnar fracture or synovial sepsis.

Clinical presentation

- unusual in the elbow and horses present with a variable, progressive lameness.
- joint effusion is usually not palpable.

Differential diagnosis

- OCLL
- collateral ligament injury.

Diagnosis

- intra-articular analgesia of the cubital joint localises the lameness to the elbow.
- **Radiography**, particularly a mediolateral projection will show:
  - changes consistent with OA:
    - osteophytosis involving cranioproximal margin of the radius (Fig. 4.114).
    - entheseous new bone may also be evident (Fig. 4.115).
- **Ultrasonography** may visualise periarticular osteophytes and joint effusion.

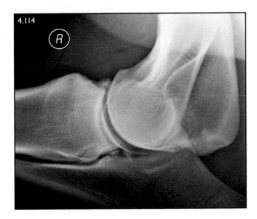

FIG. 4.114 Mediolateral radiograph of the elbow showing irregular new bone formation at the prox-imocranial aspect of the radius consistent with osteophytosis.

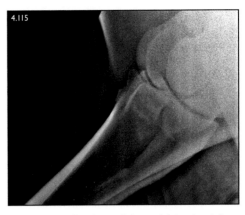

FIG. 4.115 Craniomedial–caudolateral oblique radiograph of a horse presenting with a 3-week history of right forelimb lameness localised to the elbow. This radiograph shows entheseous new bone on the proximolateral humerus. Ultrasonography revealed thickening and loss of normal fibre pattern in the lateral collateral liga-ment with capsular thickening and entheseophyte formation at the osseous margins.

- **Nuclear scintigraphy**: increased radiopharmaceutical uptake.

Management

- intra-articular medication (hyaluranon/ corticosteroids) and palliative treatment (NSAIDs).

Prognosis

- guarded for soundness.

Osseous cyst-like lesions (OCLL)

Definition/overview

- occur in the proximomedial aspect of the radius, or less commonly, distal humerus.

Aetiology/pathophysiology

- develop in a similar way to other OCLLs.
- rarely cystic lesions can develop secondary to injury to the collateral ligament.

Clinical presentation

- variable lameness which may worsen when lunged in a circle (limb on the outside):

- may worsen following upper limb flexion.
- few other localising signs.

Differential diagnosis

- elbow OA
- collateral ligament injury.

Diagnosis

- localisation to the elbow joint following diagnostic analgesia.
- **Radiography:**
  - standard radiographic projections (particularly craniocaudal views).
  - usually demonstrate a cystic lucency (Fig. 4.116).
  - evaluation of radiographic evidence of OA is important prognostically.
- **Nuclear scintigraphy:** increase radiopharmaceutical uptake in the region of the cyst.

Management

- conservatively:
  - intra-articular medication (hyaluranon/corticosteroids) can improve clinical signs.
- surgically:
  - extra-articular curettage of the cyst (Fig. 4.117):
    - with or without bone grafting or another osteo-inductive substitute.
    - may resolve clinical signs but filling of the cyst is unpredictable.
  - cortical screw can be placed across the cyst encourages the cyst to fill with bone.

Prognosis

- usually better with surgical than conservative management:
  - evidence of pre-existing joint disease will reduce outcome.
- results with screw treatment have been good.

Elbow subluxation

Definition/overview

- subluxation of the radiohumeral joint

Aetiology/pathophysiology

- usually occurs due to severe trauma/ disruption of the elbow joint and

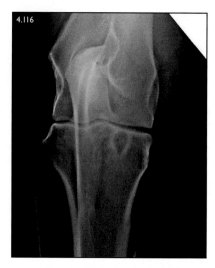

FIG. 4.116 Craniocaudal radiograph of the elbow region showing a well-demarcated osseous cyst-like lesion in the proximomedial radius surrounded by a sclerotic margin.

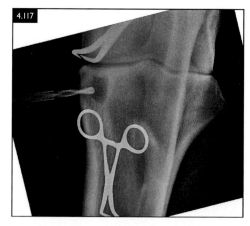

FIG. 4.117 Intraoperative craniocaudal radiograph showing an extra-articular approach to enucleation of a medially located osseous cyst-like lesion in the proximal radius.

supporting structures (e.g. collateral ligament avulsion):
  - often occur at high speed/impact.
  - also reported following recovery from general anaesthesia.
- elbow luxations often have associated fractures. ('Monteggia' fracture).

Clinical presentation

- acute severe non-weight-bearing lameness.
- marked soft tissue swelling and anatomical deformity.

Differential diagnosis

- ulnar fracture
- humeral fracture.

Diagnosis

- clinical examination.
- **Radiography:**
  - usually sufficient to provide a diagnosis of luxation.
  - difficult to achieve good radiographs in acutely distressed horse.
  - evaluate radiographs for concurrent fractures (Fig. 4.118).

Management

- fracture/luxations of the elbow usually require euthanasia.
- closed repair of elbow luxation is possible in ponies/small horses (<250 kg).

Prognosis

- poor, as marked elbow OA usually develops.

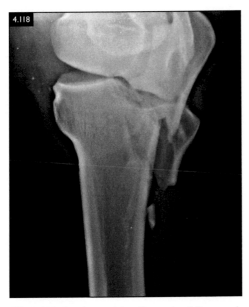

FIG. 4.118 Craniocaudal radiograph of the proximal radius of a 4-year-old colt showing comminuted proximal radial and ulnar fracture with concurrent subluxation of the radiohumeral joint.

Elbow hygroma

Definition/overview

- hygroma of the elbow or 'capped elbow' or 'shoe boil'.
- non-painful swelling over the point of the elbow (Fig. 4.119).

Aetiology/pathophysiology

- fluid-filled, acquired subcutaneous bursae:
  - caused by repetitive trauma and inflammation.
  - chronically become more fibrous than fluid-like.
- horses with a high action may catch themselves with the shoe of the ipsilateral forelimb.
- also horses that lie down regularly, particularly where bedding is inadequate.

Clinical presentation

- usually, a non-painful mass and not associated with lameness.
- if infected, more swollen, painful and oozes pus.

Differential diagnosis

- ulnar fractures
- soft tissue trauma.

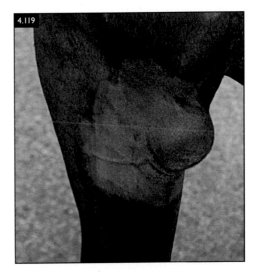

FIG. 4.119 Chronic subcutaneous bursa in the left foreleg over the point of the elbow.

Diagnosis

- clinical presentation.
- **Ultrasonography** can be used to delineate the mass, especially when contemplating removal.

Management

- most cases are cosmetic and are not treated:
  - correction of any underlying cause should be instigated.
- initial drainage and then placement of a drain may resolve some cases.
  - infection may occur, turning a cosmetic issue into a difficult to solve problem.
- *en bloc* surgical resection can be attempted but wound breakdown is common.

Prognosis

- risk of dehiscence of surgical wound is increased during general anaesthesia recovery:
  - prognosis is then unfavourable in such cases.

HUMERUS AND SHOULDER

Scapulohumeral OA

Definition/overview

- degenerative joint disease of the scapulohumeral joint.

Aetiology/pathophysiology

- secondary to joint trauma, synovial sepsis, osteochondrosis or intra-articular fracture.
- miniature ponies (e.g Shetland) associated with dysplasia/luxation.

Clinical presentation

- moderate–severe lameness.
- muscle atrophy over the shoulder region is common.
- pain on extension or flexion of the upper limb is usually obvious.
- joint effusion is rarely palpable.

Differential diagnosis

- osteochondrosis   • dysplasia
- luxation   • synovial sepsis.

Diagnosis

- localisation to the scapulohumeral joint by local analgesia.
- **Radiography** of the joint shows:
  - evidence of OA particularly at the caudal angle of the scapula and humerus (Fig. 4.120).
  - flattening of the joint surface and fragmentation may also be appreciated.
  - craniocaudal projections may be useful to assess:
    - ◆ joint congruency and presence of periarticular osteophytosis.
- **Nuclear scintigraphy** will also reveal increased uptake in the shoulder region.

Management

- most cases are relatively advanced on diagnosis.
- palliative treatment with NSAIDs or intra-articular medication may be helpful.

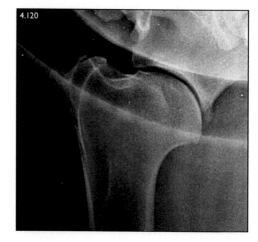

FIG. 4.120 Mediolateral radiograph of the shoulder region of a 3-year-old Shetland pony with a large osteophyte in the caudal aspect of the scapula consistent with OA.

- severe cases particularly in conjunction with dysplasia/luxation in miniature horses:
  - arthrodesis of the scapulohumeral joint has been described.

Prognosis

- guarded.

Osteochondrosis

Definition/overview

- osteochondritis dissecans (OCD) of the humeral head and glenoid.
- OCLLs of glenoid cavity of the scapula (rarely humerus).

Aetiology/pathophysiology

- less common than other locations but has a similar aetiology.

Clinical presentation

- OCD may present relatively late (yearlings or older) compared to other sites (Fig. 4.121):
  - usually moderate–severe lameness.
  - evidence of muscle atrophy and in some cases, joint effusion.
  - clubbed or boxy foot in chronic cases due to secondary contracture of the DIP joint.
  - pain on manipulation.

- OCLLs cases are generally older.
  - can present with lameness of an intermittent nature and severity.
  - clinical signs may suggest shoulder pain, but often localising signs are scant.

Differential diagnosis

- shoulder OA
- luxation
- dysplasia
- synovial sepsis.

Diagnosis

- suspected from clinical signs and signalment.
- **Radiography** of horses with shoulder OCD may show range of signs (Fig. 4.122):
  - mild flattening of the humeral head and subchondral bone lucency.
  - overt mineralised flaps
  - contrast radiography may further delineate flaps.
  - radiographic signs often underplay the severity of the condition.
  - signs of secondary OA are common.
- **Radiography** of shoulder OCLLs may be:
  - variable size.
  - located on the glenoid rim with, or without, a sclerotic rim (Fig. 4.123).
- **Ultrasonography** may be helpful in:
  - confirming joint effusion.
  - osteochondral flaps can be visualised (caudal part of the joint).
- **Nuclear scintigraphy** – increased radiopharmaceutical uptake with some OCLLs.

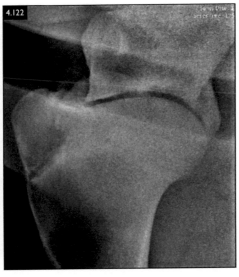

FIG. 4.121, 4.122 (4.121) Photo of a yearling Thoroughbred filly with marked atrophy over the shoulder musculature and presence of a joint effusion in the caudal aspect of the shoulder joint. (4.122) Mediolateral radiograph of the same horse showing subchondral lysis in the distal scapula and modelling of the cranial and caudal margins of the scapula consistent with osteochondrosis.

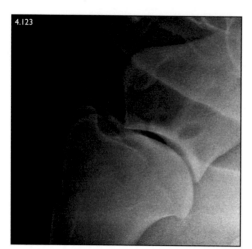

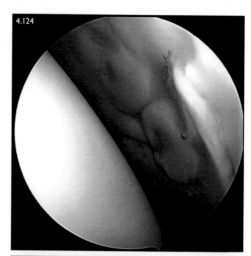

FIG. 4.123 Mediolateral radiograph of the shoulder showing a well-demarcated osseous cyst-like lesion in the distal scapula surrounded by a sclerotic rim.

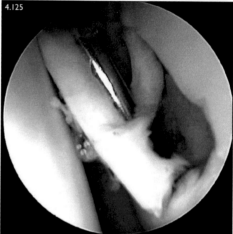

FIG. 4.124, 4.125 Arthroscopic images of the case presented in Figs. 4.121 and 4.122 showing marked irregular contour of the articular cartilage on the distal scapula (4.124). Probing of the cartilage showed lose attachment of the cartilage from underlying diseased subchondral bone (4.125)

Management

- arthroscopic surgery is advised to:
  - remove dissecting cartilage flaps and curette underlying subchondral bone (Figs. 4.124, 4.125).
  - severe cases may be present with complete lifting away of the cartilage:
    - pinning techniques have been described.
    - euthanasia in severe cases is recommended due to the rapid development of secondary joint disease.
  - debridement is possible for OCLLs but depends on the location of the cyst.
- intra-articular medication may help to resolve clinical signs associated with OCLL in the shoulder.

Prognosis

- OCD is often associated with secondary joint disease and a poor prognosis.

Dysplasia

Definition/overview

- associated with Miniature Horses and Shetland ponies.

Aetiology/pathophysiology

- malalignment of the humeral head and glenoid cavity during development of the scapulohumeral joint can result in pain and instability.

Clinical presentation

- moderate–severe forelimb lameness localised to the shoulder region.

Differential diagnosis

- shoulder OA
- luxation.

Diagnosis

- suspected based on clinical signs and signalment.
- **Radiographic** findings of:
  - shallow glenoid cavity on mediolateral radiographs.
  - measurement of the curvature of the glenoid cavity (Fig. 4.126).
  - commonly radiographic evidence of OA and occasionally signs of subluxation.

Management

- managed palliatively or through arthrodesis of the scapulohumeral joint.
- severe cases may require euthanasia.

Prognosis

- poor due to development of secondary OA.

Subluxation

Definition/overview

- subluxation of the scapulohumeral joint.

Aetiology/pathophysiology

- due to trauma or associated with dysplasia of the scapulohumeral joint.

Clinical presentation

- acute and severe lameness, often non-weight bearing.
  - chronic cases quickly develop muscle atrophy.
- palpation is resented.
- abnormal range of movement detected on manipulation.
- irregular contour to the shoulder region may be observed.

Differential diagnosis

- intra-articular fracture
- synovial sepsis.

Diagnosis

- confirmed **radiographically** (Fig. 4.127):
  - luxation usually occurs cranially and dorsally.
  - concurrent fractures may be present.
  - evidence of dysplasia may be noted.

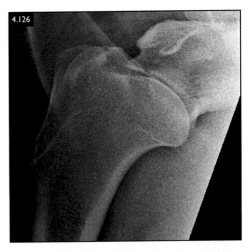

FIG. 4.126 Mediolateral radiograph of the shoulder region showing abnormal contour of the scapulohumeral articulation consistent with dysplasia. There is new bone formation at the caudal aspect of the scapula suggestive of secondary joint disease.

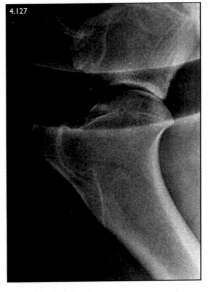

FIG. 4.127 Mediolateral radiograph of an 8-year-old miniature horse with 1-month duration of severe right forelimb lameness, showing subluxation of the scapulohumeral joint.

Management

- manual reduction under general anaesthesia is possible, particularly in smaller breeds.
  - recurrence is common.
  - scapulohumeral arthrodesis is described for management of these cases.

Prognosis

- guarded following reduction of shoulder subluxation and cases often develop OA.
- scapulohumeral joint arthrodesis cases are left with residual mechanical lameness but can carry on a normal life.

Humeral fractures

Definition/overview

- fractures of the diaphysis, deltoid tuberosity, greater tubercles and stress fractures.

Aetiology/pathophysiology

- most humeral fractures are caused by external trauma (e.g. kick or fall):
  - complete or incomplete, transverse, spiral, oblique or comminuted.
  - complete spiral or oblique fractures are the most common.
- stress fractures occur at the proximocaudal, distal cranial or caudal humerus (Fig. 4.128).
- deltoid tuberosity fractures are generally due to external trauma.

Clinical presentation

- acute severe lameness:
  - complete humeral fractures show instability and loss of limb function.
- stress fractures may show moderate lameness but often few other clinical signs.
- deltoid tuberosity fractures usually have:
  - evidence of external trauma (e.g. wound).
  - pain/crepitus over the affected region but able to bear weight (Fig. 4.129).

Differential diagnosis

- ulnar and scapular fractures
- radial or brachial plexus neuropathy.

Diagnosis

- clinical signs may be sufficient for a diagnosis of a complete humeral fracture.
- **Radiography** is important to characterise the fracture:
  - may be difficult to achieve (Figs. 4.130, 4.131).

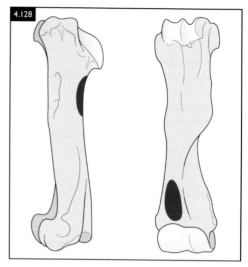

FIG. 4.128 Schematic drawing of the equine humerus showing the common locations of stress fractures.

FIG. 4.129 Swelling over the upper forearm and shoulder region due to a non-displaced fracture of the proximal humerus.

  - specific oblique views (e.g. cranioproximal/craniodistal) may be required.
  - periosteal new bone formation with chronic incomplete stress fractures.

- **Ultrasonography** may be useful in some cases.
- **Nuclear scintigraphy** can help detect stress fractures at their predilection sites.

Management

- complete humeral fracture cases should be euthanased.
- incomplete stress fractures may be managed conservatively with rest and alterations to training regimes.
- deltoid tuberosity fractures can be managed by local debridement of fragments if loose but respond well to conservative management.
- non-surgical management or removal/internal fixation of fractures of the greater tubercle have been described.

Prognosis

- hopeless for complete humeral fractures in the adult horse.
- incomplete fractures or stress fractures can respond well to conservative management:
  - return to work undertaken carefully to avoid acute propagation of the fracture.
- deltoid tuberosity fractures and fractures of the greater tubercle have a good prognosis for return to function.

Scapular fractures

Definition/overview

- fractures of the supraglenoid tubercle, neck, body or spine, and stress fractures.

Aetiology/pathophysiology

- most fractures of the scapula are due to a traumatic episode (e.g. high-speed impact with a fence or kick).
- stress fractures are due to non-adaptive remodelling following intense training:
  - occur at the distal end of the scapular spine.
- supraglenoid tubercle fractures occur more often in young horses:
  - through the original physis.
  - avulse due to the attachment of the biceps brachii.

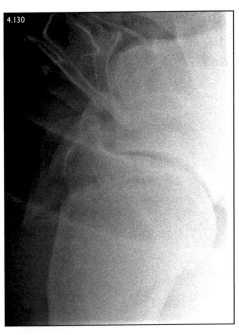

FIG. 4.130 Mediolateral radiograph of a 5-year-old cob with acute right forelimb lameness following a fall, showing a displaced comminuted fracture involving the humeral tubercles, extending into the scapulohumeral joint.

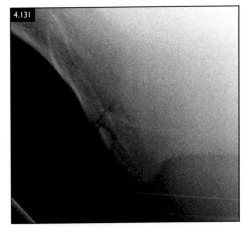

FIG. 4.131 Cranioproximal/caudodistal radiograph of the proximal humerus showing an incomplete, non-displaced fracture of the deltoid tuberosity following a kick injury to the region.

Clinical presentation

- usually moderately–severely lame and may lose function of the stay apparatus:
  - confusing the presentation with an ulnar fracture or radial neuropathy.
- may be evidence of a wound or external trauma.

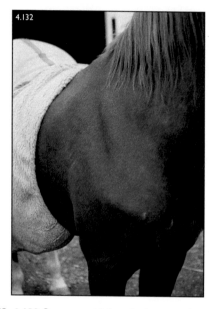

FIG. 4.132 Supra- and infraspinatus muscle atrophy in an Arab stallion that was kicked on the point of the right shoulder while covering a mare and developed a suprascapular nerve paralysis.

- moderate swelling is usually present, and pain/crepitus demonstrated on palpation.
- chronic fractures involving the neck of the scapula may show:
  - neurogenic atrophy of the supraspinatus/infraspinatus muscles.
  - due to damage to the suprascapular nerve (Fig. 4.132).
- stress fractures may be bilateral and present as lameness with pain on direct palpation in some cases:
  - complete stress fracture will present as an acute-onset severe lameness.

Differential diagnosis

- humeral fracture
- synovial sepsis
- luxation.
- radial or suprascapular neuropathy or brachial plexus injury.

Diagnosis

- often confirmed **radiographically:**
  - evaluation of the body and spine of the scapula can be difficult.
- **Nuclear scintigraphy** will show increased radiopharmaceutical uptake in the injured regions and in the sites of stress fractures.
- **Ultrasonography** may be useful with fractures of the spine (Fig. 4.133 and 4.134).

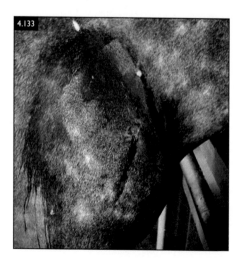

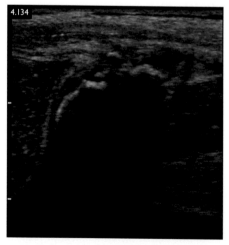

FIG. 4.133, 4.134 (4.133) Photograph of a discharging tract over the distal spine of the scapula following a kick injury. (4.134) Ultrasonography shows irregularity and bone fragmentation consistent with a fracture. A 6 cm section of bone was removed under standing sedation and local anaesthesia.

Management

- conservative management of incomplete, non-articular fractures includes rest and controlled exercise.
- wounds may lead to a sequestra formation and removal of affected fragments can be achieved under ultrasonographic guidance.
- supraglenoid tubercle fractures can be managed conservatively or surgically.
  - former often associated with ongoing pain and development of OA (Fig. 4.135).
  - surgical management includes removal or internal fixation:
    - partial tenotomy of the biceps brachii may be required.
- internal fixation of fractures of the scapular body has been described.
- complete stress fracture cases are usually euthanased.

Prognosis

- guarded in intra-articular fractures due to the development of secondary joint disease.
- non-articular fractures have a good outcome.

Intertubercular (bicipital) bursitis

Definition/overview

- inflammation of the bursa underlying the origin of the biceps brachii from the supraglenoid tubercle as it passes over the proximal humerus.

Aetiology/pathophysiology

- trauma to the region may result in bursitis.
- entheseopathy of the origin of the tendon can result from chronic tearing of the biceps brachii.

Clinical presentation

- mild to moderate lameness that often worsens with work.
- clinical examination may reveal pain over the upper humerus, with resentment following retraction of the limb.
- bursal effusion is variable and often difficult to palpate (Fig. 4.136).
- proximal limb muscle loss may be evident in chronic cases.

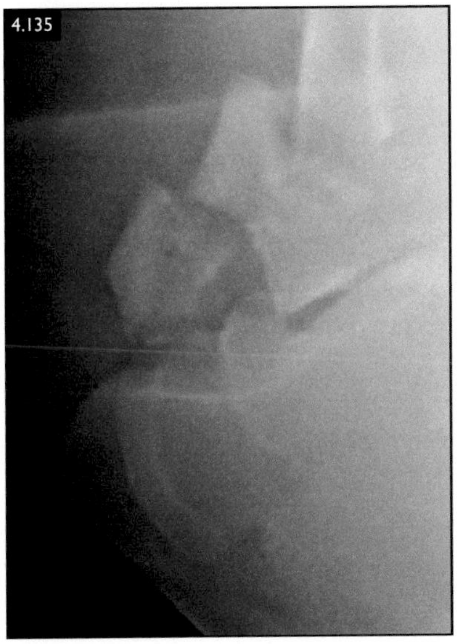

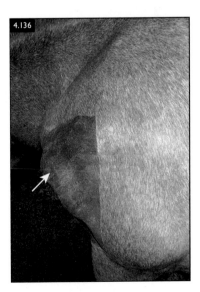

FIG. 4.135 Mediolateral radiograph of a 7-year-old Thoroughbred gelding following a fall the previous day, showing a comminuted displaced fracture of the supraglenoid tubercle.

FIG. 4.136 An Arab gelding with swelling and pain on palpation over the cranial aspect of the shoulder region following a fall. There is both subcutaneous and bicipital distension (arrow).

Differential diagnosis

- synovial sepsis
- scapular and humeral fracture
- scapulohumeral joint disease.

Diagnosis

- suspected from clinical signs.
- intrasynovial analgesia may be required to confirm the source of the lameness.
  - in up to 20% of horses the **bicipital bursa and shoulder joint communicate.**
- **Radiography** may reveal modelling of the supraglenoid tubercle with entheseopathy of the bicep tendon (Fig. 4.137).

- **Ultrasonography** is helpful in examining the tendon, bursa and the surface of the intertubercular groove of the humerus (Fig. 4.138).

Management

- managed through intrathecal medication (e.g. hyaluranon/corticosteroids).
- entheseopathy may respond to extracorporeal shock-wave therapy.
- diagnostic tenoscopy may be required in protracted cases.

Prognosis

- generally fair although can be refractive to treatment in chronic cases.

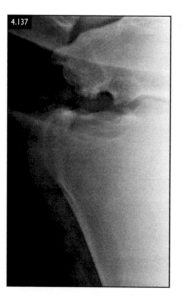
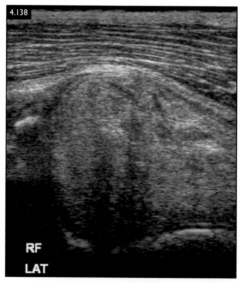

FIG. 4.137, 4.138 (4.137) Mediolateral radiograph of a 7-year-old Thoroughbred-cross horse showing chronic entheseopathy of the insertion of the biceps brachii on the supraglenoid tubercle. (4.138) A transverse ultrasonogram shows enlargement and reduced echogenicity of the lateral lobe of the biceps brachii consistent with tendonitis.

The Hindlimb

Osteoarthritis of the distal tarsal joints (bone spavin)

Definition/overview

- Osteoarthritis (OA) of the tarsometatarsal (TMT) and distal intertarsal (centrodistal) (DIT) joints.
  - less frequently, the talocalcaneal–centroquatral (proximal intertarsal (PIT)) joint.
  - most common cause of lameness involving the hindlimbs of mature horses.
- low motion joints act primarily as shock absorbers with limited horizontal movement.
- also occurs in young horses as 'Juvenile spavin'.
- incidence has little relationship to type of horse or to type/amount of work it has undertaken.

Aetiology/pathophysiology

- normal 'use trauma' to joints.
- abnormal conformation such as 'cow' or 'sickle hock'.
- excessive compression and rotational forces of the tarsal bones during jumps or stops.
- excessive tension of the ligaments over the dorsal aspect of the hock.
- other factors include:
  - incomplete ossification and collapse of the third and central tarsal bones (Fig. 5.1).
  - septic arthritis, fractures of the cuboidal tarsal bones, and osteochondrosis (OCD).
- Icelandic horses appear to have a hereditary component.

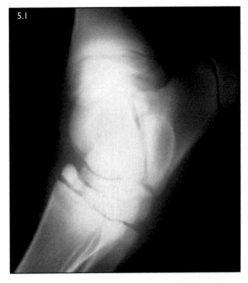

5.1

FIG. 5.1 Lateromedial radiograph of the hock in a young foal demonstrating collapse of the central and third tarsal bones.

Clinical presentation

- gradual onset of lameness or poor performance.
- variety of clinical signs ranging from:
  - subtle change in performance without obvious lameness.
  - moderate or severe lameness.
- mild cases may warm out of the lameness and improve after a short period of work.
- severe cases tend to worsen with exercise.
- generally, lameness will worsen with hard work and improve with rest.
- some horses will appear 'stiff' when restricted to the stable.
- gait alterations are not specific for bone spavin but include:

- reduction in the height of the foot flight arc and toe dragging.
  - ♦ scuff or square-off of the toe of the hind feet (not specific).
  - ○ shortening of the cranial phase of the stride.
  - ○ axial swing of the foot during protraction and landing on the lateral wall.
  - ○ asymmetric movement of the tuber coxae.
- lameness may be accentuated when worked on a circle:
  - ○ usually, worse on the inside leg.
- flexion of the limb (spavin test) usually increases the degree of lameness (not specific).
- thickening of the periarticular soft tissues occurs occasionally in chronic cases:
  - ○ enlargement on the medial aspect of the hock (Fig. 5.2).
- muscle wastage, especially of the gluteals (not specific).

Differential diagnosis

- Occult spavin
- OCD
- Proximal suspensory ligament desmitis.
- talocalcaneal OA
- cunean tendon bursitis.
- other cause of hind lameness such as stifle problems.

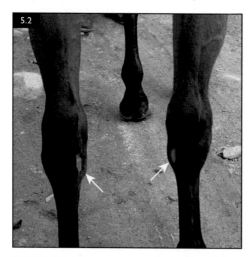

FIG. 5.2 This horse was presented with a history of chronic bilateral hindlimb lameness that was diagnosed after radiographs and intra-articular analgesia as bone spavin. Note the distal medial enlargements in both hocks due to soft tissue filling over the spavin region (arrows).

Diagnosis
Intra-articular analgesia of the small tarsal joints.

- important to confirm the source of pain.
- tarsometatarsal and distal intertarsal joints physically communicate in <40% of horses:
  - ○ local anaesthetics do freely communicate between the lower two joints.
  - ○ some clinicians recommend injecting both joints separately.
  - ○ others inject the TMT joint first and assess response before attempting the DIT.
  - ○ assess response 10 minutes after injection, and again at 40–60 minutes.
    - ♦ improvement in degree of lameness of >50% is considered positive.
    - ♦ bilateral cases exhibit an increased lameness in the contralateral hindlimb.
    - ♦ failure to improve lameness does not rule out the condition.
- tibial and peroneal nerve block may be required in some cases to improve the lameness.
- Deep branch lateral plantar nerve block can also affect the distal tarsal joints.
- local anaesthetic solution within the TMT joint can affect the proximal metatarsal region, including the proximal part of the suspensory ligament.
- **do not assess the effects of local analgesia on the response to a flexion test.**

Radiography

- four views of the tarsus of each hindlimb:
  - ○ centre the beam accurately at the level of the joints.
  - ○ signs are first evident on the dorsomedial/dorsolateral aspects of joints and include (Figs. 5.3–5.5):
    - ♦ periarticular osteophyte formation.
    - ♦ localised subchondral bone lysis.
    - ♦ narrowing of the joint space.
    - ♦ sclerosis of the distal and third tarsal bones.
  - ○ poor correlation between severity of lameness and radiographic changes.

Nuclear scintigraphy

- may be used:

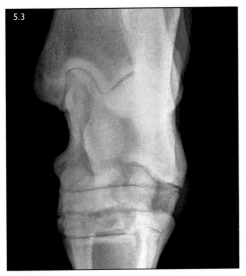

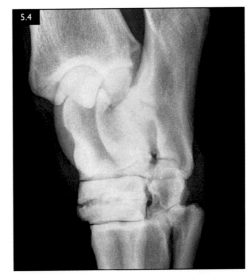

FIGS. 5.3, 5.4 Dorsoplantar (5.3) and dorsolateral–plantaromedial (5.4) oblique radiographs of a horse with OA (spavin) of the distal intertarsal joint (centrodistal).

o where no radiographic abnormalities are present.
o when intra-articular blocks are difficult to perform.
o horses presented for performance problems rather than lameness.

Management

- multiple treatments are available and determining which is used is dependent on:
  o clinical and radiographic findings
  o use of the horse.
  o response to previous treatments
  o financial constraints.
- **mild or no radiographic change** are best treated with:
  o intra-articular medication:
    ♦ triamcinolone (4–10 mg/joint, total dose of 18–20 mg in a 500 kg horse).
    ♦ methylprednisolone acetate (20–40 mg/joint).
    ♦ with or without hyaluronan.
  o systemic NSAID such as phenylbutazone.
  o lameness decreases initially but many horses suffer recurrence (variable period).
  o prolonged rest is not beneficial:
    ♦ box rest for 24–72 hours and then gradually returned to work.

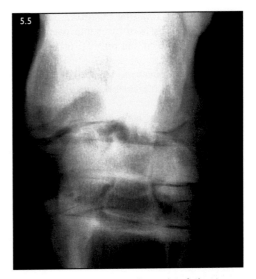

FIG. 5.5 Dorsoplantar radiograph of the tarsus showing osteoarthritic changes of both the centrodistal and talocalcaneal–centroquatral joints. Note the marked bone lysis in the upper joint.

    ♦ change in work programme (type/degree) can decrease the degree of lameness.
- bisphosphonate drugs include tiludronate (intravenous) and clodronate (intramuscular):
  o studies suggest a mild but positive response in some cases.

- severe **radiographic changes** or failure to respond to conservative management justifies surgical intervention.
- **surgical arthrodesis** can be performed in several ways:
  - simple drilling across the joint with or without the addition of cancellous bone graft.
  - use of T-plates or Bagby baskets filled with cancellous bone placed across the joints.
  - prognosis for soundness with these techniques is 60–75%.
- **chemical arthrodesis** taking between 9 and 18 months can be achieved by:
  - intra-articular monoiodoacetic acid (MIA):
    - ◆ under general anaesthesia, with a success rate of 40–80%.
    - ◆ rarely used due to post-injection pain.
  - intra-articular 70% ethanol:
    - ◆ under standing sedation.
    - ◆ up to 80% of cases improve in the short term but only 50-60% longer term.
- cunean tendonectomy is rarely used as it is unlikely to restore soundness in the long term.
- localised partial neurectomy of the branches of the tibial nerve and neurectomy of the deep peroneal nerve supplying the small tarsal joints can be performed:
  - success rate of 65% but rarely used.
  - requires prolonged general anaesthetic time, extensive soft tissue dissection, and effective for only 1–2 years because of reinnervation.

Prognosis

- varies from guarded to poor depending on:
  - joint(s) affected
  - severity and chronicity of articular changes.
  - secondary causes of lameness or back pain.
  - type of work carried out by the horse
  - response to treatment.
- horses with a good response to intra-articular medication often continue in work for some time with the aid of ongoing medication.

- poor response to intra-articular medication presents a greater challenge and will most likely require an arthrodesis.

Tarsocrural synovitis (bog spavin)

Definition/overview

- 'Bog spavin' refers to excessive distension of the tarsocrural (TC) joint with synovial fluid subsequent to acute or chronic synovitis.

Aetiology/pathophysiology

- common causes of 'bog spavin' include:
  - OCD (see page 35)
  - OA   ○ trauma   ○ sepsis.
  - haemoarthrosis
  - intra-articular fractures
  - joint capsule/ligament sprain.
  - continuous cartilage microtrauma due to poor conformation (straight hocks).
  - more evident after a period of box rest, particularly in older, well-used jumping and dressage horses:
    - ◆ should disappear once exercise is initiated.
    - ◆ Idiopathic synovitis is used to describe distension without lameness or inciting pathology.
- lameness is not always present, particularly in OCD cases.

Clinical presentation

- history and clinical signs will vary with the aetiology.
- excessive effusion of the TC joint is most apparent in:
  - dorsomedial and plantarolateral pouches of the joint (Fig. 5.6).
  - OCD cases occurs from 6–24 months of age:
    - ◆ often bilateral, and associated with a variable, frequently mild lameness.
  - effusion may occur suddenly, sometimes in association with excessive exercise.
- lameness may occur later in life due to the onset of OA subsequent to the chronic presence of an undiagnosed OCD lesion.
- Idiopathic synovitis cases are sound.

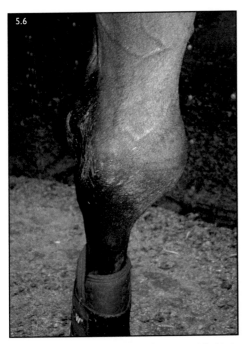

FIG. 5.6 Tarsocrural synovitis (bog spavin). Note the large synovial effusion of the joint, especially dorsomedially.

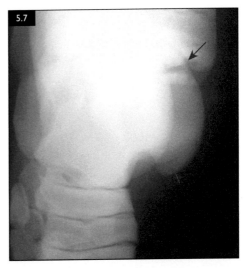

FIG. 5.7 Plantarolateral–dorsomedial oblique radiograph of the tarsus of a Thoroughbred foal with TC joint distension. There are OCD lesions of the intermediate ridge of the distal tibia (arrow) and the distal lateral trochlear ridge of the talus (arrowhead).

- fractures and sepsis cases are often extremely lame:
  - local heat, pain, and swelling may be identified, particularly following trauma.
- flexion of the hock may induce lameness or exacerbate it in some horses.

Differential diagnosis

- OCD
- non-displaced fracture
- soft tissue injury
- OA
- joint sepsis
- nutritional deficiencies.

Diagnosis

- all cases that are lame need careful assessment, including intra-articular analgesia.
- arthrocentesis in chronic cases commonly unremarkable:
  - acute cases, haemarthrosis or increased WBC count and total protein may be present.
- **Radiography** should include four standard views:
  - OCD lesions include (Fig. 5.7):

- most commonly, fragmentation of the distal intermediate ridge of the tibia.
  - these lesions can be seen incidentally in older horses.
- lesions of the distal lateral trochlear ridge of the talus.
  - less common, but of greater clinical significance.
  - variably sized, lytic lesions, causing a greater degree of lameness.
  - medial talar ridge is rarely affected.
- small bony fragments axially on the medial and lateral malleoli of the tibia (Fig. 5.8) which are uncommon.
  - standard views fail to demonstrate any bony abnormality:
- flexed lateromedial
- skyline views.
- dorsal 15° lateral–plantaromedial oblique (medial malleolar lesions).
- **Ultrasonography** can be very useful in identifying:
  - silent radiographic lesions.
  - lesions of the periarticular soft tissues, including the collateral ligaments.

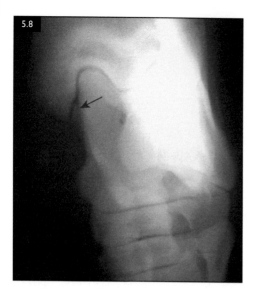

FIG. 5.8 Dorsoplantar radiograph of a yearling crossbred horse that presented with moderate TC joint distension and mild lameness. The lameness was localised to the joint by intra-articular analgesia and this radiograph reveals a medial malleolar OCD lesion just visible as a small defect and separate osseous fragment on the articular margin of the malleolus (arrow).

- absence of radiological or ultrasonographic abnormalities:
  - advanced imaging e.g. scintigraphy, CT, and MRI, or arthroscopic examination.

Management

- no treatment is required if an inciting cause is not identified, and the horse is sound.
- treatment may be attempted if the owner finds the condition cosmetically disturbing:
  - drainage of synovial fluid via arthrocentesis.
  - intra-articular medication with corticosteroids and/or hyaluronan.
  - pressure bandaging of the hock, and restricted exercise.
  - may prove successful but short term only.
- lame horses that improve on intra-articular analgesia of the TC joint should be:
  - thoroughly investigated to identify the core pathology, including arthroscopy.

- young horses, where OCD is the main cause of bog spavin can be treated conservatively:
  - rest, systemic NSAIDs, and intra-articular medications including sodium hyaluronate and corticosteroids.
- surgical arthroscopy of the joint is indicated in non-responsive cases, large lesions or increased lameness.

Prognosis

- varies with the aetiology of the condition:
  - OCD is determined by the size and position of the lesions, whether they are bilateral, the use of the animal, and whether secondary OA occurs.
    - guarded for conservative treatment of all clinically significant lesions.
    - good for surgical treatment in sport horses of distal intermediate ridge and malleolar lesions.
    - guarded for trochlear ridge lesions.
    - racehorses have slightly reduced prognosis.

Fractures and luxations

Definition/overview

- fractures of the hock include:
  - medial and lateral malleolus of the distal tibia (Fig. 5.9).

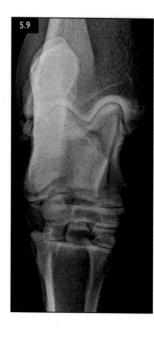

FIG. 5.9 Dorsoplantar radiograph of the tarsus showing a comminuted and slightly distally displaced fracture of the lateral malleolus of the tibia.

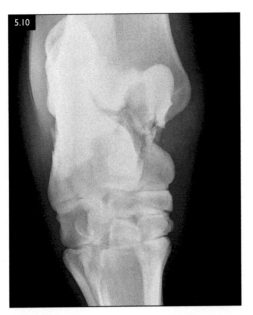

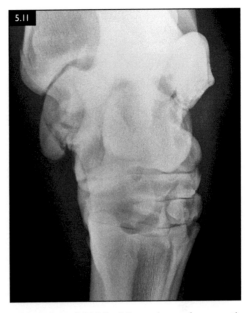

FIGS. 5.10, 5.11 Dorsoplantar (5.10) and plantaromedial/dorsolateral (5.11) oblique views of a comminuted fracture of the talus sustained following a kick.

- o talus (Figs. 5.10, 5.11) and its trochlear ridges (Fig. 5.12).
- o calcaneus (Fig. 5.13) and the cuboidal tarsal bones (Fig. 5.14).
- o osteochondral fractures of the talus.
- o incomplete or complete (uncommon) articular fractures of the dorsoproximolateral aspect of the third metatarsal bone.
- • luxation or subluxation at the level of the following joints:
  - o with or without a concurrent fracture of the hock.
  - o tarsocrural.
  - o proximal intertarsal (talocalcaneal–centroquatral) (Figs. 5.15, 5.16).
  - o tarsometatarsal (Fig. 5.17).
  - o luxation of the distal intertarsal joint (centrodistal) has not been reported.

Aetiology/pathophysiology

- • fractures and luxations are uncommon, because of dense soft tissue support.
- • usually of traumatic origin:
  - o external trauma
  - o rotation of the hock.
  - o accidents involving trapping of the distal limb in a fixed object.
- • significant disruption of the supporting structures.

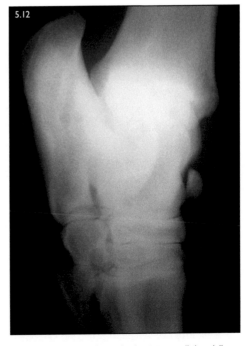

FIG. 5.12 Dorsolateral–plantaromedial oblique radiographic view of a tarsus showing a displaced fragment of the medial trochlear ridge of the talus sustained following a kick to the hock.

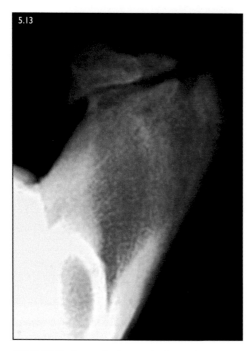

FIG. 5.13 Radiograph demonstrating a transverse fracture of the calcaneus in an adult horse following a kick injury.

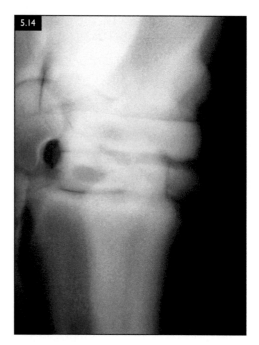

FIG. 5.14 Dorsolateral–plantaromedial radiograph of the tarsus showing a slightly displaced slab fracture of the third tarsal bone sustained after an excessive period of exercise during the convalescent period post tarsal arthrodesis.

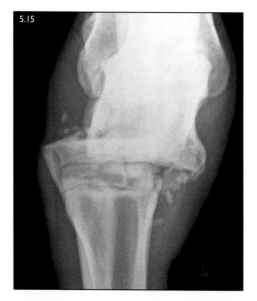

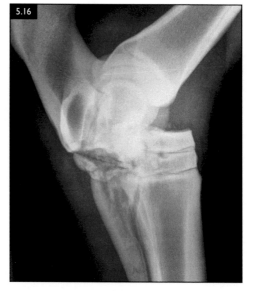

FIG. 5.15, 5.16 Dorsoplantar (5.15) and lateromedial (5.16) radiographs showing luxation of the proximal intertarsal joint following entanglement of the limb in a horse walker.

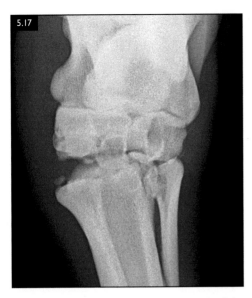

FIG. 5.17 Stressed dorsoplantar radiograph of the hock demonstrating widening of the tarsometatarsal joint following rupture of the medial collateral ligament and subluxation.

- fractures of the central and third tarsal bones are uncommon:
  - mainly seen in racehorses, especially Standardbred trotters.
  - occasionally seen secondary to previous ankylosis of the small tarsal joints.
- association between wedge-shaped conformation of the third tarsal bone and slab fractures in mature racehorses.
- fractures of the sustentaculum tali:
  - sequel to external trauma.
  - may be accompanied by sepsis of the tarsal sheath +/- osteomyelitis of bone.

Clinical presentation

- skin perforation may accompany external trauma.
- lame on the affected limb varying from:
  - mild to non-weight-bearing according to the extent of damage to the structures within, and around, the hock.
  - fracture of the small tarsal bones leads to lameness which may diminish after rest but will return when work commences.
  - mild chronic lameness may precede an acute onset of severe lameness.
- bilateral fracture of the third tarsal bone may present for poor performance.
- palpation may reveal heat, soft tissue swelling, synovial effusion, crepitation, and pain.
- fractures of the calcaneus have a dropped-hock appearance due to loss of function of the gastrocnemius muscle.
- luxation of the tarsal joints may present as abnormal deviation of the limb at the tarsus, with instability.

Differential diagnosis

- synovial sepsis
- OCD
- peritarsal cellulitis.
- other soft tissue injuries.
- spurs or fragments associated with the distal end of the medial trochlear ridge are a common incidental finding, and do not require any treatment.

Diagnosis

- flexion of the tarsus may exacerbate the lameness.
- **diagnostic regional analgesia contraindicated in any horse where a fracture is suspected.**
- **Radiography:**
  - most used imaging modality.
  - not all fractures are diagnosed, and special views may be necessary (Figs. 5.18, 5.19):
    - ♦ flexed lateromedial, skyline view of the calcaneus, and modified oblique views.
  - non-displaced fractures may not be apparent until demineralisation has occurred.
    - ♦ repetition of radiographs is recommended within 14–28 days from injury.
  - stress radiographs may be helpful to diagnose instability of the tarsal joints.
- **Scintigraphy** can be beneficial in horses where suspected fractures are not demonstrated radiographically (Fig. 5.20).
- **Ultrasonography** is useful to help diagnose concurrent soft tissue injury.
- **Standing MRI** (Fig. 5.21) and CT, of the hock may be useful where conventional imaging has not yielded a diagnosis.

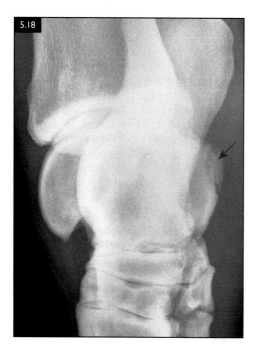

FIG. 5.18 Plantarolateral–dorsomedial oblique radiograph of a hock showing a fracture of the sustentaculum tali of the calcaneus as a result of a kick injury involving the tarsal sheath (arrow).

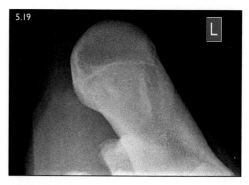

FIG. 5.19 Dorsoplantar flexed oblique radiograph of the sustentaculum tali of the calcaneus showing fragmentation and bony reaction of the medial edge following a kick to the inside of the hock.

Management

- fractures of the lateral malleolus:
  - conservative management of minimally displaced small fragments can be considered.
  - treated surgically by the arthroscopic removal of small fragments (less than 1 cm).
  - repair of larger fragments via an arthrotomy may be successful.
- fractures of the trochlear ridges:
  - small fragments should be treated by arthroscopic removal.
  - large fresh fractures may be amenable to internal fixation.
- fractures of the talus:
  - complete sagittal fractures of the talus can be repaired by internal fixation.
  - incomplete fractures can be treated conservatively.
  - severely comminuted fractures are inoperable and often lead to euthanasia.
- fractures of the calcaneus:
  - small bone fragments should be removed or fixed internally.
  - transverse fractures require internal fixation using bone plates and screws.
  - open comminuted fractures carry a grave prognosis and should be euthanased.
  - chip fractures of the sustentaculum tali should be removed via tenoscopy of the tarsal sheath.
    - ◆ those associated with wounds, tarsal sheath sepsis and/or focal osteomyelitis require more extensive resection.
      - – tenoscopy or open approach.
      - – long-term antibiotic treatment.
  - large bone fragments, where a significant portion of the flexor surface of the sustentaculum tali is affected, should be stabilised using screws.
- fractures of the small tarsal bones (Fig. 5.22):
  - slab fractures can be treated by box rest or internal fixation (slightly better success rates).
  - fragments too small or osteoarthritic changes already evident:
    - ◆ surgical drilling to facilitate arthrodesis of the joints is an option.
- luxations:
  - reduce luxation if possible (difficult to achieve, especially with TC luxation).
  - reduction is successful, place leg in a full-limb cast for 4–8 weeks.

5

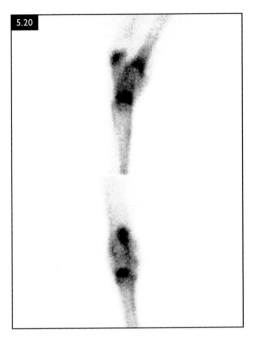

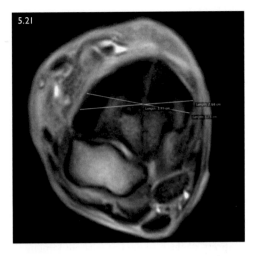

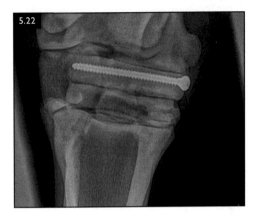

FIGS. 5.20–5.22 Bone scan (5.20) and standing low-field MR image (5.21) of the right hock of a 5-year-old eventer that presented with sudden-onset, moderately severe, right hindlimb lameness. No abnormalities were seen radiographically. The bone scan reveals a focal marked radiopharmaceutical uptake on the two views, which was confirmed on the MR image as a complete fracture of the central tarsal bone. This was repaired with a single lag screw (5.22).

Prognosis

- varies depending on the pathology and involvement of bones, joints, and soft tissues:
  - good (small fragments, especially if they can be removed).
  - hopeless where there is severe comminution and/or soft tissue trauma.

Hygroma of the tuber calcis (calcaneus) (capped hock)

Definition/overview

- bursitis or inflammatory process within a subcutaneous bursa at the point of the tuber calcis.
- may be naturally present or, more commonly, acquired.

Aetiology/pathophysiology

- chronic slow development of a bursal swelling due to repetitive trauma to the region:
  - eventually becomes cosmetically unacceptable to the owner.
  - cavity may become filled with fibrous bands and subdivided by septa.
- non-septic unless wounds occur.
- subcutaneous bursa communicates with intertendinous calcaneal and gastrocnemius bursae in approximately 40% of horses.

Clinical presentation

- rarely lame, unless septic or mechanical interference.
- localised, fluctuant swelling present over the point of the tuber calcis (Fig. 5.23):

- o size and thickness are dependent on the stage of the condition.
- pain on palpation may be evident in acute cases, but in chronic cases it is non-painful.

Differential diagnosis

- Gastrocnemius tendonitis and associated bursitis
- sprain of the long plantar ligament (Curb)
- tarsal sheath distension (Thoroughpin)
- luxation of the SDFT.

Diagnosis

- clinical presentation of a swelling over the point of the tuber calcis without lameness.
- Ultrasonography:
  - o demonstrate hygroma and contents, and to differentiate it from other conditions.
- positive contrast radiography can be useful if there is confusion regarding the anatomic structures involved,.

Management

- prevention of further trauma by using hock boots is recommended.
- fluid drained from the swelling, corticosteroids injected, and a pressure bandage applied:

- o results are often disappointing and short-lived.
- surgical resection should be avoided due to the high incidence of wound breakdown:
  - o only indicated in show horses.

Prognosis

- good for soundness but guarded for cosmesis.

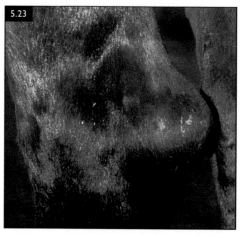

FIG. 5.23 Photograph of a show horse with hygroma of the tuber calcis.

TIBIA

Fractures

Definition/overview

- proximal physeal fractures, fissure fractures, and diaphyseal displaced or non-displaced fractures, as well as stress-related fractures.
- distal physeal fractures are rare.

Aetiology/pathophysiology

- stress fractures of the tibia:
  - o stress accumulation and microfracture at the level of the caudal or caudolateral cortex of the tibia.
  - o usually occur in 2-year-old Thoroughbreds in training.
  - o commonly unilateral.

- physeal and diaphyseal fractures are usually of traumatic origin:
  - o kick or fall.
- proximal physeal fractures are most common in foals:
  - o following a kick or stepped on by the mare.
  - o originate medially and propagate laterally and then vertically.
    - ♦ commonly form a Salter–Harris type II fracture.
  - o may be minimally displaced initially, but usually progressive displacement occurs.
- diaphyseal fractures are usually displaced and occur at all ages:
  - o spiral in configuration and/or comminuted.

Clinical presentation

- stress fractures commonly present with acute lameness:
  - improves greatly, but only temporarily, after a few days of rest.
- severe non-weight-bearing lameness may indicate:
  - incomplete fissure fracture or a complete displaced or non-displaced fracture.
- abduction and instability of the affected limb at the level of the tibia:
  - fracture is displaced and/or comminuted.
- wounds may be present at the injured site:
  - medially in complete fractures due to poor soft tissue coverage and sharp fracture ends.

Differential diagnosis

- Pelvic stress fractures.
- stress-related injury of the distal aspect of the third metatarsal bone.
- fracture of any other bone
- septic arthritis.

Diagnosis

- history, clinical signs and findings on palpation.
  - stress fracture cases may be painful on deep palpation or percussion of the medial aspect of the tibia, but this is unreliable.
  - typical history for stress fractures in racehorses is acute lameness, which improves with rest, and recurs after work.
- diagnostic analgesia should be avoided in acute lameness where a fracture is suspected.
- **Radiographs** are suitable for imaging of all displaced fractures:
  - stress, non-displaced or fissure fractures may initially be radiographically silent.
  - only become evident after 10–14 days.
- **Scintigraphy** is very helpful in all radiographically silent fractures:
  - increased radiopharmaceutical uptake in the affected cortex.
  - usually, a single area, but occasionally multiple sites or bilaterally (Figs. 5.24, 5.25).

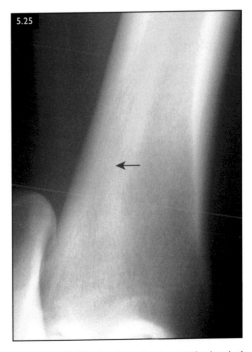

FIGS. 5.24, 5.25 Gamma scintigraphy (5.24) and plain radiograph (5.25) of a stress fracture at the level of the distal third of the tibia in a racehorse. Note the increased sclerosis evident on the radiograph (arrow).

Management/prognosis

- stress fractures should be:
  - box rested for 1 month.
  - box rest with controlled walking exercise for 2 months before resuming training.
  - prognosis for return to racing is good.
  - introduction of training too early predisposes to recurrence.
- incomplete fissure or non-displaced diaphyseal fractures should be:
  - treated conservatively with box rest and cross-tying or sling for the first 2 weeks.
- complete fractures:
  - adequate immobilisation prior to referral to avoid opening of the fracture.
  - displaced fractures may be internally fixated using bone plates and screws.
    - ♦ only recommended for horses weighing <325 kg.
    - ♦ humane destruction recommended for heavier horses (hopeless prognosis).
- proximal physeal fractures in foals:
  - managed conservatively if minimal displacement is present.
  - displaced should be treated by internal fixation by a variety of techniques.
  - excellent prognosis for life and fair for future soundness.

STIFLE

Patella

Definition/overview

- conditions associated with the patella include upward fixation of the patella, fractures, fragmentation of the apex, and luxation.

Aetiology/pathophysiology

- medial patellar ligament locks the patella and its medial cartilaginous extension above the medial trochlear ridge of the femur as part of the stay apparatus of the hindlimb.
- Upward fixation of the patella is caused by failure to unlock this mechanism.
  - more common in young horses and ponies (may be hereditary in some breeds).
  - straight hindlimb conformation, poor body and muscle conditioning, and weak thigh musculature predisposes horses to this condition.
  - also following coxofemoral luxation, trauma to the stifle region, or sudden box rest in older and/or fit horses.
  - acute and permanent fixation is usually unilateral.
  - intermittent fixation can be bilateral.
  - some horses suffer from a milder condition in which there is delayed

release of the patella leading to milder lameness and poor performance.
- fractures of the patella occur because of an external injury:
  - kick or hitting a fixed object while jumping.
  - medial pole of patella most common.
    - ♦ variety of other fracture configurations can occur.
  - fragmentation of the apex of the patella:
    - ♦ closely correlated to desmotomy of the medial patellar ligament.
- Lateral luxation of the patella:
  - usually inherited and most common in miniature-breed foals.
  - reported in other breeds, possibly due to trauma.
  - usually accompanied by hypoplasia of the lateral trochlear ridge.
    - ♦ can occur with normal conformation.
- medial luxation is usually due to trauma to the region.
  - must overcome the larger medial trochlear ridge.

Clinical presentation

- lameness varies between mild and severe depending on the condition affecting the patella.

- flexion commonly exacerbates the degree of lameness:
  - some horses may resent this manipulation.
- Upward fixation of the patella causes typical gait abnormalities characterised by:
  - affected hindlimb extended caudally and dragged while walking.
  - usually evident during rest or at the start of movement (Fig. 5.26).
  - may be temporarily relieved after a few strides or remain for a prolonged duration.
  - release after an intermittent locking involves a snap followed by rapid hyperflexion of the stifle and hock.
  - when manifested as 'delayed release' or 'catching' of the patella:
    - sudden jerk observed during flexion of the stifle as the horse moves.
- foals affected with luxation of the patella:
  - unable to extend the stifle
  - stand in a crouching position.
  - less severe cases have a stiff gait and refusal to flex the affected limb.
- effusion of the femoropatellar joint is common with:
  - fractures of the patella, fragmentation of the apex of the patella, and all chronic conditions that may promote OA.
- fragmentation of the apex of the patella after desmotomy of the medial patellar ligament may have excessive fibrous tissue at the surgical site and resent flexion of the affected limb.

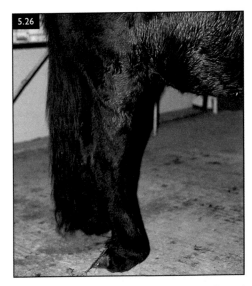

FIG. 5.26 This Dale pony has an upwardly fixated patella in the right hindleg, which extends the hock and stifle, causing it to stand with the distal limb joints flexed.

- palpation of foals with patellar luxation reveals the position of the patella, the possibility for relocation, and if the luxation is intermittent or permanent.
- cranioproximal/craniodistal (skyline) view of the patella should be included if a fracture is suspected (Figs. 5.27, 5.28).
- ultrasonography is useful for demonstrating excessive scar tissue after desmotomy.

Differential diagnosis

- Stringhalt
- fibrous or ossifying myopathy
- OCD   • septic arthritis.
- meniscal and/or cruciate ligament tear.
- desmitis of the patellar ligaments.
- fracture of any other bony component of the stifle
- low-grade hindlimb ataxia.

Diagnosis

- based on clinical signs, palpation, and radiographic findings, including:
  - lateromedial, caudocranial, and caudocranial oblique radiographs of the stifle.

Management/prognosis

- Upward fixation of the patella should initially be treated conservatively:
  - increase muscle mass over the hindquarters through exercise and improved nutrition.
  - stifle discomfort may prevent an increase in muscle mass and can be alleviated with:
    - medium term – oral NSAIDs.
    - intra-articular corticosteroids in the femoropatellar joint.
  - condition spontaneously resolves in many young animals.
  - failure of conservative management, or severe clinical signs warrants surgical intervention:
  - medial patellar ligament desmotomy:

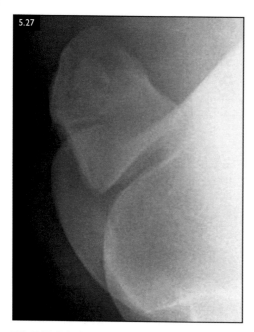

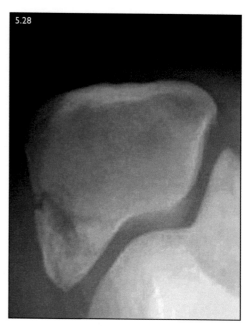

FIG. 5.27, 5.28 Lateromedial (5.27) and cranioproximal–craniodistal oblique (skyline) (5.28) radiographs of the stifle demonstrating a sagittal fracture of the medial pole of the patella. Note that the fracture is difficult to identify on the lateromedial view.

- ♦ most easily performed in the standing horse under sedation and local anaesthesia when the ligament is under tension.
- ♦ recommended to rest the horse in a box for 2 months postoperatively with gentle walking in hand during the second month.
- ♦ horses expected to return to normal use (some develop a slightly restricted gait).
- ♦ complications reported include distal patellar fragmentation and severe lameness after surgery (usually too early return to work) (Fig. 5.29).
- ○ medial patellar ligament splitting (desmoplasty) percutaneously:
  - ♦ under ultrasonographic guidance
  - ♦ standing sedation or under GA.
  - ♦ as effective as desmotomy and without serious complications.
  - ♦ less likely to lead to a restricted gait in the longer term.
- ○ counterirritant agents injected into the patellar ligaments is preferred by some

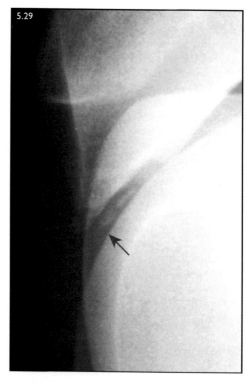

FIG. 5.29 Magnified lateral view radiograph of the distal patella of a horse that developed distal patellar fragmentation (arrow) post medial patellar desmotomy.

clinicians with little scientific evidence of its effectiveness.

  ◆ inadvertent intra-articular injection of a counterirritant could prove disastrous.

- management of a fractured patella is determined by the fracture configuration:
  ○ small non-displaced, non-articular fractures can be treated conservatively with box rest, and the prognosis for soundness is good.
  ○ medial pole fractures may be amenable to arthroscopic removal of up to 30% of the patella with a good prognosis for future athletic function.
  ○ larger fragments are harder to treat:
    ◆ patellar instability may ensue if the attachment of the medial patellar ligament is completely disrupted.
  ○ internal fixation is required for all other fracture configurations:
    ◆ quadriceps muscle contraction may promote breakdown.
  ○ comminuted fractures have a poor prognosis.
- patellar luxation is difficult to treat and may not lead to athletic soundness:
  ○ lateral release and medial imbrication surgical techniques have been used.
  ○ breeding from affected animals should be avoided.

Osteochondrosis of the femoropatellar joint

Definition/overview

- Femoropatellar joint is one of the most common sites for OCD in the horse:
  ○ particularly in the Warmblood and Thoroughbred (see page 35).
- usually presents in actively growing large individuals when foals or weanlings:
  ○ may remain clinically silent until the animal starts active work.
- may predispose to the early onset of OA of this joint.

Clinical presentation

- usually presents before 2 years of age (particularly between 4 and 14 months) in excessively growing individuals.

- may remain asymptomatic in early life but presents later:
  ○ work programme increases between 3 and 6 years of age.
  ○ excessive wear and tear leads to OA of the joint.
- usually obvious femoropatellar joint distension (Fig. 5.30), often bilateral, but possibly asymmetrical in degree.
- lameness can vary:
  ○ mild bilateral stiffness to a marked unilateral lameness.
  ○ especially lame where there are large lesions of the lateral trochlear ridge and possible lateral patella instability.
  ○ may improve with rest but returns with increased exercise.

Differential diagnosis

- traumatic lesions of the stifle joints and patella.
- Subchondral bone cysts of the medial femoral condyle.
- sepsis of the femoropatellar joint.
- older animals include other causes of OA and inflammation of the femoropatellar joint.

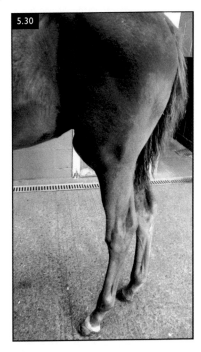

FIG. 5.30 Extensive synovial effusion of the left hindlimb femoropatellar joint in a 6-month-old Thoroughbred with stifle OCD.

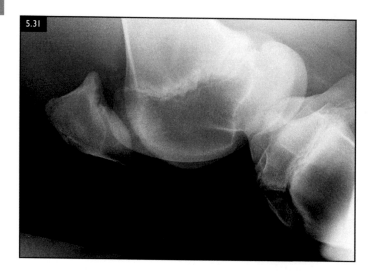

FIG. 5.31 Lateromedial radiograph of a Swedish Warmblood foal with stifle distension and lameness. Note the upper third lateral trochlear ridge and upper subpatellar OCD lesions.

Diagnosis

- clinical history and presenting signs are highly suspicious.
- confirmation requires **radiographs of the stifle**, preferably of both limbs.
  - lateromedial and caudal 30° lateral–craniomedial oblique views are the most useful.
  - most common site for lesions is the lateral trochlear ridge of the femur.
  - occasionally lesions restricted to the articular surface of the patella (Fig. 5.31).
  - medial trochlear ridge and the trochlear groove lesions are rare.
  - significant radiographic lesions include:
    - marked flattening of the mid- to upper-third of the lateral ridge.
    - irregularity of the contour of the ridge.
    - one or more subchondral lucent defects in the ridge +/– surrounding sclerosis.
    - osseous density fragments within ridge defects or free in the joint distally (Fig. 5.32).
    - irregularity of the articular surface of the patella towards the apex +/– surrounding sclerosis.
    - evidence of osteophytic development in older horses with secondary OA.
- **Ultrasonography** will identify joint distension and localise trochlear ridge cartilage lesions (see Figs. 2.22, 2.23 – see pages 36 and 37).

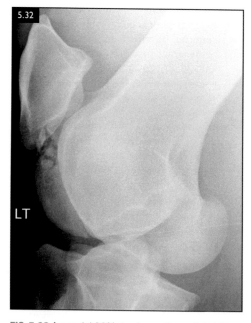

FIG. 5.32 A caudal 30° lateral–craniomedial oblique radiograph of the left stifle of a 6-year-old eventer. Osteochondral fragments can be seen associated with the upper third of the lateral trochlear ridge of the femur.

  - provides the means to measure the size of any lesion (useful prognostically) (<2 cm; 2–4 cm; >4 cm)
- Arthroscopy of the joint allows identification of all lesions and assessment of the extent of cartilage damage (Fig. 5.33).

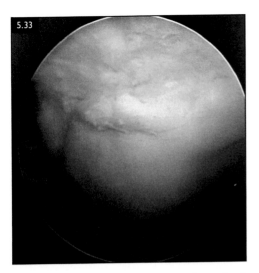

FIG. 5.33 Arthroscopic view of the underneath of the patella (upper part of image) with an OCD lesion visible in the distal half. The trochlear groove is ventral.

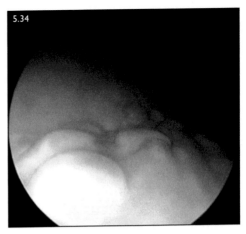

FIG. 5.34 Arthroscopic view of the upper half of the medial side of the lateral trochlear ridge showing a large area of abnormal and loose cartilage associated with an OCD lesion.

Management

- early or mild cases conservative treatment with:
  - rest, systemic NSAIDs, and possibly intra-articular medications may be adequate to return the animal to soundness in the short term.
  - some cases may develop problems later in life due to the onset of early joint disease and OA.
- surgical intervention using arthroscopy is recommended for most cases but particularly:
  - those that are lame
  - not responded to conservative treatment.
  - large lesions
  - where an athletic career is envisaged (Fig. 5.34).
- horses with OA are treated conservatively in most cases.

Prognosis

- depends on the extent and severity of the lesions, at what age they are detected, and whether one or both limbs are involved.
- generally guarded to fair for return to athletic soundness for lesions requiring surgical intervention and guarded to poor where OA is evident:
  - size of the lesion measured arthroscopically is important.

  - significantly improved prognosis if lesion less than 2 cm in length.
  - large lesions of the trochlear ridge in foals where stability of the patellar is affected carry a poor prognosis.
  - foals with stifle OCD rarely reach high performance levels.

Subchondral osseous cyst-like lesions of the medial femoral condyle

Definition/overview

- medial femoral condyle is the most common site for osseous cyst-like lesions (OCLL), often referred to simply as bone cysts:
  - lesions tend to form on the weight-bearing aspect of the condyle.
  - affects the overlying articular cartilage, leading to synovitis and lameness.
- may hasten joint degeneration and lead to OA.

Clinical presentation

- usually recognised in young animals (<4 years), particularly at the beginning of ridden exercise.
- may remain asymptomatic until later in life, when joint degeneration, or an

increase in athletic demands, leads to a more pronounced synovitis and discomfort.
- lameness is usually mild or moderate and acute:
  - older animals may be insidious in onset.
- effusion of the medial femorotibial joint is usually present but is not always detected.
- may be positive response to upper hindlimb flexion.

Differential diagnosis

- femoropatellar joint OCD
- meniscal tears.
- chondromalacia of the medial femoral condyle
- older animals OA.

Diagnosis

- clinical signs.
- diagnostic regional and intrasynovial analgesia:
  - essential in isolating the site of pain and confirming clinical significance.
    - ♦ **Note** sufficient time should be allowed for the local anaesthetic to diffuse into the cyst (e.g. up to 30–60 minutes).
    - ♦ partial improvement (50%) warrants radiographic evaluation.
- **Radiographic** examination of the stifle should demonstrate:
  - radiolucent, flask-shaped lesion of subchondral bone of medial femoral condyle.
  - most easily seen on a caudocranial and flexed lateromedial views (Fig. 5.35).
  - contralateral stifle should always be imaged as lesions can be bilateral.
  - cysts vary in size from small indentations to a large radiolucent lucency depending on stage of development.
  - often oval or rounded in shape, usually confluent with the articular margin, and may have a sclerotic peripheral margin.
- **Ultrasonography** of the articular surface of the medial femoral condyle can be achieved if the limb is held in a flexed position.
  - confirms a defect in the subchondral bone and articular cartilage.

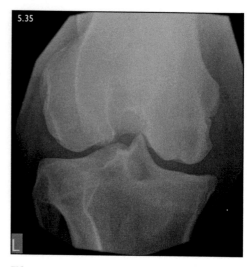

FIG. 5.35 A caudocranial radiograph of the left stifle of a 6-year-old Welsh gelding. A depression in the subchondral bone of the medial femoral condyle is seen to communicate with a flask-shaped radiolucency within the condyle itself. This is an OCLL of the medial femoral condyle.

- useful in defining the size of the opening to the cyst which can affect prognosis.
- arthroscopy of the joint allows identification of the cyst opening and treatment.

Management

- conservative treatment with rest, systemic NSAIDs, and possibly intra-articular medication (corticosteroids) may be adequate to return the animal to soundness in the short term.
- direct treatment of the cyst is required to:
  - provide long-term resolution of the lameness associated with the lesion.
  - reduce the likelihood of the development of OA.
- reduction of inflammatory mediator production by the cyst lining:
  - corticosteroids injected directly into the cyst lining under ultrasonographic or arthroscopic guidance under general anaesthesia.
  - latter is more reliable.
- arthroscopic debridement of the cysts under general anaesthesia:
  - primary treatment.

o those animals in which corticosteroid medication has proved unsuccessful.
- resurfacing the medial femoral condyle with mosaic arthroplasty and chondrocytes or stem cells in fibrin glue are successful but currently not widely available.
- cortical or resorbable bone screw placed across the cyst under radiographic and arthroscopic guidance.
  o drilling through the cyst and the ongoing presence of the screw appear to stabilise the cyst and allow it to fill with hard bone.

Prognosis

- depends on the age of the horse and the size of the defect in the joint surface.
- arthroscopic assessment of lesion size has proved to be a reliable guide to prognosis, with lesions of 15 mm diameter or smaller more likely to have a positive outcome.
- 50% success rate with conservative treatment of medial femoral condylar subchondral bone cysts has been reported.
- arthroscopic treatment:
  o 70% success rate reported in horses under 3 years old.
  o much lower success rate is reported in horses over 4 years old (35%).
- younger horses where corticosteroids are injected directly into the cystic lining, seem to have a favourable prognosis.
- 75% of 20 cases of medial femoral condyle cysts subjected to a screw technique were sound at 120 days postoperatively.
- accompanying DJD and bilateral cysts worsens the prognosis.
- OA at the time of treatment leads to a significantly poorer prognosis.

Soft tissue injuries of the stifle joint

Definition/overview

- stifle is a complex structure, and all components are prone to injury:
  o array of ligaments maintaining normal movement.
  o two large fibrocartilaginous menisci.

o large areas of joint cartilage.
- soft tissue injuries of the stifle include:
  o meniscal tears.
  o lesions of the cruciate, meniscal, or collateral ligaments.
  o desmitis of the patellar ligaments.
- penetrating injury into the synovial compartments can result in septic arthritis.
- trauma to the articular cartilage is frequently encountered.

Aetiology/pathophysiology

- injuries of the meniscus and associated ligaments are probably caused by a combination of crushing forces with rotation of the tibia and flexion or extension of the stifle.
- cruciate injuries are possibly the result of a combination of torsional and compressive forces when sudden turning of the horse occurs while the stifle is flexed.
- medial collateral ligament is more commonly affected than the lateral one.
- ligament and meniscal injuries are most often seen in adult horses and a combination of soft tissue injuries may be present.
- desmitis of the patellar ligaments, usually the middle one:
  o rare and occurs most commonly in jumping horses.
  o also occur in horses with intermittent upward fixation of the patella or as a sequela to desmotomy of the medial patellar ligament.
- blunt trauma to the region can induce a variety of injuries.

Clinical presentation

- affected horses are usually lame:
  o degree of lameness varies from subtle to severe depending on the type, severity, and duration of the injury.
- effusion of any of the stifle joint compartments may be palpable, and signs of local inflammation may be present as well.
- stifle joint in a horse is divided into three compartments:
  o femoropatellar joint, and the lateral and medial femorotibial joints.

- commonly the femoropatellar and medial femorotibial joints communicate:
  - these joints are most often affected by disease and most likely to be effused.
- upper limb flexion usually exacerbates the lameness and may be resented.
- intra-articular analgesia should greatly improve the degree of lameness with injuries involving any of the joint components:
  - only a partial improvement in some cases.
- evidence of external injury or wounds may occasionally be present.
- severe injury to the supporting ligaments of the joint (collateral and cruciate ligaments) may lead to instability and severe lameness.
  - intra-articular analgesia of the stifle joints in these cases can be difficult to interpret.

Differential diagnosis

- OCD
- septic arthritis
- rupture of the origin of the long digital extensor tendon and peroneus tertius.
- articular fractures
- femoral fractures.

Diagnosis

- clinical findings at the level of the stifle may point to this region as the potential site of pain, but careful examination of the rest of the limb is mandatory.

- when performing intra-articular analgesia, it is important to anaesthetise all three compartments to localise the pain.
  - local analgesia should be avoided if lameness is severe.
- **Radiography** is useful to identify concurrent bony injury.
- **Ultrasonography** is useful for imaging many soft tissue conditions of the stifle region (Figs. 5.36–5.39):
  - some lesions, particularly those of the cruciate ligaments, are difficult to visualise.
  - useful to identify lesions of the cartilage of the trochlear ridges of the distal femur and medial femoral condyle.
- **Diagnostic arthroscopy** is indicated if no abnormalities are detected on any imaging modality and the lameness is localised to the stifle (Fig. 5.40).
- **MRI** of the stifle can be performed under general anaesthesia and may provide further useful information to identify soft tissue lesions.

Management

- mild soft tissue injuries of the stifle may be treated with a combination of rest and intra-articular medication:
  - corticosteroids frequently used (5–10 mg triamcinolone acetonide per affected joint compartment).
  - **avoid an excessive total dose.**

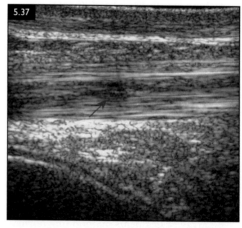

FIGS. 5.36, 5.37 Transverse (5.36) and longitudinal (5.37) ultrasonographic images demonstrating heterogenicity and disruption of fibre pattern indicative of desmitis of the middle patellar ligament (arrows).

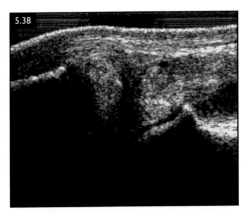

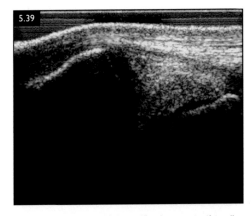

FIGS. 5.38, 5.39 Longitudinal ultrasonographic image of the medial aspect of the stifle demonstrating disruption of the medial meniscus, indicative of a tear (5.38). Note the normal ultrasonographic appearance of the contralateral meniscus (5.39).

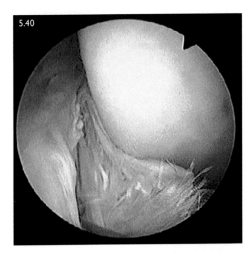

FIG. 5.40 Arthroscopic image of the left lateral femorotibial joint of a 12-year-old dressage horse in which a tear of the cranial cruciate ligament was found.

- careful rehabilitation programme aimed at improving muscling around the stifle with correct patellar tracking reliant on the muscle groups of the lateral thigh.
- arthroscopic evaluation of the joint is indicated:
  - poor response to intra-articular medication.
  - soft tissue lesion that may benefit from debridement is suspected.
  - severe unilateral lameness suggests a significant soft tissue lesion.
- desmitis of the patellar ligaments or damaged collateral ligaments should be treated with rest and controlled exercise, and the progress monitored ultrasonographically:
  - some cases have been treated with shock-wave therapy, PRP or stem cells.

Prognosis

- guarded for soundness in horses with meniscal tears, and poor in horses with moderate to severe cruciate ligament injuries.
- concomitant injuries, particularly involving the articular cartilage, further deteriorate the prognosis.
- meniscal tears can be graded from 1 to 3 depending on severity:
  - return to athletic function:
    - 63% of horses affected by grade 1 tears.
    - 56% of horses affected by grade 2 tears.
    - 6% of horses affected by grade 3 tears.
- collateral ligament injury – prognosis for return to athletic exercise is poor.
- some horses with desmitis of the patellar ligament may not return to their previous level of exercise.

Chondromalacia of the medial femoral condyle

Definition/overview

- abnormal, soft, fissured, poorly adherent cartilage over the medial femoral condyle

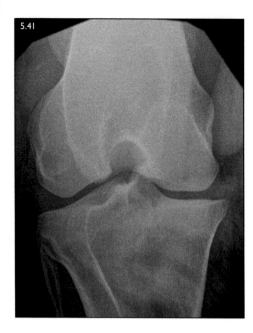

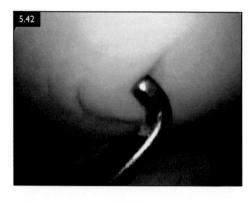

FIGS. 5.41, 5.42 (5.41) Caudocranial radiograph of the left stifle of a 6-year-old gelding that had suffered diffuse chondromalacia of the medial femoral condyle. The medial femoral condyle appears flattened on this projection. (5.42) Arthroscopic image of the left medial femoral condyle of the same horse showing cracked and poorly adherent cartilage.

is well recognised by arthroscopic surgeons:
- ○ seen in isolation or in combination with other injuries.
- ○ still a debate about whether chondromalacia is a primary cause of lameness.

Aetiology/pathophysiology

- aetiology as a primary condition is currently unknown.
- when seen in combination with soft tissue pathology:
  - ○ assumed to have either arisen from the same injurious incident.
  - ○ occurred subsequent to an altered joint environment.

Clinical presentation

- overt unilateral or bilateral lameness.
- when bilateral may present as poor performance.
- sudden or insidious in onset.
- more apparent in ridden work.
- palpable effusion of the medial femorotibial joint is common.

Differential diagnosis

- soft tissue injuries of the medial femorotibial joint.

- OCD.
- OCLL of the medial femoral condyle.

Diagnosis

- positive response expected to diagnostic analgesia of the medial femorotibial joint:
  - ○ increased lameness in the contralateral limb indicative of a positive response.
    - ♦ helpful where lameness is subtle and bilateral.
- **Radiography** is useful to identify concurrent bony pathology:
  - ○ flattened or concave medial femoral condyle on a caudocranial projection is common (Fig. 5.41).
- **Ultrasonography** is likely to demonstrate:
  - ○ effusion and synovial proliferation medially.
  - ○ abnormal cartilage lining the condyle with the limb in a flexed position.
- **Diagnostic arthroscopy** is currently the only way to reliably assess articular cartilage of the medial femoral condyle (Fig. 5.42) and exclude concurrent soft tissue injuries.

Management

- mild lameness can be treated with intra-articular corticosteroid medication (5–10 mg triamcinolone acetonide) injected into the medial femorotibial joint:

o avoid in animals where corticosteroids are contraindicated.
o careful rest and rehabilitation programme.
- arthroscopic evaluation of the joint is indicated:
  o intra-articular medication failed to resolve lameness.
  o more severe lameness means a concurrent soft tissue lesion is suspected.
  o damaged cartilage may require removal.
  o post-operative rest and careful graduated rehabilitation is essential.

Prognosis
- guarded for future athletic function where chondromalacia is a primary lesion.
- concurrent soft tissue injuries reduce return to athletic soundness to guarded to poor.

Osteoarthritis of the stifle joints

Definition/overview
- not uncommon and medial femorotibial joint is most frequently affected.

Aetiology/pathophysiology
- usually occurs secondary to other joint pathology:
  o where this is not recognised may be part of 'wear and tear' pathology.

Clinical presentation
- often older, active, working animals.
- low-grade, bilateral lameness, often exacerbated by upper limb flexion, is common.
  o may be insidious or sudden in onset.
- palpable, mild to moderate effusion of the femoropatellar and medial femorotibial joints.
- muscle atrophy of the affected limbs.

Differential diagnosis
- Chondromalacia.
- distal tarsal OA
- OCD.

- soft tissue injuries of medial femorotibial joint.
- OCLL of medial femoral condyle.

Diagnosis
- positive response to diagnostic analgesia of the affected joint compartments:
  o lameness present elsewhere may mean additional joint blocks are necessary to render the horse completely sound.
- **Radiographs** required to identify the periarticular osteophytes indicative of OA (Fig. 5.43):
  o complete radiographic series to fully characterise the joint disease that is present.
- **Ultrasonography** should confirm the presence of osteophytes at the joint margins:
  o may indicate marked synovial hypertrophy typical of chronic joint disease.
- **Diagnostic arthroscopy** may be warranted if clinical findings are suggestive of concurrent soft tissue or cartilage pathology.

Management
- intra-articular medication such as:
  o corticosteroid (5–10 mg triamcinolone acetonide into the medial femorotibial joint).
  o Interleukin-1 receptor antagonist protein (IRAP), PRP or hydrogels.
  o brief period of rest followed by a rehabilitation programme.

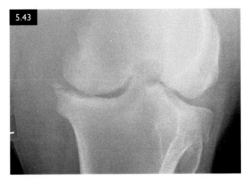

FIG. 5.43 Caudocranial radiograph of the stifle of a horse with long-term stifle lameness. Note the medial tibial and femoral condylar new bone proliferation due to severe OA of the medial femorotibial joint.

- more advanced cases may require concurrent oral non-steroidal anti-inflammatory drugs to adequately control discomfort.
- rider's expectations and the work that is required of the horse may need to be adjusted.

Prognosis

- fair prognosis for the management of clinical signs if identified at relatively early stage:
  - joint degeneration will progress at a rate proportional to the workload.
- advanced cases can be very difficult to treat, often leading to retirement or a significant reduction in workload.

Fractures affecting the stifle.

Definition/overview

- these include those of the:
  - patella (see above) o tibial tuberosity
  - trochlear ridges of the femur.
  - femoral condyles
  - intercondylar eminences of the tibia.

Aetiology/pathophysiology

- fractures of the above (excluding the intercondylar eminences) are most common in horses engaged in jumping/hunting activity, after hitting a hard object while jumping.
  - may also occur following a kick injury or a foreign body penetration.
- fractures of the lateral trochlear ridge and, less commonly, the medial trochlear ridge, may occur either alone or concurrently with patellar fractures.
- fractures of the tibial tuberosity vary in size, may be displaced by the pull of the patellar ligaments (avulsion fracture) and occasionally are intra-articular.
- fracture of the intercondylar eminence of the tibia are thought to be a sequela to lateral impingement of the medial condyle on the eminence rather than an avulsion.

Clinical presentation

- lameness, with the chronicity and the location of the fracture affecting the degree.

- flexion of the limb is very painful, and, in some injuries, joint instability may be felt.
- acutely the stifle region may be generally swollen and painful:
  - over time, joint effusion, often very pronounced, is palpable.
- evidence of external injury or wounds may be present.

Differential diagnosis

- OCD   • septic arthritis
- femoral fractures.
- rupture of the origin of the long digital extensor tendon and peroneus tertius.

Diagnosis

- clinical findings, but careful examination of the rest of the limb is mandatory.
- where the cause of lameness is unclear, intra-articular analgesia may be useful:
  - only partial improvement in lameness may be evident in some cases.
  - should be avoided in severely lame cases.
- **Radiography** and **ultrasonography** may identify the injury (Figs. 5.44–5.48).
  - additional views such as skyline of the patella, flexed lateromedial, and oblique views at different angles are sometimes required.
- **Scintigraphy** is indicated in acutely lame horses when a fracture is suspected but the location is not identified.
- CT and MRI of the stifle under general anaesthesia may provide further information on fracture location and configuration.

Management/prognosis

- treatment for tibial tuberosity fractures is determined by the size of the fragment, the degree of displacement, and involvement of the joint.
  - minimally or non-displaced fragments may be treated conservatively with box rest.
    - ♦ cross-tie these horses to minimise proximal displacement of the fractured fragment by the pull of the quadriceps muscles via the patellar ligament.
  - small, displaced fragments are best removed.

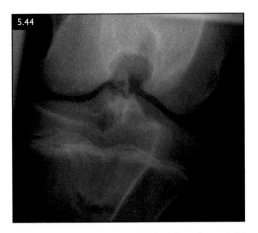

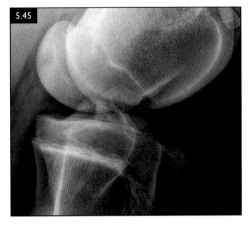

FIGS. 5.44, 5.45 Caudocranial (5.44) and caudolateral–craniomedial oblique (5.45) radiographs showing a fracture of the medial intercondylar eminence of the tibia sustained spontaneously during exercise.

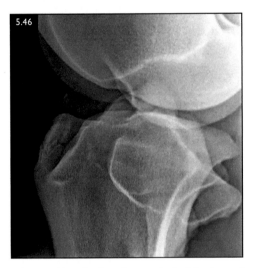

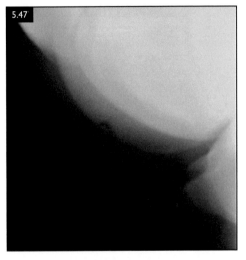

FIG. 5.46 Non-displaced comminuted fracture of the upper part of the tibial tuberosity caused by a kick to the cranial stifle region.

FIG. 5.47 A deliberately under-exposed lateral radiograph of the femoropatellar joint to highlight a small chip fracture off the cranial aspect of the lateral trochlear ridge due to a kick injury.

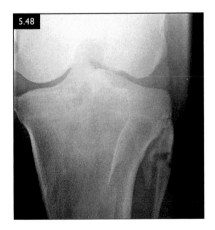

FIG. 5.48 Caudocranial radiograph of the stifle reveals a healing 3-month-old fracture of the proximal fibula caused by a direct trauma.

○ large fragments should be fixed internally using a bone plate and screws.
○ prognosis for tibial tuberosity fractures is generally good.

- intra-articular fractures should be assessed, and possibly treated, arthroscopically.
  ○ intercondylar eminence fractures are best treated with arthroscopy:
    ♦ removal of the fragment if small.
    ♦ internal fixation with larger fragments.
  ○ small fracture fragments of the trochlear ridges of femur and femoral condyles:
    ♦ often traumatic in origin.
    ♦ sometimes accompanied but other injuries such as fracture of the patella.
    ♦ best treated by arthroscopic removal.
    ♦ prognosis for future athletic function depends on size and location of the fracture.
      – small bone fragments removed with minimal disruption of the articular surface carry a fair prognosis.
      – injuries that affect large areas of the joint surface, affect ligamentous attachment or lead to joint or patellar instability carry a much poorer prognosis.

Calcinosis circumscripta

Definition/overview

- uncommon condition, more often seen in young horses.
- lesions are usually in the subcutaneous tissue close to joint capsule or tendon sheaths.
- most common in the stifle region:
  ○ also described in the hock, carpus, neck, and shoulder.

Aetiology/pathophysiology

- lateral aspect of the stifle, close to the fibula.
- cause is unknown.
- dystrophic mineralisation of subcutaneous tissue, with evidence of inflammation:
  ○ trauma may be the initiating factor.

Clinical presentation

- generally presented for an unsightly mass.
  ○ often slowly increasing in size.
- lesion is a firm and well-circumscribed mass closely attached to the underlying tissues:
  ○ no involvement of the skin.
- occasional horses may have a gait abnormality.

Differential diagnosis

- tumoural calcinosis is a different metastasic condition associated with calcification of multiple periarticular sites.
- osteochondromatosis.
- mast cell tumour.

Diagnosis

- clinical findings are not specific.
- **Radiography** reveals a localised soft tissue swelling and amorphous to well-circumscribed accumulations of granular mineral opacity (Figs. 5.49, 5.50).
- **Ultrasonography** is helpful to rule out other causes of soft tissue swelling.

Management

- most horses are best not treated.
- surgical excision is indicated only if the lesion causes clinical complications, or the owner insists on removal for cosmetic reasons.
  ○ careful delicate excision is required to avoid penetrating the adjacent femorotibial joint capsule.
  ○ recurrence has not been reported.

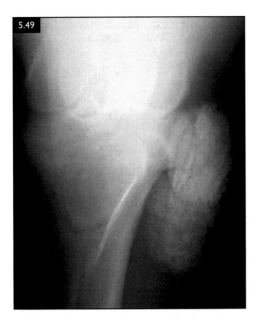

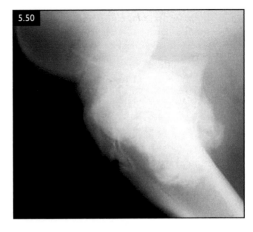

FIGS. 5.49, 5.50 Caudocranial (5.49) and lateromedial (5.50) radiographs of a horse with calcinosis circumscripta located on the lateral aspect of the stifle.

FEMUR

Fractures

Definition/overview

- fractures may occur at any level of the bone:
  - head, neck, and greater trochanter.
  - mid-diaphysis or the distal aspect of the femur.

Aetiology/pathophysiology

- found at any age but they are most common in very young horses:
  - fractures of the femoral head and neck occur almost exclusively in foals.
  - after flipping over backwards or severely abducting the hindlimbs.
- other types of femoral fractures occur most commonly after external trauma:
  - kick, a severe fall, or in young foals after being trodden on by their dam.
- bad induction or recovery from general anaesthesia may lead to femoral fractures in adults.
- diaphyseal fractures are commonly comminuted, spiral, or oblique in shape.
- distal physeal fractures are usually Salter–Harris type II with the metaphyseal component caudally.

- extensive surrounding musculature protects femoral fractures from becoming open:
  - muscle contraction promotes significant overriding of the fracture ends.

Clinical presentation

- non-weight-bearing lame in most cases:
  - young foals with a proximal physeal fracture may partially weight bear.
- crepitation and rotational instability, while manipulating the limb:
  - haemorrhage and swelling in the muscle may make it difficult to elicit crepitation.
- external rotation of the limb (continuous pull of the gluteal muscles) is common.
- swelling and severe oedema are present in all diaphyseal and distal physeal fractures:
  - mild in minimally displaced distal physeal fractures.
  - not readily apparent in acute proximal fractures.
- overriding of the fracture ends shortens distance between greater trochanter and patella:

- appearance of upward fixation of the patella.
- generally, a shorter limb with a higher hock compared with the contralateral limb.
- patella often feels loose and can be manipulated sideways.
- major blood vessels, usually the femoral artery, may be severed by the sharp fracture ends:
  - clinical signs of acute blood loss or fatal haemorrhage.

Differential diagnosis

- coxofemoral luxation
- upward fixation of the patella
- muscle tear.
- fractures of the patella or tibia
- pelvic fractures.

Diagnosis

- clinical presentation and signs, and findings on palpation and manipulation of the limb:
  - auscultation with a stethoscope over the femoral region may identify crepitus.

- proximal physeal fractures are the most difficult to diagnose clinically.
  - rectal palpation (larger animals) may reveal crepitation during limb manipulation.
- soft tissue swelling may be identified on the medial aspect of the thigh.
- **Radiography** is required for definitive diagnosis in all fractures (Fig. 5.51):
  - diagnostic images can usually only be obtained in adults in a standing position for:
    - distal aspect of the femur.
    - oblique views of the coxofemoral joint.
  - all other cases require general anaesthesia (Fig. 5.52).
    - contraindicated if a displaced fracture is suspected.
    - small foals, heavy sedation may suffice.
- **Ultrasonography** may visualise surface of the fractured bone and confirm displacement.
- **Scintigraphy** is useful for diagnosing fractures of the greater or third trochanters (Figs. 5.53, 5.54).
  - may also be confirmed ultrasonographically.

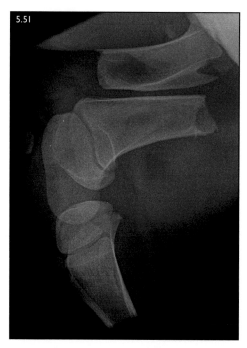

FIG. 5.51 Lateromedial radiograph showing a complete diaphyseal fracture of the right femur in a two-week old, 24kg, foal.

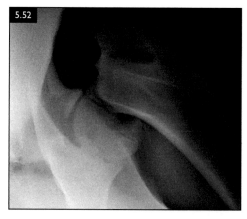

FIG. 5.52 Ventrodorsal radiograph of the proximal femur and coxofemoral joint taken under GA and confirming a displaced fracture of the femoral neck in a young horse. (Photo courtesy Henk van der Veen).

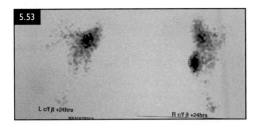

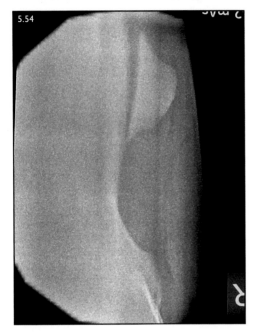

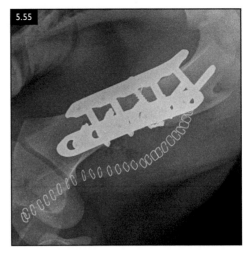

FIG. 5.55 The femoral fracture of the horse in 5.51 was successfully repaired with two locking compression plates.

FIGS. 5.53, 5.54 (5.53) Scintigraphic images from the upper femur and coxofemoral joint of the left and right hindlimbs of a horse with an acute onset right hind lameness. Note the marked IRU overlying the third trochanter of the right hindlimb. (5.54). Craniolateral/caudomedial radiograph of the upper femur of the right hindlimb of the same horse. Note the mildly displaced fractured third trochanter.

Management

- treatment depends on a variety of factors:
  - location, type and severity of the fracture.
  - degree of soft tissue damage.
  - temperament of the horse.
  - financial constraints.

- **most important factor to consider is the size of the affected animal:**
  - euthanasia is warranted for horses heavier than 200 kg with a displaced fracture.
  - successful repair of diaphyseal or proximal physeal fractures only in foals and small ponies with full fracture healing expected in only 50% of cases. (Fig. 5.55).
  - distal physeal fractures have been successfully treated surgically in yearlings.
- conservative management of a minimally displaced distal physeal fracture may be successful.
- surgical repair of proximal physeal fractures is possible:
  - guarded prognosis as significant complications such as OA of the coxofemoral joint, unstable fixation, and necrosis of the femoral head are common.
  - conservative management is unlikely to result in a pain free animal.
- fractured greater or third trochanter is best treated by box rest.

COXOFEMORAL (HIP) JOINT.

Injuries and joint disease

Definition/overview

- rarely a cause of lameness in the horse.
- conditions involving this joint include:
  - fractures of the femoral head and neck.
  - fracture involving the coxofemoral joint.
  - coxofemoral joint subluxation and luxation.
  - partial tear and rupture of the round (teres) ligament.
  - OA, OCD and hip dysplasia.

Aetiology/pathophysiology

- fractures and damage to the round ligament result from trauma, commonly after a fall:
  - coxofemoral joint is predominantly maintained in position by round ligament.
  - ligament rupture/partial tear without joint dislocation (rare) results in instability and/or subluxation of the coxofemoral joint and rapid onset of OA.
- luxation results from:
  - damage to supporting ligaments.
  - unstable fracture of the ilial shaft.
  - articular fracture of the acetabulum following a fall.
  - violent overextension or severe abduction/adduction of the limb.
  - after application of a full-limb cast.
  - complication or co-pathology of upward fixation of the patella.
  - femur commonly displaces in a craniodorsal direction:
    - ♦ greater trochanter of the femur may appear higher on the limb.
- Hip dysplasia is rare but well recognised:
  - possibly hereditary in Norwegian Dole ponies.
  - usually bilateral with malformation of the acetabulum and head/neck of the femur.
  - instability, subluxation, and OA.
- OCD and OCLL are rare in the coxofemoral joint in young horses:
  - predispose to early onset of OA.

Clinical presentation

- acute lameness following trauma.
- very positive to upper hindlimb flexion.
- may be resentful of bearing weight on the affected limb for any length of time.
- dislocated hip joint or ruptured round ligament without luxation, are severely lame.
  - affected limb is rotated with stifle and toe pointed out, and hock turned medially (Fig. 5.56).
  - hock of the dislocated limb appears higher in comparison to contralateral limb.
  - limb shortening decreases the cranial phase of the stride.
  - crepitation and pain when the limb is manipulated.
- partial tears of the round ligament may have less distinct clinical signs:
  - may resent abduction of the limb.
- chronic coxofemoral joint cases develop significant gluteal atrophy.
- OA has no distinct clinical signs other than moderate to severe lameness:
  - usually unilateral but occasionally bilateral.

Differential diagnosis

- Upward fixation of the patella.
- Fracture of the pelvis
- trochanteric bursitis.

Diagnosis

- history of trauma, severe lameness, and clinical findings (mainly outward rotation of whole limb) is suggestive of hip luxation or rupture of the round ligament.
- **Radiographs** of the coxofemoral joint (smaller horses) can be obtained in a standing position:
  - luxation of the joint or fractures can be identified (Fig. 5.57).
  - GA is necessary to confirm joint pathology in larger horses, (Fig. 5.58).
    - ♦ OA, OCD, OCLL and hip dysplasia.
    - ♦ **attendant risks during recovery from the anaesthesia.**
      - – heavy sedation in foals.

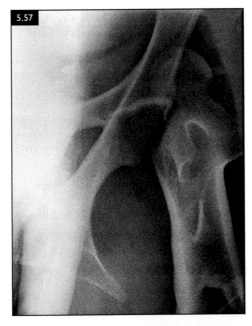

FIG. 5.56, 5.57 (5.56) Luxation of the right coxofemoral joint in a pony following upward fixation of the patella. Note the pelvic asymmetry and outward rotation of the right hind toe. (5.57) Ventrodorsal radiograph of the hip region of the same pony demonstrating luxation of the coxofemoral joint.

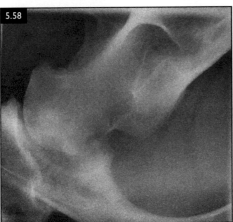

FIG. 5.58 Ventrodorsal radiograph, taken under GA, of the coxofemoral joint of a horse with OA of the joint. Note the margin osteophytosis on all aspects of the acetabular rim and the femoral head. (Photo courtesy Henk van der Veen).

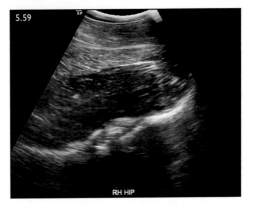

FIG. 5.59 Ultrasonographic image of the coxofemoral joint of a horse with OA. The white line to the right is the femoral neck and the line to the left is the acetabular rim. The hyperechoic white fragmentation centrally is osteophytosis at the margin of the femoral head (right) and acetabular rim (left). (Photo courtesy Henk van der Veen).

- intra-articular analgesia of the joint can be performed to localise source of pain:
  - under ultrasonographic guidance.
  - usually only partially improves the lameness.
- **Ultrasonography** of the acetabulum and femoral head is possible and can identify:
  - subluxation or luxation
  - femoral neck or head fractures.
  - periarticular changes of OA (Fig. 5.59).

- **Scintigraphy** is useful if a fracture is suspected, or intra-articular block results are equivocal.
- **Diagnostic arthroscopy** can be performed, but for horses weighing >300 kg, special long instruments are required.
- round ligament injury often only definitively diagnosed at post-mortem examination.

Management

- closed reduction of a coxofemoral luxation under general anaesthesia in acute cases only:
  - acetabulum fills with granulation tissue preventing relocation in chronic cases.
  - joint may dislocate again either during recovery or within a few days.
  - various surgical procedures for open reduction can be attempted, but the prognosis for athletic function is poor.

- treatment for rupture of the round ligament and OA:
  - arthroscopic debridement and medication of the joint with corticosteroids.
  - guarded to poor prognosis for OA and often hopeless for ligament damage.
- For hip dysplasia, OCD and OCLL the predisposition to OA at an early age means the prognosis for athletic function is poor.

GENERALISED ORTHOPAEDIC DISEASES

Immune-mediated polysynovitis

Definition/overview

- relatively uncommon with most reported cases in foals.

Aetiology/pathophysiology

- caused by deposition of immune complexes in the synovium and complement activation:
  - sequela to a primary inflammatory focus such as:
    - pneumonia, infected umbilicus, or a peripheral abscess.
- *Rhodococcus equi* is the most reported primary infective agent:
  - also following equine herpesvirus-4 and streptococcal infections.
- more than one joint is frequently involved.

Clinical presentation

- effusion of one or more joints, gait stiffness, and low-grade lameness.
- relevant clinical signs if the source of the primary infection is still active.

Differential diagnosis

- infectious arthritis
- idiopathic arthritis
- Lyme disease.

Diagnosis

- clinical presentation of multiple joint effusion with minimal lameness is suggestive.
- analysis of synovial fluid collected aseptically from affected joints reveals:
  - WBC count $<20 \times 10^9$/l, with mixed healthy neutrophils and mononuclear cells.

Management

- identify primary cause and treat accordingly.
- condition is self-limiting, and recovery is expected within a few weeks.
- restrict to box rest and possible treatment with systemic chondroprotective drugs.
- corticosteroid therapy may be contraindicated in the face of a bacterial infection.

Lyme disease (borreliosis)

Definition/overview

- confirmed clinical cases are rare:
  - certain parts of the world (northeast USA and parts of UK) cases are seen regularly.
- many horses in tick-infested areas have a positive antibody titre for *Borrelia burgdorferi* without clinical disease.
- disease has a significant zoonotic risk in humans bitten by infected ticks in endemic areas.

Aetiology/pathophysiology

- causative agent is a spirochete, *Borrelia burgdorferi*:
  - transmitted to the horse through a bite of an infected tick of the *Ixodes* family.

- ticks have a 2-year, three-stage life cycle and they can become infected during any stage:
  - feeding on mammalian hosts.
    - white-footed mouse in the USA and possibly deer in the UK.
  - stage of the life cycle infecting the horse is unknown.
- not all ticks are infected with the spirochete:
  - infection varies by tick species and geographic region.
- infective agent non-specifically activates various cells of the immune system:
  - production of proinflammatory mediators.
  - localise in joints and result in chronic arthritis.
  - may infiltrate nerve roots, spinal cord and brain.

Clinical presentation

- lameness and stiffness in one/more limbs, and low-grade fever are the most common clinical signs:
  - lameness is often caused by arthritis.
  - may involve more than one joint and can become chronic.
- other clinical signs include chronic weight loss, swollen joints, muscle tenderness, and anterior uveitis.
- neck stiffness, neurological signs and forelimb lameness if the bacteria penetrate the CNS.

Differential diagnosis

- OA   • OCD
- Polysaccharide storage myopathy
- Immune-mediated polysynovitis
- chronic intermittent rhabdomyolysis
- Equine protozoal myelitis
- any condition causing neck stiffness, forelimb lameness or neurological signs.

Diagnosis

- difficult, and presumptive diagnosis is commonly based on:
  - history, clinical signs, housing in an endemic area, and ruling out of other causes.
- ELISA titres and positive Western blot testing are reported to be diagnostic in some cases.

- multiplex assay allows acute and chronic infection to be distinguished:
  - allows antibody titres from natural infection and vaccination to be differentiated.
- PCR assay of cerebrospinal fluid for *B. burgdorferi* DNA is indicative of infection in the nervous system.

Management

- treatment by tetracycline (6.6 mg/kg i/v q24h) or doxycycline (10 mg/kg p/o q12h).
- ceftiofur (2–4 mg/kg i/m q12h) has also been used.
- duration of treatment up to 30 days, but the rationale for this is empirical.
- clinical signs in suspected cases reported to resolve within 1 week of treatment.
- supportive treatment includes chondroprotective agents and/or NSAIDs.
- prevention of the disease includes:
  - daily grooming.
  - application of tick repellents containing permethrin.
  - pasture management and care.
- equine-approved vaccine is not yet commercially available:
  - vaccination with canine recombinant OspA vaccine or whole-cell vaccines are at least partially protective.

Musculoskeletal nutritional deficiencies

Overview

- nutrition has a major role in the proper development and function of the musculoskeletal system.
- imbalanced nutrition affects horses of all ages and disciplines, starting from the embryo, through the foal and growing horse stage, to the adult.
- conditions that may occur because of these deficiencies include:
  - Nutritional secondary hyperparathyroidism.
  - OCD and/or any of the DODs.
  - Exertional rhabdomyolysis and White muscle disease.
- nutrients suggested to affect the musculoskeletal system when deficient include:

- electrolytes such as sodium, potassium, calcium, and magnesium.
- minerals such as calcium and phosphorus.
- trace minerals (copper, zinc, and selenium).
- vitamins (A, D, and E).
- protein   ○ digestible energy.
- actual digested level of nutrients is important plus the relative amounts and the interrelations between them.
- diets high in phosphorus and/or low in calcium, with a phosphorus/calcium ratio of 3:1 or more, may develop **Nutritional secondary hyperparathyroidism**:
  - feeding excessive amounts of wheat bran or other grains (high phosphorus/low calcium).
  - various pastures contain high contents of oxalate which binds calcium:
    - hyperphosphataemia stimulates parathyroid hormone (PTH).
    - inhibits the synthesis of the active form of vitamin D in the kidney.
    - osteoclastic activity is increased:
      - excessive bone resorption and bone loss.
  - clinical signs include:
    - symmetrical enlargement of facial bones
    - upper respiratory noise.
    - lameness
    - young horses may develop physitis and limb deformities.
    - bone resorption around the lamina dura of the molars and premolars:
      - pain and masticatory problems, and in severe cases loosened teeth.
  - clinical laboratory findings may include:
    - mild hypocalcaemia and hyperphosphataemia (may be within reference ranges).
    - fractional urinary clearance:
      - calcium (normal/low) and phosphorus (normal/high).
    - serum PTH level increased.
  - treatment requires:
    - correct ration – increase calcium and reduce phosphorus intake.
    - restricting access to plants containing oxalates.

- unless fractures are present, lameness usually disappears within 6 weeks.
- inadequate levels of protein in a mare's diet can cause hypothyroidism in the foetus:
  - may persist postnatally and affect normal development.
  - incomplete ossification of carpal/tarsal cuboidal bones, causing angular limb deformities (ALDs).

Fluoride toxicosis

Definition/overview

- rare in horses:
  - skeletal abnormality caused by chronic fluorine intake above the critical levels.
- acute and chronic forms have been reported, with the latter more common.

Aetiology/pathophysiology

- common sources of fluoride include:
  - contaminated forage from nearby industrial plants.
  - drinking water containing excessive amounts of fluoride.
  - feed supplements with high fluoride concentration.
  - vegetation grown on soils rich in fluoride.
- fluoride accumulates in the bone and teeth throughout the horse's life (cumulative poison).
  - almost fully absorbed from the GI tract.
  - accumulates in calcified developing tissues (i.e. bones and teeth):
    - young growing animals are more susceptible.
  - defective mineralisation of teeth.
  - pathophysiology of bone damage is not clear, but includes:
    - production of abnormal bone.
    - accelerated remodelling, and, occasionally, accelerated bone resorption.

Clinical presentation

- exostosis formation, especially on the third MC/MT bones:
  - mandible, ribs, and in sites of tendon insertion and periarticular tissues.

- may have stiff gait, intermittent lameness, and a dry and roughened coat.
- poor growth and weight loss, and abnormal bones that are susceptible to fractures.
- teeth may appear mottled and discoloured brown, and with hypoplastic enamel.
  o teeth may be missing.

Differential diagnosis

- Hypertrophic osteopathy
- Nutritional secondary hyperthyroidism
- osteomyelitis. • OA • septic arthritis.

Diagnosis

- based mainly on clinical signs and history of possible exposure.
- fluoride in urine and bone can be determined, and analysis of water and feed is advisable.
- **Radiographs** may reveal abnormal bone appearance such as thickening and increased density of the bones.

Management

- reduction of the toxic effects of fluoride can be attempted by using substances such as:
  o aluminium sulphate, aluminium chloride, and calcium carbonate.
- dietary restriction of fluoride-containing substances.
- removing animal from the source of fluoride.

Prognosis

- poor if intermittent lameness is present.
- teeth never recover from the discolouration.

Hypertrophic osteopathy

Definition/overview

- bilateral symmetric proliferation of vascular connective tissue and subperiosteal bone of the distal extremities.
- referred to as Marie's disease, hypertrophic osteodystrophy, and hypertrophic pulmonary osteoperiostitis.

Aetiology/pathophysiology

- pathophysiology not clearly understood.
- associated with thoracic pathology such as:
  o lung abscessation, tuberculosis, neoplasia, pulmonary infarction, and rib fractures.
  o also abdominal metastases, and possibly related to pregnancy.

Clinical presentation

- often presented for investigation of the enlarged bones:
  o commonly progressive deterioration over a prolonged period.
  o distal limbs may be symmetrically or asymmetrically swollen and warm.
  o restriction and pain on palpation and manipulation.
  o stiff gait and commonly reluctance to move.
- signs related to the primary lesion such as:
  o cough, nasal discharge, or chronic weight loss may precede bone enlargement.

Differential diagnosis

- Fluorosis
- Nutritional secondary hyperthyroidism.
- primary underlying disease intrathoracic or intraabdominal mass (neoplasia, abscess).

Diagnosis

- based on clinical signs and radiographic findings.
- **Radiography** commonly reveals a palisade pattern of periosteal new bone formation parallel to the cortices of the long bones (Fig. 5.60):
  o new bone may be close to the joint margins.
  o articular surfaces are rarely involved.
- **Bone scintigraphy** may show an increased uptake in the distal limb, even before radiographic changes.
- haematology and biochemistry may reveal abnormalities related to the underlying cause.
- identify the primary condition.

Management

- primary lesion should be treated if possible:
  - bone lesions may subside following resolution of the primary problem.
- no underlying disease is identified:
  - symptomatic treatment, mainly with NSAIDs.
- small proportion of horses, the condition may resolve spontaneously.

Prognosis

- primary lesion is not identified or treated:
  - prognosis is poor.
  - debilitating nature of the condition may warrant euthanasia.
- primary condition is identified and treated – prognosis is better but still guarded.

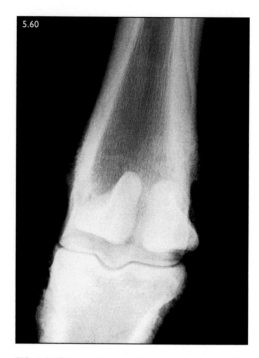

FIG. 5.60 Dorsopalmar radiograph of a horse with hypertrophic osteopathy. Note the new bone formation at the level of the third metacarpal bone and proximal phalanx.

The Axial Skeleton – The Neck

Congenital abnormalities

Definition/overview

- present at birth but may be overlooked in some horses and become clinically apparent later in life.
- common congenital abnormalities are:
  - occipitoatlantoaxial malformation (OAAM):
    - most often in Arabs (possible genetic predisposition) and rarely other breeds.
  - congenital fusions of other vertebrae.
  - meningocoele.

Clinical presentation

- OAAM is usually associated with:
  - visible and palpable deformity of the cranial aspect of the neck.
  - cranial neck stiffness:
    - may result in abnormal posture to graze (Fig. 6.1).
  - with or without ataxia.
- congenital fusion of other vertebrae may remain undetected unless the horse is evaluated radiographically.
- meningocoele is associated with both forelimb and hindlimb ataxia:
  - primary lesion at the cervicothoracic junction.
  - horse may stand with an abnormal forelimb posture.

Diagnosis

- based on **radiographic** examination (Fig. 6.2).
  - laterolateral images are usually adequate for diagnosis.
  - ventrodorsal images of the occiput, atlas and axis:
    - asymmetry of the deformities.

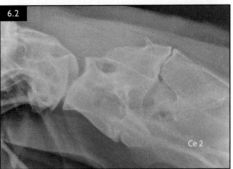

FIGS. 6.1, 6.2 (6.1) A 6-year-old Warmblood showjumper gelding adopting a rather abnormal posture in order to lower its head to graze. (6.2) Laterolateral radiograph of the occiput and first two cervical vertebrae of the horse in 6.1. Cranial is to the left. There is congenital occipitoatlanto-axial malformation. Note the truncated atlas and the duplicated odontoid pegs. Ce 2 = second cervical vertebra. (Photos courtesy Ana Stela Fonseca)

Management

- no treatment or management is available for any of these conditions.

DOI: 10.1201/9781003369226-6

Cervical vertebral malformation (see *Concise Textbook of Equine Medicine & Surgery Book 5, Neurology chapter*)

Definition/overview

- acquired variants in the shape and alignment of the cervical vertebrae:
  - can result in compression of the cervical spinal cord and ataxia.

Aetiology/pathophysiology

- unclear but there is a relationship with osteochondrosis and propensity for rapid growth.
- possible genetic predisposition.

Clinical presentation

- ataxia and weakness of variable severity.
- age of onset is highly variable:
  - clinical signs commonly occur at 1 to 4 years of age.
- may superficially appear to be loose, extravagantly moving horses.
- usually bilaterally symmetrical with hindlimbs invariably more severely affected.
- mildly affected horses may appear:
  - croup high in downward transitions from trot to walk.
  - bouncy gait in transitions.
  - occasional circumduction of outside hindlimb when turned in small circles.
- more severe cases show frequent circumduction of the hindlimbs, weakness and both forelimb and hindlimb dysmetria.

Differential diagnosis

- other possible causes of ataxia and weakness including:
  - neuroaxonal dystrophy
  - equine protozoal myelitis (asymmetrical clinical signs).
  - traumatic causes of spinal cord compression
  - equine herpesvirus 1 infection.
  - spinal cord compression associated with osteoarthritis (OA) of the caudal cervical articular process joints.
  - primary brain lesion.

Diagnosis

- **radiographic** examination of the cervical vertebrae using true laterolateral images.
- abnormalities include:
  - narrowing of the dorsoventral sagittal diameter of the vertebral canal.
  - extension of the dorsal lamina of a vertebra.
  - malalignment (subluxation) of adjacent vertebrae.
    - other changes are shown in Fig. 6.3.
  - semi-quantitative methods of assessment of vertebral canal dimensions:
    - intra- and intersagittal ratios have been utilised.
    - limited sensitivity and specificity.
  - myelography may assist in determining the site/s of spinal cord compression.
    - false-negative and false-positive results can occur.

Management

- observed in young foals:
  - strict dietary management (reduced energy input) may improve clinical signs.
- selected patients:
  - surgical fusion of vertebrae at site of spinal cord compression may improve clinical signs.

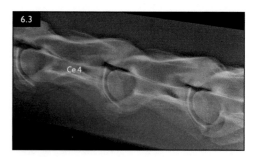

FIG. 6.3 Laterolateral radiograph of the fourth (Ce 4) to sixth cervical vertebrae of a yearling Thoroughbred with mild hindlimb ataxia. Cranial is to the left. The pedicles of each vertebra are short, the articular process joints are 'low slung' and there is enlargement of the dorsal aspect of the caudal epiphysis of each vertebra ('ski jump'). There is mild enlargement of the articular process joints between the fifth and sixth cervical vertebrae. The vertebral canal of the fifth and sixth cervical vertebrae is slightly wedge shaped. The ventral part of each physis has not yet closed.

Soft tissue injuries

Definition/overview

- variety of acquired soft tissue injuries of the neck:
  o nuchal ligament, the nuchal bursa, muscles and the jugular vein.

Aetiology/pathophysiology

- traumatic injuries, either acute or repetitive wear and tear.
- occasionally, ill-fitting tack, especially in horses used for driving, can cause muscle dysfunction, pain and reluctance to work, or overt forelimb lameness.

Clinical presentation/ diagnosis/management

- **nuchal ligament** inserts on the caudal aspect of the occiput:
  o signs of entheseopathy (new bone on caudal aspect of occiput):
    ◆ common clinical observation, often of no clinical significance (Fig. 6.4).
  o relatively rarely, a horse may have an abnormal head and neck posture when ridden:
    ◆ improved by local infiltration of local anaesthetic solution around the entheseous new bone.

- ◆ more commonly the cause of the clinical signs is secondary to forelimb/hindlimb lameness, oral problem, or another cause of neck pain.
- **Nuchal bursitis** is an unusual cause of swelling in the poll region, with or without associated neck stiffness:
  o may or may not be localised dystrophic mineralisation (Fig. 6.5).
  o diagnosis is usually confirmed using ultrasonography.
  o some horses respond to conservative management but in others surgical treatment is required.
- **Brachiocephalicus muscle soreness** is often identified in association with forelimb lameness elsewhere in the limb (e.g. foot pain or proximal suspensory desmitis):
  o primary strain injury can also occur:
    ◆ usually associated with gait abnormality.
      – most readily detected at the walk or when the horse is ridden.
      – shortened cranial phase of the stride on the affected side.
      – head and neck are raised as the affected limb is protracted.
  o physiotherapy treatment usually resolves the clinical signs.

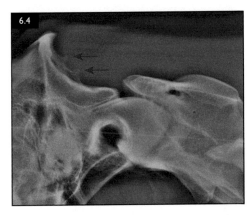

FIG. 6.4 Laterolateral radiograph of the occiput and first cervical vertebra of a 12-year-old advanced event horse. Cranial is to the left. There is entheseous new bone on the caudal aspect of the occiput (arrows), an incidental finding in this horse.

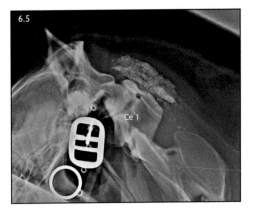

FIG. 6.5 Laterolateral radiograph of the occiput and first two cervical vertebrae of a 14-year-old Icelandic pony mare. Cranial is to the left. There is extensive mineralisation dorsal to the first cervical vertebra (Ce 1). Ultrasonography revealed marked fluid distension of the nuchal bursa with multifocal hyperechogenic foci.

- occasionally, a horse may sustain an acute **strain/tear of the pectoral muscles:**
  - usually due to a fall.
  - muscle tearing and subsequent haemorrhage is extremely painful:
    - severe lameness, distress, signs mimicking colic, and difficulties in examining the horse.
  - usually palpable swelling of the muscle.
- **muscle abscess** is usually a direct consequence of a previous intramuscular injection:
  - neck swelling, pain and stiffness (Fig. 6.6).
  - ultrasonography usually reveals a thick-walled unicameral or multicameral structure filled with anechoic purulent material (Fig. 6.7).
  - surgical drainage is sometimes successful:
    - radical *en bloc* resection of the entire abscess may be required.
- **Jugular vein thrombophlebitis** is usually the result of intravenous medication given via a needle or catheter, or an adverse reaction to a catheter:
  - localised soft tissue swelling around the vein, thickening of the wall of the vein and accumulation of echogenic material within the vein (Fig. 6.8).
  - if severe, the vessel may become transiently or permanently occluded:
    - obvious palpable thickening and firmness of the vein.
    - no associated heat, pain or neck stiffness.
  - infectious thrombophlebitis is usually the result of circulating bacteria invading a thrombus:
    - associated heat, pain and neck stiffness.
    - diagnosis can be confirmed using ultrasonography.
    - systemic antimicrobial therapy and local injection of antimicrobial drugs into the infected thrombus may result in complete resolution.
    - occasionally, surgical removal of the vein is required.
- **Rupture of the ligament of the dens** is a rare injury:
  - results in dorsal subluxation of the dens, sometimes associated with an audible click.

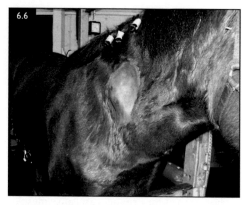

FIG. 6.6 A mid right-sided neck swelling due to an abscess in the muscles following a vaccination injection.

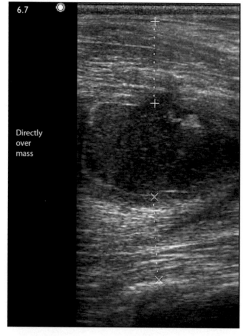

Directly over mass

FIG. 6.7 Ultrasonographic image of a neck muscle abscess showing the circumscribed abscess containing speckled hyperechoic foci of pus. Note the hyperechoic bony structure deep to the abscess, which is the edge of the cervical vertebral column.

  - head carried stiffly in extended position relative to the neck.
  - vertebral canal is wide at this level and spinal cord may be unaffected with no ataxia.
  - prognosis is guarded for return to full athletic function.

- **Acquired torticollis** (Fig. 6.9) is believed to be the result of cerebrospinal *Parelaphostrongylus tenuis* migration:
  - affected horses develop an obvious twist in the neck with no associated radiological or ultrasonographic abnormalities.

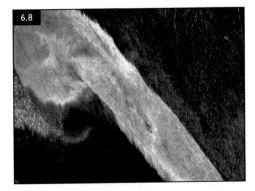

FIG. 6.8 A left-sided jugular vein thrombophlebitis following an intravenous catheter placed for a general anaesthesia. Note the enlarged, thickened and corded vein.

FIG. 6.9 Dorsal image of a 3-year-old Mangalarga Marchador with acquired torticollis of 3 months' duration. There is convexity of the neck to the left. *Parelaphostrongylus tenuis* migration in the spinal cord has been described as a cause.

- particularly in **driving horses**, using a collar or breast plate may result in forelimb gait abnormalities or unwillingness to work:
  - may only be reproduced when the horse is driven.
  - may not be possible to identify focal areas of pain in the neck region.
  - diagnosis is usually dependent on exclusion of other potential causes of pain and by changing the tack.

Fractures (see also *Concise Textbook of Equine Medicine & Surgery Book 5, Neurology chapter*)

Definition/overview

- occur in many configurations and may be displaced or non-displaced.

Aetiology/pathophysiology

- traumatically induced, often as the result of a fall or a collision with a solid object.
- secondary to OA of the caudal cervical articular process joints.

Clinical presentation

- depends on:
  - site of the fracture and its configuration.
  - whether one or more vertebrae are involved.
- may or may not be visible deformity of the neck (Fig. 6.10).
- usually neck pain and stiffness.
- dermatomal sweating can occur.
- catastrophic fracture may result in spinal cord trauma and respiratory arrest.
- occasionally associated forelimb lameness.

Differential diagnosis

- neck abscess
- jugular vein septic thrombophlebitis
- severe muscle injury.
- meningitis/meningoencephalitis.

Diagnosis

- **Radiography** is required to identify most fractures and their laterality.
  - laterolateral (Fig. 6.11) and lateroventral–laterodorsal oblique images.

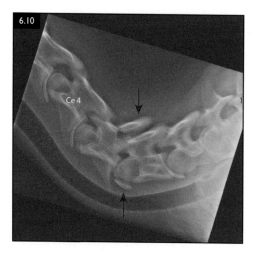

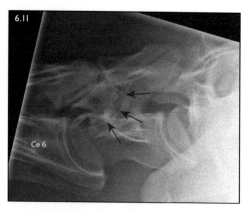

FIG. 6.10 Laterolateral radiograph of the third cervical to second thoracic vertebrae of a 7-month-old pony that had collided with a fence the previous night. Cranial is to the left. There is marked distortion of the shape of the neck. There are displaced fractures of the dorsal lamina, caudal articular process and caudal aspect of the vertebral body of the fifth cervical vertebra (arrows). There is mild subluxation between the third and fourth cervical vertebrae and the seventh cervical and first thoracic vertebrae. Ce 4 = fourth cervical vertebra.

FIG. 6.11 Laterolateral radiograph of the sixth cervical (Ce 6) to first thoracic vertebrae of a 7-year-old Irish Sports Horse eventer. The horse had a sudden onset of severe ataxia 3 days previously, but retrospectively had been difficult to turn for several months. Cranial is to the left. There is considerable asymmetric enlargement of the articular process joints between the sixth and seventh cervical vertebrae, with ventral buttressing. The vertebral canal of the seventh cervical vertebra is wedge shaped. There is a fracture extending from the right articular process joint through the pedicle of the seventh cervical vertebra (arrows). There is mild dorsal displacement of the head of the first thoracic vertebra.

- o ventrodorsal images give additional information in selected cases.
- complete appraisal of fracture configuration may require CT.

Management and prognosis

- in absence of neurological signs, manage conservatively:
  - o reasonable prognosis for return to full athletic function.
- acute neurological signs:
  - o may progressively improve in the first week after diagnosis where there is no further trauma to the spinal cord.
  - o persistence of neurological signs warrants a guarded prognosis.
- callus associated with fracture repair may result in later onset ataxia, but this is unusual.
- complex fractures which result in visible neck deformity – more guarded prognosis.

Osteoarthritis of the caudal cervical articular process joints

Definition/overview

- post-mortem studies reveal highest prevalence of OA of the cervical articular process joints occurred between the third and fourth cervical vertebrae.
- clinically significant lesions most common from the fifth cervical to first thoracic vertebrae.

Aetiology/pathophysiology

- high mobility of the caudal aspect of the neck may predispose to OA.
- Osteochondrosis may be a factor in some horses.
- congenital variations in the symmetry of conformation of caudal cervical vertebrae and associated muscles.

Clinical presentation

- variety of clinical signs may occur including:
  - neck stiffness (Figs. 6.12–6.14).
  - neck pain, local muscle atrophy, dermatomal sweating (Figs. 6.15, 6.16):
    - neck pain and stiffness are not particularly common.
    - primary presenting clinical problem with pain is sometimes alteration in behaviour (e.g. bolting, rearing, stopping and refusing to go forwards).
    - localised muscle atrophy in the caudal neck region and sometimes focal pain on palpation.
    - marked pain may reflect fracture of an articular process.
  - **Locking neck syndrome** is characterised by:
    - sudden onset low neck posture, unwillingness to move and sometimes marked pain (Fig. 6.17).
    - may be focal patchy sweating on one side in the caudal neck region.
    - clinical signs usually resolve:
      - spontaneously within 24 hours.
      - following neck manipulation.

FIG. 6.13 Assessment of neck flexibility using food in a 10-year-old Warmblood showjumper with advanced osteoarthritis of the caudal cervical articular process joints and marked neck stiffness, which compromised performance. The horse could not flex its neck any further and rotated its head sideways. Note that the horse has also positioned the right forelimb in front of the left forelimb and is partially flexing the carpus and distal limb joints.

FIG. 6.12 Assessment of neck flexibility using food in a normal horse. The horse is positioned against a fence so that it cannot move its body away. The forelimbs are both fully load bearing, and the horse has flexed the neck laterally, maintaining the head more or less vertical.

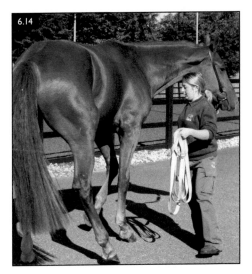

FIG. 6.14 Turning in small circles, this horse is holding its neck very stiffly.

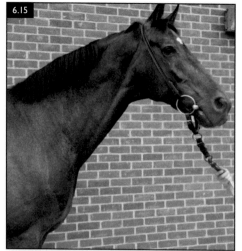

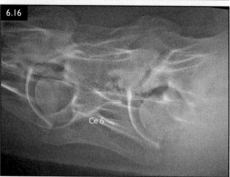

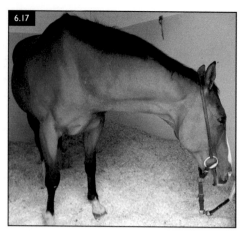

FIG. 6.17 A 6-year-old Warmblood gelding with its neck in a fixed low position bent to the left, a posture typical of the neck locking syndrome. The horse had patchy sweating on the right side at the level of the articulation between the sixth and seventh cervical vertebrae. The horse had moderate enlargement of the articular process joints between the fifth and sixth cervical vertebrae and marked enlargement of the articular process joints between the sixth and seventh cervical vertebrae, with ventral buttressing (osteoarthritis).

FIGS. 6.15, 6.16 (6.15) A 13-year-old Warmblood dressage horse gelding with chronic neck stiffness, resistant behaviour when ridden on the left rein and dermatomal sweating on the right side. (6.16) Laterolateral radiograph of the fifth cervical to first thoracic vertebrae of the horse in 6.15. Cranial is to the left. There is considerable asymmetric enlargement of the articular process joints between the fifth and sixth (Ce 6) and sixth and seventh cervical vertebrae, with ventral buttressing. Oblique images revealed that the osteoarthritic changes were worst on the right side.

- ♦ speculated that the syndrome reflects transient nerve root compression:
  - – evidence of chronic nerve root pathology has been found.
- ○ **Dermatomal sweating:**
  - ♦ result of neural compromise.
  - ♦ may be accompanied by some change in hair coat colour.

- ♦ result of impingement on the vertebral nerve or its branches, which provide autonomic fibres locally.
- ♦ articular process joint modelling must be abaxial to affect these branches.
- ○ **forelimb lameness** due to nerve root compression:
  - ♦ unusual but well recognised.
  - ♦ lameness can be highly variable in presence and severity.
  - ♦ lameness may only be present when ridden or worsens at this time:
    - – may be influenced by whether ridden to a contact or on a longer rein.
  - ♦ lameness may be accentuated by local analgesic techniques performed in the lame limb.
  - ♦ bilaterally short-stepping forelimb gait.
  - ♦ hopping-type forelimb gait, only seen when ridden (Fig. 6.18).
    - – may be noted in a subset of horses with nerve root compression.

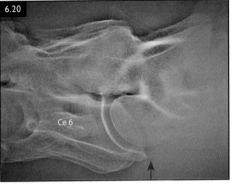

FIG 6.18 A 5-year-old Irish Sports Horse that exhibited a right forelimb lameness only when ridden and which was characterised by a hopping-type gait. The horse is trotting. There is advanced diagonal placement of the left hindlimb. The left forelimb has left the ground but the right hindlimb remains weight bearing. The entire forehand is elevated. The horse is above the bit, with the ears back and a glazed expression reflecting pain. No nerve block improved the lameness.

FIGS. 6.19, 6.20 (6.19) Posture typical of root signature (pain associated with nerve root compression). This 12-year-old advanced event horse had episodic severe left forelimb lameness during flat work and would stop suddenly, adopting this posture. During these episodes the horse exhibited severe pain on palpation and muscle tension over the sixth and seventh cervical vertebrae on the left side. (6.20) Laterolateral radiograph of the sixth (Ce 6) and seventh cervical vertebrae of the horse in 6.19. Cranial is to the left. There is marked enlargement of the articular process joints between the sixth and seventh cervical vertebrae, with ventral buttressing. There is a spur on the caudoventral aspect of the vertebral body of the sixth cervical vertebra (arrow). The small spinous process on the craniodorsal aspect of the seventh cervical vertebra is a normal variant.

- rarely, episodic signs of lower motor neuron dysfunction (stumbling).
- compromised forelimb function due to lower motor neuron dysfunction:
  - root signature posture (Figs. 6.19, 6.20).
  - hyperaesthesia episodic neck 'locking' (Fig. 6.17).
  - ataxia and weakness.
- in severe lameness, affected horses are usually:
  - less willing to go forwards.
  - change their facial features, reflecting pain.
  - ears back, glazed expression in eyes, mouth open (Figs. 6.21, 6.18).
  - may tilt the head consistently in one direction (Fig. 6.21).

Differential diagnosis

- other neck lesions such as discospondylitis, fracture, neoplasia, bone necrosis and osteomyelitis have similar clinical signs but are much less common.

- neck stiffness when ridden may be secondary to forelimb or hindlimb lameness:
  - diagnostic analgesia will remove the lameness and the secondary neck signs.
- oral pain may create neck stiffness.

FIG. 6.21 A 6-year-old Warmblood showjumper in a canter on the left rein. There is a marked tilt of the neck and head with the nose to the right and the poll to the left. The horse's ears are erect and divergent, and the eyes have a glazed expression reflecting pain. In trot the horse exhibited a mild left forelimb lameness, especially on the right rein, which was accentuated markedly after median and ulnar nerve blocks. Intra-articular analgesia of the elbow and shoulder joints did not improve the lameness or the head posture.

Diagnosis

- laterolateral and lateroventral–laterodorsal **radiographic** oblique images from right to left and left to right permits the best evaluation of the articular process joints:
  - many normal mature horses have some degree of enlargement and modelling of the caudal cervical articular process joints, often biaxial.
  - greater the degree of modelling, enlargement, OA and/or subluxation (Figs. 6.20, 6.22, 6.23), the more likely to be associated with clinical signs.
    - ◆ **some nerve root compression cases have no detectable radiological abnormality in the joints.**
  - significance of OA changes requires the exclusion of other potential causes of the clinical signs.

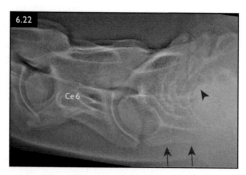

FIG. 6.22 Laterolateral radiograph of the fifth cervical to first thoracic vertebrae of the horse in 6.21. Cranial is to the left. The vertebrae have short pedicles and low-slung articular process joints. The articular process joints of the sixth (Ce 6) and seventh cervical and seventh cervical and first thoracic vertebrae are moderately and massively enlarged, respectively, with ventral buttressing. The vertebral canals of both the sixth and seventh cervical vertebrae are wedge shaped. There is mild subluxation of the first thoracic vertebra (arrowhead). Both ventral processes are transposed from the sixth to the seventh cervical vertebra (arrows).

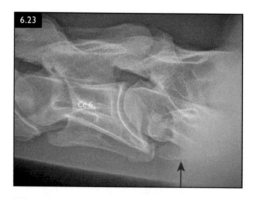

FIG. 6.23 Laterolateral radiograph of the sixth cervical to first thoracic vertebrae of the horse in 6.18. Cranial is to the left. There is mild subluxation of the seventh cervical vertebra. The vertebral foramen of the seventh cervical vertebra is wedge shaped. There is enlargement of the caudodorsal aspect of the epiphysis of the sixth cervical vertebra (a 'ski jump') (Ce 6). The articular process joints between the seventh cervical and first thoracic vertebrae are enlarged, with reduction in size of the intervertebral foramen. One ventral process has been transposed from the sixth to the seventh cervical vertebra (arrow).

- intra-articular analgesia performed under ultrasonographic guidance can be used.
  - **risk of inducing transient paresis.**
- CT, either standing or under general anaesthesia, is now available for neck cases.
- Myelography is used in horses with ataxia and weakness if surgical intervention is being considered.

Management

- short-term improvement in neck pain, stiffness and forelimb lameness may be achieved by intra-articular corticosteroid medication of the articular process joints:
  - often radiological abnormalities of at least two adjacent articular process joints are present.
    - ◆ identification of the most significant joints is challenging.
    - ◆ common practice to treat more than one joint per side.
    - ◆ clinical signs may be unilateral or bilateral:
      - – medication of the left and right sides is often indicated.
      - – bilateral pathology is common and optimal function of both is required.
- locking neck syndrome is usually sporadic and treatment is difficult to justify because clinical efficacy is impossible to judge.
  - dermatomal sweating often persists long-term.

Discospondylosis

Definition/overview

- degenerative changes of intervertebral discs are common post-mortem observations:
  - strong overlying dorsal longitudinal ligament usually prevents disc material protruding into the vertebral canal.
- occasionally, clinical signs referable to an intercentral articulation and degeneration of the intervertebral disc are recognised (Fig. 6.24).

Aetiology/pathophysiology

- aseptic degeneration, traumatically induced injury and infectious discospondylitis have been recognised.

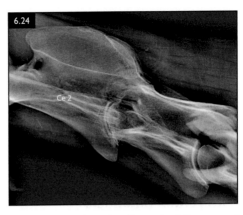

FIG. 6.24 Laterolateral radiograph of the second to fourth third cervical vertebrae of a 5-year-old Warmblood potential dressage horse with neck stiffness and reluctance to turn. Cranial is to the left. There is marked narrowing of the intercentral joint space between the second (Ce 2) and third cervical vertebrae consistent with discospondylosis. The outline of the head of the third cervical vertebra is a little irregular. The articular process joints between the second and third cervical vertebrae are also abnormal.

Clinical presentation

- clinical signs are related to the region of the neck involved.
- neck stiffness and pain, with or without difficulties in lowering the head and neck to graze, are the most common features.
- may or may not be associated neurological signs.

Diagnosis

- based on good-quality true laterolateral **radiographs:**
  - marked loss of the intercentral joint space (Fig. 6.24).
  - with, or without, changes in the caudal end plate of the more cranial vertebra and the head of the more caudal vertebra.

Management

- surgical fusion could be considered.

Vertebral osteomyelitis

Definition/overview

- infection of one or more cervical vertebrae is rare.

Aetiology/pathophysiology

- may be secondary to *Rhodococcus equi* infection in the foal.
- avian tuberculosis has been recognised in adult horses.
- extension into the cervical vertebrae has also occurred from neck abscesses post intramuscular injections.

Clinical presentation

- neck stiffness due to bone pain.

Differential diagnosis

- Multiple myeloma and other neoplastic conditions.

Diagnosis

- based on radiological, haematological and bone biopsy examinations:
  - usually lucent zones within one or more vertebrae:
    - may be surrounded by a rim of more radiopaque bone (Fig. 6.25).
  - may be an increase in serum amyloid A and fibrinogen concentrations.

Management

- difficult to manage successfully:
  - long-term appropriate antimicrobial therapy is essential.
- prognosis is guarded to poor.

Subluxation

Definition/overview

- misalignment of two adjacent vertebrae in the sagittal plane, or subluxation, results in a change in orientation of the vertebral canal from cranial to caudal:
  - sagittal diameter of the canal is large:
    - may be no associated clinical signs.
    - incidental radiological abnormality (Fig. 6.26).
  - smaller vertebral canal:
    - potential for spinal cord compression and ataxia and weakness.
- lesions most common between the third and fourth cervical vertebrae.
- less commonly, in the caudal neck region involving the fifth/sixth, sixth/seventh or seventh cervical/first thoracic vertebrae.

Aetiology/pathophysiology

- congenital or acquired and result in spinal cord compression or cervical nerve root injury.

Clinical presentation

- most horses present with a variable degree of ataxia and weakness:
  - between third and fourth cervical vertebrae predominantly seen in young horses.

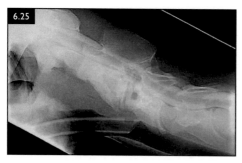

FIG. 6.25 Laterolateral radiograph and myelogram of a young foal with a septic bone abscess within the cranial aspect of the third cervical vertebra. Note the collapse of the third vertebra and the caudal part of the second vertebra, plus narrowing of the ventral column of contrast medium immediately dorsal to the affected intercentral joint. (Photo courtesy Jane Boswell)

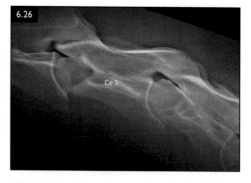

FIG. 6.26 Laterolateral radiograph of the second to fifth cervical vertebrae of a 13-year-old Grand Prix dressage horse. Cranial is to the left. There is subluxation of the third (Ce 3) and fourth cervical vertebrae, which was an incidental finding. Each vertebra has short pedicles, so the articular process joints are low slung, the dorsoventral sagittal diameter of the vertebral canal is small and the intervertebral foramina are small.

- loss of performance (e.g. a breeding stallion in covering mares) or forelimb lameness:
  - mature horses often with caudal neck lesions.

Differential diagnosis

- any cause of ataxia and weakness can cause similar clinical signs.

Diagnosis

- laterolateral radiographic images are essential for diagnosis (Fig. 6.27).

Management

- conservative management is usually unsuccessful.
- surgical fusion can be considered.

Multiple myeloma and other neoplastic lesions

Definition/overview

- neoplastic lesions are rare in the cervical region.
- primary bone lesions include multiple myeloma, osteosarcoma and lymphosarcoma.

Aetiology/pathophysiology

- Myeloma is a myeloproliferative disorder that can cause radiolucent lesions in any bone, including the cervical vertebrae.

Clinical presentation

- neck stiffness due to bone pain.
- ataxia may occur if a tumour involves the vertebral canal.
- involvement of nerve roots may result in forelimb lameness.
- other systemic clinical signs may be present depending on the nature of the primary tumour.

Differential diagnosis

- osteomyelitis.

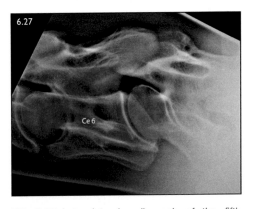

FIG. 6.27 Laterolateral radiograph of the fifth cervical to first thoracic vertebrae of 9-year-old Warmblood gelding with poor performance. Cranial is to the left. The horse exhibited mild hindlimb ataxia. There is subluxation of the sixth (Ce 6) and seventh cervical vertebrae. The head of the seventh cervical vertebra is displaced dorsally. The vertebral canal of the seventh cervical vertebra is wedge shaped. There is enlargement of the caudodorsal aspect of the epiphysis of the sixth cervical vertebra (a 'ski jump'). There is asymmetrical enlargement of the articular process joints of the sixth and seventh cervical vertebrae.

Diagnosis

- based on radiological, haematological and bone biopsy examinations.
  - usually clearly demarcated lucent zones within one or more vertebrae, usually without a rim of more radiopaque bone (Fig. 6.28).
  - haematological abnormalities associated with myeloma include anaemia, leucocytosis, neutrophilia and lymphocytosis:
    - total protein concentration is very elevated.
    - protein electrophoresis shows a monoclonal peak in the gamma region.
- further information may be acquired by CT or MRI of the neck.

Management

- no treatment and grave prognosis.

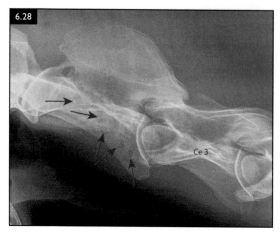

6.28

Ce 3

FIG. 6.28 Laterolateral radiograph of the occiput and first three cervical vertebrae of an aged pony with recent onset ataxia. Cranial is to the left. There is multifocal osteolysis in the second cervical vertebra (arrows). The differential diagnosis was neoplasia or osteomyelitis. Post-mortem examination confirmed the presence of lymphosarcoma. There was an extradural mass on the left side of the spinal cord. (Photo courtesy Lucy Meehan)

The Axial Skeleton – Thoracolumbar Region

Back anatomy and function

Axial skeleton

- horse has 18 thoracic and 6 lumbar vertebrae.
- size and shape of vertebrae changes gradually from cranial to caudal based on their function within the back.
- primary ossification is normally complete shortly after birth:
  - separate centres of ossification can remain present for many years.
    - epiphyses of the lumbar vertebral bodies may take between 5 and 7 years to close (Fig. 6.29).

- extremities of transverse and dorsal spinous processes (9–14 years or even later Fig. 6.30).
- each vertebral body is separated from its neighbour via a single fibrocartilage intervertebral disc.
  - articulates via left and right articular processes:
  - form a facet joint on each side (Figs. 6.31–6.33).
- all vertebrae undergo flexion and extension in the sagittal plane during locomotion:
  - range of motion increases, moving caudally through the thoracolumbar area, occurring maximally in the caudal lumbar area.

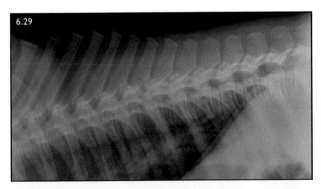

FIG. 6.29 Lateral radiograph of the caudal thoracic region of a 3-week-old foal. The bones are formed but epiphyses remain open in the vertebral bodies and the dorsal tips of the DSPs are not ossified.

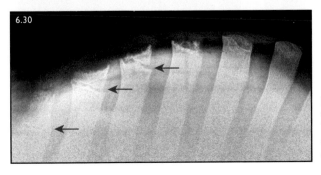

FIG. 6.30 Lateral radiograph of the withers in a 7-year-old Thoroughbred. The epiphyses in the dorsal tips of the dorsal spinous processes can remain open for up to 14 years (arrows).

Soft tissue structures

- several ligaments link the bones together:
  - supraspinous and interspinous ligaments functionally link the dorsal spinous processes (DSPs) of the vertebrae (Fig. 6.34).
  - dorsal and ventral longitudinal ligaments run along the inside of the floor and the outside, of the vertebral column, respectively:
    - functionally opposing the supra- and interspinous ligaments.
  - ligamentum flavum links adjacent vertebral body laminae.

- intrinsic muscles of the thoracolumbar region are divided into those above (epaxial) and those below (hypaxial) the transverse processes:
  - dorsal epaxial muscles consist of:
    - longissimus group axially and iliocostalis abaxially.
    - easily visualised and palpated in the back.
  - underneath are the multifidus muscles which sit next to midline against the DSPs.
    - provide stabilisation and proprioception.

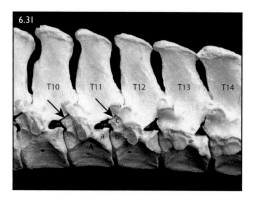

FIG. 6.31 Mid-thoracic vertebrae (T10–T14) from a normal mature horse. Note the narrow spaces between the tips of the processes, which point cranially. A = vertebral body of T11; a = caudal costal facet; b = cranial costal facet; c = mamillary process; arrows = dorsal intervertebral joint space.

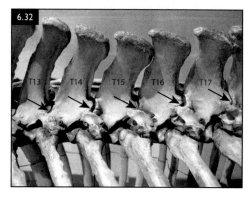

FIG. 6.32 Mid-thoracic vertebrae (T13–T17) from a clinically normal mature horse. Note the mild osteoarthritic changes in the facet joints. Arrows = dorsal intervertebral (facet) joint space.

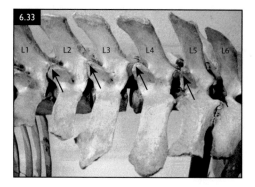

FIG. 6.33 Lumbar vertebrae (L1–L6) from a normal mature horse. Note the larger horizontally oriented dorsal intervertebral (facet) joint spaces (black arrows).

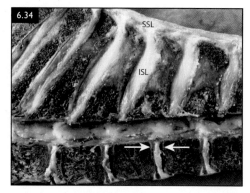

FIG. 6.34 Sagittal plane section of the thoracic spine of a young pony with spinal cord removed. SSL = supraspinous ligament; ISL = interspinous ligament. Note the vertebral body epiphyses remain open and visible (arrows).

o psosas, iliacus and quadratus lumborum muscles are in the hypaxial group acting to ventroflex the spine.
- equine spinal function has been likened to the archer's bow:
  o vertebrae and dorsal soft tissue structures form the bow.
  o muscles and ligaments of the ventral abdominal wall form the string.
    ♦ huge mechanical advantage over the epaxial muscles allowing them to ventrally flex the spinal column.
    ♦ frequently weak and dysfunctional in horses with chronic back problems.
      – addressing this is an important part of the treatment plan of such cases.
- ligaments and adjacent fascial planes are richly endowed with multiple types of sensory nerve receptor as well as bare nerve endings.

Spinal deformities

Definition/overview
- osseous abnormalities of the vertebral column are quite common:
  o 20–36% of examined Thoroughbreds having deviations from the standard vertebral formula in the thoracolumbar or lumbosacral region.
  o only rare cases with severe deformity may interfere with the athletic use.
- excessive curvature of the thoracolumbar spine can be ventral (lordosis) (Fig. 6.35), dorsal (kyphosis) (Fig. 6.36) or lateral (scoliosis) (Fig. 6.37).

Aetiology/pathophysiology
- unclear aetiology including congenital and developmental causes.
- severe curvatures are usually due to an anomaly in one or several vertebrae.
- Lordosis is relatively common in geriatric animals, especially when retired from regular ridden exercise.
  o broodmares with increasing age and repeated pregnancies.

Clinical presentation
- affected animals are usually pain free despite obvious vertebral column deviation.

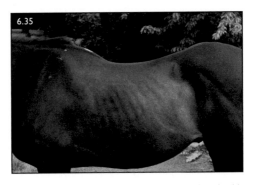

FIG. 6.35 A late middle-aged Thoroughbred with lordosis of the back. The horse had no clinical signs of back pain.

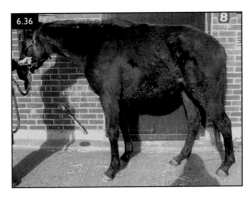

FIG. 6.36 A 3-year-old Thoroughbred filly with kyphosis.

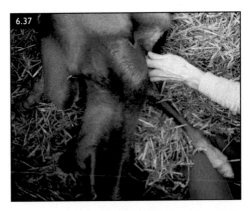

FIG. 6.37 Neonatal Thoroughbred foal born with congenital axial and appendicular abnormalities including a lateral scoliosis, most evident at the level of the withers.

- fitting tack correctly can be difficult, leading to pinching and secondary back pain.
- neurological gait abnormalities are not usually seen.
- stride length may be restricted.
- progressive lordosis can affect young horses after the age of puberty before skeletal maturity occurs, between the age of 2 and 7 years.
  - typically, there is no known cause.
  - affected individuals often suffer back pain if ridden due to DSP impingement.

Differential diagnosis

- gradually progressive developmental curvatures must be distinguished from mild curvatures of the vertebral column caused by asymmetric muscle contraction:
  - usually painful to palpation and will respond to analgesic medication.
- fractures of the thoracic or lumbar vertebrae may cause lordosis or kyphosis:
  - history and clinical presentation differ markedly.
- severe epaxial muscle wastage can give a similar appearance to kyphosis due to increased prominence of the DSPs.

Diagnosis

- examination and inspection.
- **Radiography** useful to identify bony abnormalities and other concurrent pathology (Fig. 6.38).

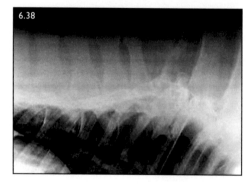

FIG. 6.38 Laterolateral radiograph of a foal with a marked deformity of its thoracic spine. Note the major disruption of the normal alignment of the thoracic vertebral column at T12/13.

Management

- no reports of attempted correction of congenital anomalies of the spine.
- careful fitting of the saddle is essential to minimise secondary back pain in ridden animals.
- owners may wish to breed from affected animals and as the inheritance of these anomalies is unknown, giving advice on this matter is difficult.

Back pain

Definition/overview

- back pain is a challenging diagnostic and therapeutic syndrome.
- most of the conditions described in this chapter can cause primary back pain.
- secondary back pain is more common than primary back pain:
  - most often through compensation for insidious limb lameness, especially hindlimbs.

Aetiology/pathophysiology

- acute trauma can lead to back pain via nociceptive sensory nerves distributed in the soft tissues of the back:
  - negative feedback loops within the lower levels of the CNS can lead to variations in pain perception.
  - expectation of pain is known to increase its perception ('cold backed' horse that anticipates pain when saddled or backed)
- chronic stimulation and nerve damage (neuropathy) can lead to nervous system dysfunction:
  - inappropriate sensation of pain (termed neuralgia).
  - abnormally heightened sensation (termed allodynia).
  - observed in equine back pain, whereby the lightest brush of fingers over the back elicits an electric shock-type pain response.
- secondary overuse of thoracolumbar epaxial muscles can develop as a response to lameness, so-called compensatory or secondary back pain:
  - muscles become tight and painful to palpate due to an abnormal workload.

- anti-inflammatory mode of action of NSAIDs as painkillers means that they have little or no effect on primary back problems caused by muscle pain or on neuropathic pain:
  o using NSAIDs to distinguish 'untreatable behavioural' back problems from those horses with finite pathology can be misleading.

Clinical presentation

Introduction

- assessment of a back pain case should include the following steps:
  o clinical history.
  o passive physical examination, including assessment and palpation of the whole musculoskeletal system.
  o dynamic physical examination, including walk, trot and canter.
  o further tests and investigations:
    ♦ neurological examination.
    ♦ oral examination.
    ♦ ridden test.
- aim to identify whether pain is present in the back and/or elsewhere in the musculoskeletal system.
- next decide on the most appropriate means of identifying the problem.

Clinical History

- back pain has many manifestations.
  o symptoms vary hugely between individuals and within the same individual over time.
  o symptoms vary from mild loss of performance through to extreme and dangerous avoidance behaviour, with bucking, rearing and bolting.
  o owners and riders may not associate them with problems until symptoms worsen, instead finding various alternative explanations.
  o most horses with finite orthopaedic causes of back pain exhibit behavioural and schooling abnormalities.
  o horses owned and ridden by inexperienced riders may also exhibit these abnormalities in the absence of pathology.
- loss of performance, evidence of distress or avoidance behaviour before or during work may be the first sign.

- failure to engage the core muscles of the back.
  o frustrated owners complain that:
    ♦ they cannot get their horses to develop muscles of the topline.
    ♦ will not work in an outline.
    ♦ work excessively on the forehand.
- symptoms can develop insidiously.
- horse may continue working with intermittent rest or repeated musculoskeletal manipulative treatments.
- highly skilled riders can mask or improve a horse's exhibition of pain.
- owners may not identify actual back pain until relatively late in the process:
  o reporting physical therapist's findings.
  o struggling to approach or groom the horse's back.
- horses can be totally normal until a sudden onset of severe topical pain and guarding in the dorsal thoracolumbar area.
- breed differences can occur, both in the symptoms of pain and in the response to examination.
- minority of horses with back pain become dangerous to handle.

Clinical examination

- whole horse should be examined:
  o concurrent lameness and pelvic, neck or even dental problems are common.
  o clinically silent back pathology can become symptomatic when pain develops elsewhere in the body or limbs.
  o recognising and successfully treating the distant pathology sometimes causes back pain to resolve in such horses.
- using NSAIDs for short periods can provide useful information to help distinguish horses with compensatory and primary causes of back pain.
  o e.g. lame horses with distal tarsal pain and compensatory (secondary) lumbar epaxial tenderness often resolve the back pain fairly quickly after commencing oral NSAID treatment.
  o horses with kissing spines (primary) are unlikely to resolve back pain as quickly or as profoundly.
- gastric ulcers occur commonly both as a primary problem and secondary to

orthopaedic problems, including back pain:

- o gastroscopy allows diagnosis and monitoring.
- o treatment of both the ulcers and orthopaedic problems is essential for success.

Differential diagnosis

- primary muscle pain from tack or riding technique
- trauma • facet joint OA.
- impingement of DSPs
- supraspinous desmitis
- ventral spondylosis.
- more than one condition can be present.
- secondary muscle pain from:
  - o hindlimb or forelimb lameness
  - o neck pain • pelvic pathology.
- dental pain, gastric ulcers or other forms of gastrointestinal pain.

Diagnosis

- detailed passive and dynamic assessment of the musculoskeletal system.
- diagnostic analgesia is indicated to prove the connection between clinical and imaging findings, but this is only possible in limited areas.
- diagnostic therapy is widely used as a practical alternative for some conditions in the back.
- clinical and imaging abnormalities correlate poorly:
  - o **radiography** and **ultrasonography** are important complimentary imaging modalities for epaxial structures, but both have limitations.

- o radiation scatter progressively deteriorates the image quality, with increasing size.
- o thoracic and cranial lumbar DSPs can be imaged using mobile units with laterolateral projections (Fig. 6.39).
- o vertebral bodies and more caudal lumbar DSPs may be obscured in adult horses and ponies.
- o ventrolateral 20°–dorsolateral oblique views of the ipsilateral caudal thoracic facet joints is more challenging.
  - ♦ requires a focused grid or aluminium wedge filter and much higher power (Figs. 6.40, 6.41).
  - ♦ lumbar facet joints are rarely possible in any animal larger than a pony and requires optimal equipment and technique.
- o **ultrasonography** provides good visualisation of the dorsal spinal structures and epaxial muscles, but no hypaxial structures can be seen.
  - ♦ supraspinous ligament and tops of the DSPs can be viewed in good detail with a high-frequency linear tendon probe (Fig. 6.42).
  - ♦ dorsal and transverse spinous processes, along with the facet joints, dorsal lamina and deeper epaxial muscles:
    - – require a low-frequency curvilinear probe (Fig. 6.43).
- **gamma scintigraphy** is commonly used for the investigation of back cases:
  - o thoracolumbar area most useful for acute bone injury, such as following trauma, or in detection of stress or other fractures (Fig. 6.44).

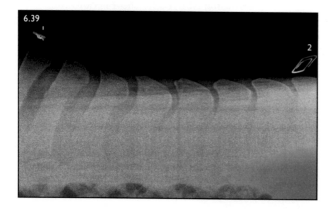

FIG. 6.39 Laterolateral radiograph of the caudal thoracic region. This view is the mainstay of radiographic imaging of this area.

- ○ poorer specificity for chronic conditions, increasing the risk of false-negative results.
- use of thermography to investigate back problems is controversial and not routine.

- pressure algometry can be used to study the dynamic interaction between the horse's back, rider and tack.
- examination of the saddle/tack and its fit should be carried out by a suitably qualified person in all cases of back pain.

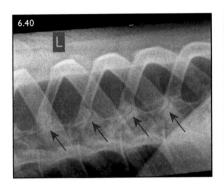

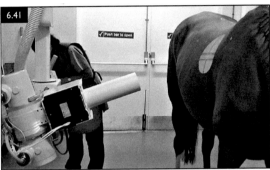

FIGS. 6.40, 6.41 (6.40) Ventrolateral 20°–dorsolateral oblique of the same area of the horse in 1.483. This view allows the ipsilateral facet joints to be visualised very clearly (arrows). (6.41) The radiograph is obtained using this oblique projection and the use of a cone.

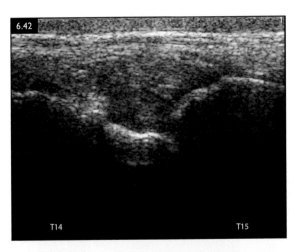

FIG. 6.42 Longitudinal ultrasound image obtained on midline using a 13.3 MHz linear probe between T14 and T15. The supraspinous ligament and Y-shaped upper part of the interspinous region can clearly be seen, but the interspinous ligament appears black due to the orientation of the fibres (see 6.34).

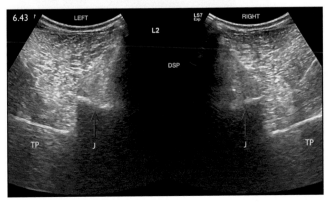

FIG. 6.43 Transverse parasagittal ultrasound images of the lumbar facet joints at L2 using a 4 MHz curvilinear probe. J = facet joint space; TP = transverse process. Note that the joint is in line with the arrow.

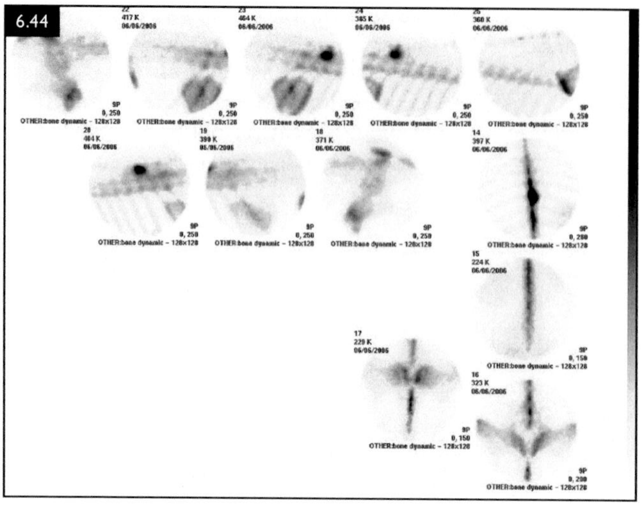

6.44

FIG. 6.44 Bone scan of the thoracolumbar spine and pelvis of a horse presenting with acute back pain. Note the intense focal radiopharmaceutical uptake between T15 and T16 in all views, suggesting acute bone pathology. Radiographs confirmed a fracture of the DSP.

Management

- symptomatic treatment coupled with active rehabilitation is the cornerstone of the management of back pain:
  - specific strengthening exercises and stretching is beneficial, but many owners lack the skills and time required to carry it out properly.
    - use of postural training aids is widespread.
    - encourages the horse to work with a low head carriage, pushing from behind and building support musculature around the back.
  - time off ridden work is often needed, depending on the cause of the back pain.
  - physiotherapy techniques can reduce local muscle spasm and compensatory tension elsewhere.
  - water treadmills can also be helpful:
    - water depth should be no higher than the shoulder to prevent an excessively high head carriage and compensatory dipped back posture.
  - swimming is contraindicated for the same reason.
- any concurrent orthopaedic or lameness problem requires specific treatment:
  - foot pain is common in horses and can rapidly induce back dysfunction.
    - optimising foot balance and support is a helpful therapeutic step.

- acupuncture may provide analgesia for some thoracolumbar conditions:
  - electroacupuncture at higher frequencies (15–100 Hz) can provide greater stimulation than plain needle acupuncture (Fig. 6.45).
- successful treatment and rehabilitation require a collaborative approach involving physical therapists, farriers and veterinarians.
- heavy riders, as well as equine obesity, are both increasing issues:
  - studies have concluded that horses should carry an absolute maximum of 20% of their lean body weight, including tack.

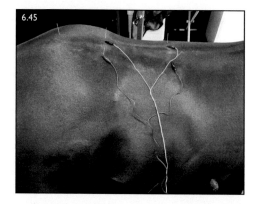

6.45

FIG. 6.45 Electroacupuncture being carried out on a horse with lumbar pain.

Soft tissue trauma

Definition/overview

- thoracolumbar region soft tissues are subject to traumatic injury:
  - affected structures include skin, muscles (epaxial and hypaxial), ligaments, nerves, and blood vessels.

Aetiology/pathophysiology

- acute one-off trauma, such as a fall, casting or road traffic accident.
- chronic repetitive wear and tear through dysfunction, compensation or iatrogenic factors such as ridden exercise with a maximally flexed neck and an excessively low head carriage.
- poorly fitting tack is a common causes of muscle pinching and pain.
- ligament, nerve or even bone injury can occur depending on the severity of the trauma:
  - muscle or ligament trauma may cause rupture or detachment from an origin or insertion.
    - new bone may form at a ligament origin or insertion site (entheseopathy).
    - may, or may not, lead to chronic pain (Fig. 6.46).
  - lesser injuries are more common.
- bony pathology of the vertebrae, such as overlapping DSPs, may result in secondary damage to muscles and ligaments.
- supraspinous bursa, lying between the supraspinous ligament and the tips of the cranial thoracic DSPs at the withers, can become infected (Fig. 6.47):
  - following a wound at this site ('fistulous withers')
  - less commonly, subsequent to chronic infection by *Brucella abortus* or *Actinomyces bovis*.
- haematomas may occur in the vicinity of the withers and can be exceptionally painful.
- penetrating injuries can introduce foreign material and bacteria into the epaxial muscles, causing infectious myositis (Fig. 6.48).

Clinical presentation

- acute back trauma typically induces focal swelling, heat, crepitus and avoidance behaviour.

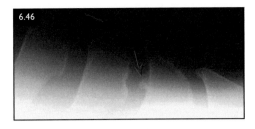

FIG. 6.46 Radiograph of several DSP tips in the cranial thoracic region. There is a spur of bone (arrow) on the cranial aspect of the DSP of T11 which was insignificant.

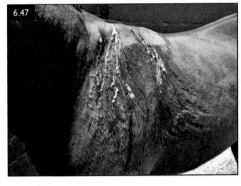

FIG. 6.47 Fistulous withers. Multiple draining tracts are present from the infected supraspinous bursa.

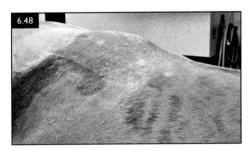

FIG. 6.48 Abscess in the lumbar epaxial muscles caused by a puncture wound. Swelling, heat and extreme tenderness were present.

- asymmetry in the size, texture and response to palpation of epaxial muscles may be appreciable.
- ill-fitting saddles may cause direct soft tissue trauma:
  - may manifest as hair loss (Fig. 6.49), focal muscle atrophy, white hairs, painful blisters or even ulcers (Fig. 6.50).
  - can become chronic, with painful scar tissue formation, termed saddle galls.

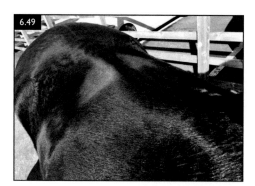

FIG. 6.49 Poor fitting tack can lead to saddle slip, excessive rubbing and hair loss.

FIG. 6.50 With chronicity, tack can rub right through the skin, causing open painful saddle sores as in this working mule.

- acute muscle trauma causes variable focal pain and spasm:
  - with chronicity, the symptoms may become more diffuse and muscle wasting more prominent.

Differential diagnosis

- abnormal function caused by poor riding technique, heavy riders and poor fitting saddles.
- lameness, particularly of the hindlimbs, can lead to asymmetry of back movement and rider.
  - may also cause the saddle to move asymmetrically, to slip and rub (termed saddle slip), thereby causing secondary back pain.
- traumatic injuries of muscles, nerves and ligaments, especially the supraspinous and interspinous ligaments.
- bone injuries, including DSP impingement, facet joint disease, stress and other fractures, and ventral spondylosis.
  - may have concurrent secondary soft tissue injuries.
- other causes of poor performance, especially limb lameness.
- behaviour and temperament problems.
- dental problems.
- more rarely: other forms of visceral pain, such as nephrolithiasis, cyclic ovarian activity, gastric ulcers and mesorchium tension; aortic or iliac artery thrombosis.

Diagnosis

- acute superficial trauma visual findings.
- ultrasound and radiographic assessment.

- injection of small volumes of local anaesthetic into identified pathology can help ascertain their relevance.
  - care should be taken in how this is interpreted.
  - anticipation and compensation for focal pain may not resolve.
  - contraindicated for deeper perispinal structures due to the risk of spinal injection.
- **gamma scintigraphy** provides a means of locating chronic perivertebral pathology:
  - unless the soft tissue damage has changed local bone physiology.
  - false negatives will be encountered.
- diagnostic therapy can provide an effective alternative:
  - crucial to ensure total understanding and cooperation from the owner before initiating therapy.

Management

- superficial ulcers and rubs usually respond quickly to basic wound management.
- deeper saddle galls, especially those nearer the midline, are sometimes most easily managed by surgical excision of scar tissue, carried out standing under local anaesthesia.
- saddles and tack should be checked carefully to ensure a good fit before riding resumes.
- damaged ligament or muscle can be accurately located:
  - targeted treatment is possible:
    - ♦ corticosteroids +/- pitcher plant extract (sarapin/saragyl) injected directly into ligaments using

ultrasound guidance have been reported to be of use.

- ◆ helpful symptomatically where inflammation is present.
- ◆ excessive corticosteroid use has been associated with ligament mineralisation (Fig. 6.51).
- focused extracorporeal or radial-pulsed shock-wave therapy provides an alternative targeted analgesic treatment for ligament injuries.
- mesotherapy involves injections of small quantities of a combination of corticosteroid, local anaesthetic and a myorelaxant subcutaneously:
  - ○ into the dermatomes corresponding to the proposed sites of pathology over the thoracolumbar region.
  - ○ popular in some parts of Europe.
- judging treatment success is best monitored via the restoration of normal function and symptomatic resolution of pain, or the behavioural symptoms of it:
  - ○ reversal of muscle atrophy seen in chronic injuries.
- ultrasound can be used to monitor healing.

Fractures

Definition/overview

- incidence of complete fractures of the thoracic, lumbar or sacral vertebrae is low (<2%).

Aetiology/pathophysiology

- acute vertebral fractures are traumatic, usually caused by heavy falls or road traffic collision.
- fractures occur in three basic locations:
  - ○ DSPs, vertebral facet joints and vertebral bodies.
  - ○ fractures of the DSPs are the most common at the withers (T2–T9).
    - ◆ caused by backwards falls when rearing in young animals.
    - ◆ usually displaced due to the ventral pull of the supraspinous ligament.
    - ◆ can become infected if associated with an open wound (Figs. 6.52, 6.53).
    - ◆ can also involve the supraspinous bursa, allowing a fistula to form if infection persists (see Fig. 6.47).

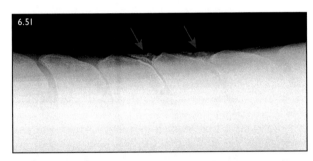

FIG. 6.51 This horse had repeated corticosteroid injections in the interspinous ligament over a period of several years. Note the mineralisation in and around the top of the interspinous space (arrows).

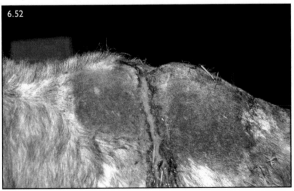

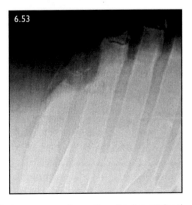

FIGS. 6.52, 6.53 (6.52) Chronic purulent discharge from the withers of a horse several months after it sustained a wound to the area during a fall. (6.53) Laterolateral radiograph of the withers of the same horse. Note that the dorsal part of the DSP of T3 is missing, with evidence of osteomyelitis in the surrounding part of the process. Also note the overriding fractures of the DSPs of T4 and T6, which occurred at the original traumatic episode.

○ fractures of the vertebral bodies may result in acute and severe neurological signs due to spinal cord compression:
  ♦ hindlimb paresis or paralysis with normal forelimb function is suggestive of a lesion caudal to T2.
○ vertebral lamina at the base of the DSP:
  ♦ predilection site for incomplete stress fractures and subsequent remodelling.
    – flat racing Thoroughbreds.
  ♦ result from repeated mechanical overloading.
  ♦ may be responsible for poor performance or other signs of back pain.
○ in foals, vertebral end plate fractures or separations may occasionally be seen following trauma during or soon after parturition (Fig. 6.54).

Clinical presentation

• depends on the location (within the back and individual vertebra) and type of fracture:
  ○ stress fractures of the vertebral lamina may show only vague and low-grade signs of back pain or poor performance.
  ○ severely displaced fracture caudal to T2 will result in hindlimb paralysis.
  ○ complete but minimally displaced fracture may still allow the horse to move without overt discomfort:
    ♦ hindlimb paresis, ataxia and severe focal muscle spasm and guarding.
    ♦ subsequently, may be local or generalised epaxial muscle atrophy:
      – chronic disuse or neurological dysfunction of lower motor neuron pathways (Fig. 6.55).
  ○ fractured articular facets may lead to a reflex scoliosis orientated towards the fracture or present later in life with OA pathology of the joint/s (see Osteoarthritis).
  ○ fractured DSPs usually have swelling, heat, pain and guarding.
    ♦ most apparent in the withers (Fig. 6.56).
    ♦ crepitus may be noted if the horse allows physical palpation or manipulation.

FIG. 6.54 Sagittal section through several thoracic vertebrae of a neonatal foal that was unable to rise following parturition. There is separation at one of the vertebral cranial physes.

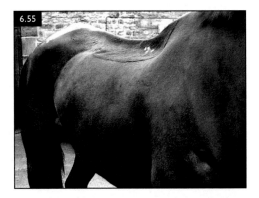

FIG. 6.55 Chronic and profound epaxial muscle atrophy of the right side of the back of a horse that was involved in a traumatic episode several years earlier. This had been associated with severe back pain, muscle guarding and spasm, and rapid onset of muscle wastage.

  ♦ displaced DSP fragments in the withers remain attached to the supraspinous ligament and conjoined nuchal ligament.
    – may present with an apparent reluctance to move the neck or not be able to lower their heads to graze.
    – once the initial soft tissue swelling has reduced, there may be long-standing visible bony deformity at the withers.

6b

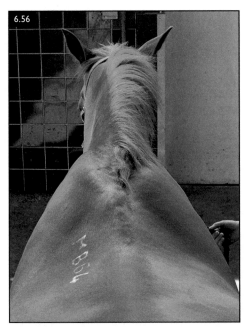

FIG. 6.56 View from above of a horse that had fallen backwards earlier that day. Note the swelling over the proximal aspect of the left scapula. Radiography confirmed fractures of the DSPs of T4–T7.

Differential diagnosis

- horses with lumbar fractures without neurological signs can resemble animals with pelvic fractures.
- if presenting signs are vague vertebral column discomfort and poor performance, then the other causes must be considered.
- discharge at the withers secondary to an infected fracture(s) should be differentiated from an infected supraspinous bursa.

Diagnosis

- diagnostic approach is the same as that outlined above for soft tissue lesions, unless there are obvious clinical signs such as ataxia, marked swelling or discharge.
- if vertebral displacement is present, a deviation of the DSP(s) may be palpable.
- DSPs can be assessed ultrasonographically and radiographically in the standing horse (Figs. 6.57–6.59).
- Scintigraphy is sensitive at detecting suspect fractures (see Fig. 6.44).

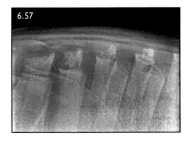
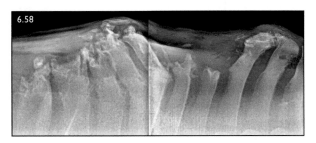

FIGS. 6.57, 6.58 (6.57) Lateral radiograph of the withers area in a Thoroughbred that had suffered a heavy fall on to the area. Multiple displaced fractured DSPs are visible. The pull of the nuchal ligament tends to pull the dorsal sections downward, making the fragments override one another. (6.58) Composite radiograph of the same horse 3 months later. The withers were externally healed, but modified tack was needed.

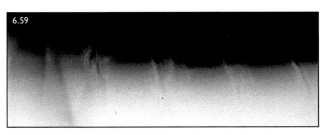

FIG. 6.59 Lateral radiograph of a single DSP fracture of T11.

Management

- absence of neurological symptoms:
  - at least 6 weeks box rest.
  - anti-inflammatory medication as required, depending on symptoms (usually around 2 weeks).
  - additional one-off dose of corticosteroid is indicated to minimise early swelling.
  - horses should be fed and watered from a height as they are often unwilling or unable to put their head to the ground.
    - ◆ increased risk of pleuropneumonia due to pooling of bronchial secretions:
      - – temperature taken daily.
      - – bloods obtained if any doubt about the horse's respiratory tract health.
- with neurological deficits:
  - systemic corticosteroid medication continued for at least three to four treatments.
  - internal fixation of fractures of the vertebral bodies in the thoracolumbar area has not been described.
  - euthanasia is indicated if hindlimb ataxia or paresis is significant and progressive despite treatment, or if pain cannot be controlled.
- cases of closed DSP fractures:
  - require good levels of analgesia early on.
  - eventually heal, allowing work to recommence from 3 to 4 months post injury.
- open fractures benefit from:
  - early surgical debridement of loose bone fragments and lavage.
  - ensuring adequate postoperative drainage in the elevated position is the principal challenge for any withers' surgery (Fig. 6.60).
  - prolonged postoperative periods of antibiotic medication, pain management, and local wound treatment.
  - repeat surgery for sequestrum removal and lavage is common.

Prognosis

- open fractures of the DSPs have a significant rate of repeated swelling,

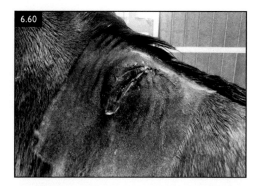

FIG. 6.60 Ensuring adequate drainage is the principal challenge associated with surgery of the withers.

draining supraspinous fistulas (fistulous withers) and sequestration:
- degree of cosmetic blemish depends on the extent of displacement.
- modified tack may be required after healing.
- articular facet joint fractures tend to heal by ankylosis:
  - residual functional restriction, OA and pain are significant risks.
  - chances of full recovery are guarded.
- acute traumatic fractures of the vertebral bodies in the back have a guarded to hopeless prognosis depending on the incidence of neurological deficits.

Spondylitis and discospondylitis

Definition/overview

- infection and osteomyelitis within vertebrae are termed spondylitis.
- if it involves the intervertebral disc, it is termed discospondylitis (Fig. 6.61).

Aetiology/pathophysiology

- osteomyelitis occurs predominantly in immune-compromised individuals:
  - foals with partial or complete failure of passive transfer of immunity.
- types of bacteria associated with sepsis in such individuals are implicated, including staphylococcal and streptococcal species, mycobacteria, *Rhodococcus* and others.

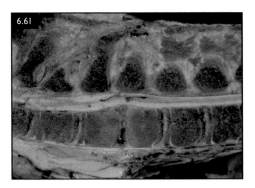

FIG. 6.61 Post-mortem specimen of a sagittal section of the thoracolumbar spine of a yearling that presented with acute localised spinal pain, pyrexia and hindlimb neurological deficits. Note the discospondylitis of the intervertebral disc and surrounding vertebral bones, with subsequent spinal cord compression.

Clinical presentation

- clinical signs predominantly relate to systemic symptoms of sepsis.
- focal spinal pain is present.
- haematology is consistent with infection:
  - leucocytosis, neutrophilia and elevated acute phase proteins.
- progressive infection may compress the spinal cord and spread into the meninges:
  - deteriorating neurological signs and meningitis.

Differential diagnosis

- sepsis, osteomyelitis and septic synovitis elsewhere in the body.
- soft tissue or bone trauma from being trodden on or kicked.

Diagnosis

- **radiography** of the painful area will usually show the characteristic signs of epiphyseal lysis and surrounding sclerosis:
  - ventral spur formation can develop between adjacent affected vertebrae.
- infection close to the spinal cord, analysis of cerebrospinal fluid:
  - evidence of inflammation and infection.
- **scintigraphy** will demonstrate high radiopharmaceutical uptake but does not necessarily differentiate this condition from trauma.

Management

- vague presenting symptoms often lead to the infection being well established before diagnosis.
- high-dose antibiotic therapy is required:
  - preferably following culture and sensitivity testing.
  - intensive supportive treatment.
- umbilical remnants should be scanned ultrasonographically to check for infection in foals.
- if neurological signs have already developed, then treatment is usually unsuccessful.

Impingement of dorsal spinous processes

Definition/overview

- also known as 'kissing spines' and 'overriding DSPs'.
- one of the most common conditions associated with thoracolumbar pain in the horse.
- space between the DSPs in the thoracic or, less commonly, the lumbar vertebral column narrow dorsally, and the opposing bone surfaces remodel.
- any space 4 mm or less on static radiographs is considered at risk of dynamic impingement.

Aetiology/pathophysiology

- cause of the condition is unclear:
  - multifactorial in origin.
- unknown why only a proportion of affected horses develop clinical back pain:
  - detected on radiographic and gross post-mortem examination in many clinically asymptomatic horses.
    - ♦ 30 to 90% of examined horses in various surveys.
- incidence is highest in the saddle region (T12–T18).
- impingement can affect between 1 and 13 spaces.
- reported in horses that have not been ridden, including young foals.
- thoracolumbar conformation may affect the degree of impingement:
  - slightly lordotic and/or short back conformation brings DSP tips closer.

- differences in shape of the DSP (hooked conformation in tip of Thoroughbreds).
- abnormal and deformed vertebra(e).
- level and type of work performed may affect likelihood of clinical pain.
- breed predisposition, with Thoroughbreds or part-Thoroughbreds widely reported to have a higher incidence.
- back pain may develop more often in horses with both impingement of DSPs and injury elsewhere in the axial and appendicular musculoskeletal system.
- local pathological processes occur at the sites of DSP impingement:
  - dorsal one-third to one-half of affected process increases its rate of remodelling:
    - radiographically seen initially as sclerosis of subperiosteal bone (Fig. 6.62).
  - occasionally followed by development of cyst-like zones of bone resorption (Fig. 6.63).
  - DSP contact may lead to pseudoarthroses and small bursae formation.

- common to find all grades of disease distributed within one individual horse.
- associated inter- and supraspinous ligaments also become damaged.
- source of pain in these horses is still unclear and varies considerably in individual cases.
  - local anaesthetic infiltrated into narrowed spaces alleviates the pain transiently:
    - implies sensory nerves in local soft tissues are a significant source of pain (Fig. 6.64).

Clinical presentation

- symptoms of back pain as discussed earlier.
- many cases present as chronic back pain with epaxial muscle wastage, intermittently responsive to medication or time off work.
- limb lameness is commonly present in such horses, hindlimbs more than the forelimbs.
- some horses become painful suddenly, e.g. following a fall.
- symptoms may develop following a period off work, apparently occurring suddenly when riding restarts.
- clinical signs correlate poorly with the number of impingements or the perceived severity of the radiographic changes.
- poor saddle fit can exacerbate the back pain.

Differential diagnosis

- soft tissue and bone trauma in back
- saddle pinch or pressure.
- vertebral facet disease (can coexist)
- supraspinous desmitis.
- ventral spondylosis.
- hindlimb lameness and foot imbalance.

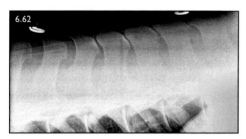

FIG. 6.62 Impingement of DSPs (kissing spines) may only be manifest as narrowing of the space, sometimes with sclerosis. This horse has four kissing spines.

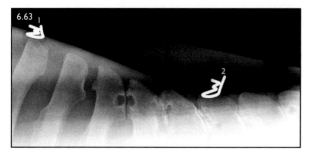

FIG. 6.63 Severe impingement of DSPs may be associated with lytic areas and DSP overlap.

FIG. 6.64 Local anaesthetic injected into narrowed, or immediately to the side of, close kissing spines often results in a marked reduction in pain and increased spinal function in affected horses.

Diagnosis

- palpation in the midline of the back over the DSPs:
  - narrowing or loss of the interspinous spaces is palpable in the thoracic region.
    - less reliable in the lumbar area.
- **radiography** is required to demonstrate the full extent of impingement (Figs. 6.62, 6.63) and other pathology elsewhere in the back:
  - focal consistent back pain in the region of radiographic DSP impingement may be adequate for a diagnosis.
  - majority of cases do not have such clear signs of focal pain:
    - further tests to prove cause and effect.
- local anaesthetic can be infiltrated into the affected interspinous spaces (Fig. 6.64) or immediately adjacent:
  - re-evaluate case straight away, preferably when ridden.
  - **potential for false positives in normal horses due to disruption of proprioceptive pathways.**
- alternatively, corticosteroids can be injected into the same affected spaces and the horse re-evaluated after 2–3 weeks.
- **scintigraphic** examination may demonstrate local increased metabolic activity:
  - **false negatives and positives are frequently encountered.**
  - **undermines value when used in isolation** (Fig. 6.65).

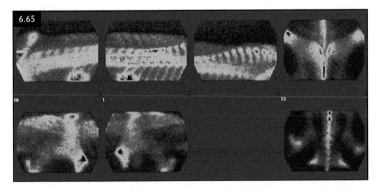

FIG. 6.65 Bone scan of the thoracolumbar spine and pelvis of a horse with back pain. Note the radiopharmaceutical uptake in several DSPs in the thoracic region and also associated with the dorsal facet joints of the thoracolumbar junction. Additional diagnostic methods will be required to ascertain the significance of these findings.

Management

- management to reduce pain and restore normal function is needed:
  - systemic or local anti-inflammatory medications.
  - local counterirritant medication.
  - bisphosphonate drugs.
  - acupuncture and mesotherapy.
  - usually in combination with some form of manipulative therapy e.g. physiotherapy, osteopathy or chiropractic.
  - published success rates following non-surgical management are very limited.
    - ◆ short-term success rates with 70–80% of affected horses experiencing resolution of symptoms for 4–6 weeks.
    - ◆ 30–40% of these horses experience recurrence of symptoms when the medication wears off.
- surgical management has similar success rates, which often persist significantly longer:
  - consequently, sometimes used in preference to medical treatment.
  - several surgical options exist:
    - ◆ under general anaesthesia or standing sedation and local analgesia.
    - ◆ most studies agree that the number of spaces involved, and the radiographic severity of signs, are unrelated to outcome.
  - involved DSPs are removed *in toto* or by wedge resection (Fig. 6.66).
    - ◆ return to soundness is expected in approximately 70% of cases.
    - ◆ delay before return to exercise is longer because of the increased surgical trauma and duration of wound healing.
    - ◆ minimally invasive resection has recently been published.
  - Interspinous ligament desmotomy is a minimally invasive alternative with lower morbidity and cost (Fig. 6.67).
    - ◆ often leads to rapid resolution of symptomatic back pain.
    - ◆ treated horses can return to riding in 4 weeks.
  - regardless of the surgical technique, a period of active non-ridden rehabilitation is considered vital before riding restarts.

FIG. 6.66 Intraoperative view of surgery to remove several thoracic DSP tips. The surgeon has removed a wedge of bone from the cranial aspect of the caudal DSP and the caudal aspect of the cranial DSP, thereby removing the bony interference at this point.

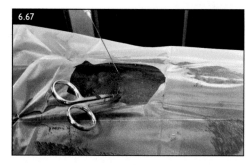

FIG. 6.67 Interspinous ligament desmotomy is a minimally invasive alternative surgical treatment for kissing spines. The spinal needle is placed in the space to aid triangulation, while Mayo scissors are used to cut the interspinous ligament in the area of impingement.

Supraspinous desmitis

Definition/overview

- supraspinous ligament is conjoined physically and functionally with the interspinous ligament.
- injuries broadly fall into two subtypes: acute trauma and chronic overuse.

Aetiology/pathophysiology

- pathogenesis of chronic supraspinous and interspinous desmitis:
  - not clearly defined.
  - nor separated from that of other pathologies, such as DSP impingement, facet disease and ventral spondylosis.

- ○ may be increased osteoclastic activity at the ligament–bone interface (Fig. 6.68).
- ○ may predispose to DSP avulsion injury or ligament tearing if loaded suddenly (Fig. 6.69).

Clinical presentation

- little or no difference in clinical presentation to the other forms of back pain.
- acute supraspinous desmitis can present as a discrete focal firm swelling over one to five DSPs in the dorsal midline.
- extensive chronic desmitis is harder to identify due to generalised thickening, potentially limiting the palpation of DSPs.

Differential diagnosis

- soft tissue and bone trauma
- saddle pinch or pressure.

- vertebral facet disease
- DSP impingement • ventral spondylosis.

Diagnosis

- laterolateral **radiographs** of the thoracolumbar region:
  - ○ avulsion fracture fragments at the insertion point of the ligament onto the DSP (Fig. 6.69).
  - ○ thickening of the soft tissues overlying the DSPs and rounded entheseophytes on the dorsal tips of the DSPs is a progressive finding with chronicity (Fig. 6.70).
- **ultrasonography** assessment of the supraspinous ligament:
  - ○ demonstrates entheseophytes more clearly.
  - ○ acute hypoechoic zones or thickening/ fibrosis present in the supraspinous ligament (Fig. 6.71).

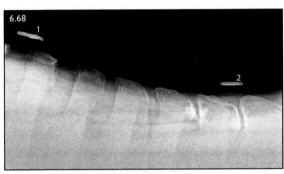

FIG. 6.68 Supra- and interspinous desmitis in a horse without DSP impingement. Increased osteoclastic activity weakens the ligament attachment, predisposing to injury. Sclerosis, lysis and entheseophyte formation can be seen in the three spaces to the right of the image.

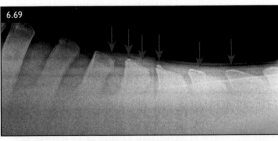

FIG. 6.69 Small avulsion bone injuries sustained after supraspinous ligament trauma (arrows).

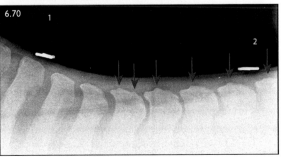

FIG. 6.70 Chronic supraspinous desmitis manifest in this horse with thickening of the soft tissues and rounded entheseophytes on the dorsal tips of the DSPs (arrows).

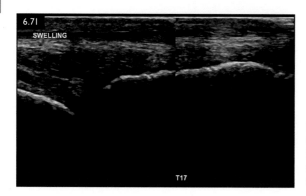

FIG. 6.71 Longitudinal ultrasound image of a discrete focal area of soft tissue swelling between T16 and T17. A hypoechoic area of supraspinous ligament is visible, consistent with injury.

- injection of local anaesthetic around the injured ligament may reduce pain dramatically.
- Interspinous ligament desmitis is not well defined:
  - radiographic changes seen may resemble those seen with impingement of DSPs.
    - sclerosis and occasional lysis in the adjacent DSP.
    - **interspinous space tends to be of normal size.**
  - increased osteoclast activity in the enthesis is readily visible on gamma scintigraphy.

Management

- acute injuries with avulsion fractures:
  - symptomatic treatment with rest and gentle field turnout is indicated.
  - long periods of rest are rarely needed.
  - physiotherapy is indicated as soon as the acute inflammation is controlled.
  - topical anti-inflammatory drugs may be beneficial given the subcutaneous position of the supraspinous ligament.
- chronic cases.
  - bisphosphonate drugs are contraindicated in acute avulsion fractures:
    - may be of benefit in chronic cases by reversing the increased osteoclast action in the enthesis and reducing pain.
  - focused or radial-pulsed extracorporeal shock-wave therapy potentially beneficial to reduce pain and encourage normal function.

- Interspinous desmotomy can resolve pain in horses with chronic reactive inter- and supraspinous entheseopathy in the absence of DSP impingement.

Osteoarthritis

Definition/overview

- occurs at the thoracic and lumbar dorsal intervertebral joints.

Aetiology/pathophysiology

- dorsal intervertebral joints are formed by the cranial and caudal facets of the articular processes in the thoracolumbar vertebral column (see Figs. 6.32, 6.33, 6.40, 6.43).
- incidence of OA in these joints is unknown:
  - recorded throughout the thoracolumbar vertebral column.
  - most common around the thoracolumbar junction.
  - may occur concurrently with impingement of the DSPs.
    - changes back movement and adversely affects facet joints, leading to OA.
- clinical signs do not correlate with the degree of pathology observed on diagnostic imaging or post-mortem examination.
- incidence of stress fractures of the vertebral lamina adjacent to the dorsal intervertebral joints is high in young Thoroughbred racehorses and positively correlated with the presence of OA.

- other types of horse, such as sport horses, OA of the facet joints is more common in:
  - middle-aged to older horses.
  - suggests chronic wear and tear as a major factor in their aetiology.

Clinical presentation

- onset of clinical signs is insidious:
  - ranges from overt vertebral column pain to poor performance.
  - no clinical signs to differentiate this condition from other back problems.

Differential diagnosis

- other conditions that can cause back pain including:
  - impingement of the DSPs
  - spondylosis.
  - ligamentous or muscular pathology
  - non-displaced fractures.

Diagnosis

- **Radiography** (see Figs. 6.40 and 6.41)
  - imaging of the thoracic facet joints is easier than in the lumbar area.
  - normal joint space appears as a radiolucent oblique line sloping from cranial to caudal at an angle of approximately 45° to the vertebral bodies.
  - sclerosis of the articular facets, subchondral lucencies and periarticular bone proliferation obscuring the joint space can be detected (Fig. 6.72).
  - concurrent impinging DSPs visible on laterolateral radiographs:
    - negative effect on the clinical signs and prognosis.
- **Ultrasonography** (see Fig. 6.43),
  - 3–4 MHz curvilinear probe in a parasagittal position.
  - facet joints much smaller and diagonally orientated in thoracic region:
    - longitudinal orientation more helpful.
    - osteophytes on the axial aspect of the joint space indicate OA.
  - joints are large and horizontally orientated in the lumbar region:
    - easiest to image in a transverse position.
    - cranial articular facet more dorsal than caudal articular facet.
    - abaxial mammillary process should be higher than the joint space.
    - OA is indicated when an osteophyte makes the joint appear higher than the mammillary process (Fig. 6.73).

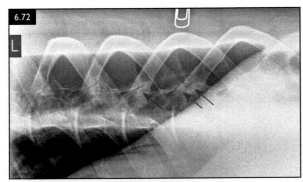

FIG. 6.72 Ventrolateral 20°–dorsolateral oblique radiograph of the caudal thoracic region of a horse with facet joint disease. Sclerosis of the articular facets, subchondral lucencies and periarticular bone proliferation obscuring the joint space can be detected (arrows). Compare with normal in 6.40.

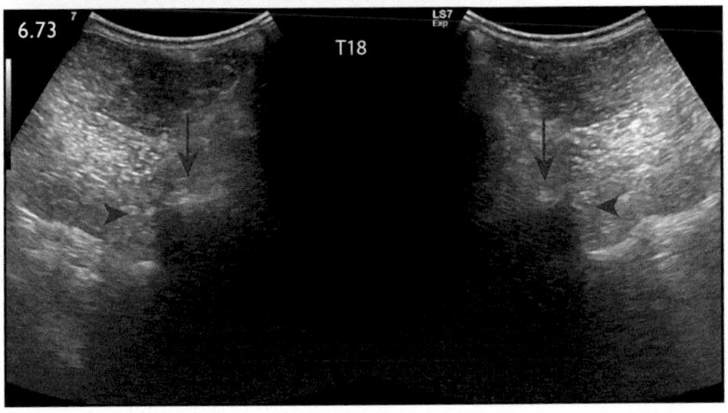

FIG. 6.73 Transverse parasagittal ultrasound images of arthritic lumbar facet joints at T18/L1 using a 4 MHz curvilinear probe. Osteophytes have formed on the joint (arrow), making this higher than the abaxial edge of the mammillary process of T18 (arrowhead). Compare with normal in 6.43.

- **Scintigraphy** (see Fig. 6.65).
  - increased bone activity in a proportion of horses:
    - increases clinical significance.
  - false negatives and error can occur if this technique is used in isolation.
- ultrasound-guided injections of the facet joints:
  - confirmation of the significance of abnormalities using diagnostic analgesia.
  - **close proximity of the spinal cord increases the risk of catastrophic spinal nerve blocking.**
  - diagnostic therapy with corticosteroid is a safer alternative:
    - improves significantly within 7 days of medication if clinically significant OA.
    - concurrent regional pathology may diminish the perceived improvement.

Management

- corticosteroids such as methylprednisolone or triamcinolone can be injected directly into the tissues overlying the joints, preferably using small volumes (e.g. 1 ml per site).
- bisphosphonate drugs are also helpful.
- previously described non-specific symptomatic methods of managing back pain.

Prognosis

- depends on:
  - number of joints affected and the severity of the pathology.
  - presence of concurrent lesions such as DSP impingement.
  - type of work being performed and the ability of the horse and rider.
  - negative prognostic signs are:
    - poor thoracolumbar flexibility and pain caused by multiple affected joints.
    - further lesions elsewhere in the vertebral column.
- resolving pain and increasing epaxial muscle strength and mobility improves results:
  - pain can recur when medication wears off.

- ignoring other risk factors for back pain (e.g. lameness) means recurrence is likely.

Spondylosis

Definition/overview

- alternatively 'spondylosis deformans'; 'ventral spondylosis'; 'vertebral spondylosis'.
- new bone that is produced on the ventral and ventrolateral surface of thoracic or lumbar vertebrae.
- intervertebral discs are not affected.

Aetiology/pathophysiology

- bone formation at the sites of soft tissue attachments on the ventral surface of the vertebrae may represent a form of entheseophytosis, due to wear and tear:
  - occurs in regions occupied by the ventral longitudinal ligament and the outer annular fibres of the intervertebral disc, ventral to the disc spaces.
  - may extend to form a solid plate of bone on the ventral surface of the thoracolumbar vertebral column.
  - lesions are most common in the region T9 to T15:
    - usually multiple sites affected, especially in severe cases.
- tends to be seen in older horses.
- asymptomatic in the majority of horses:
  - occasionally, proliferative bone can fracture, leading to acute onset of clinical signs.

Clinical presentation

- rare clinical condition (3.4% of 670 cases with back pain in one USA report).
- frequently encountered as an incidental finding:
  - dorsal thorax radiographed for other reasons.
  - post-mortem studies of back pathology.
- clinical signs are those seen in other chronic back pain conditions and may include:
  - back stiffness and poor hindlimb impulsion.

○ poor jumping ability, reluctance to flex and extend the thoracolumbar vertebral column.
- 60% of cases had concurrent osseous back pathology (complicates presentation).

Differential diagnosis

- other osseous or soft tissue pathology of the thoracolumbar vertebral column.

Diagnosis

- not possible to distinguish spondylosis from other causes of chronic thoracolumbar pain on clinical examination.
- **Radiography** will confirm the presence of new bone on the ventral aspect of the thoracic vertebrae (Fig. 6.74).

○ caudal thoracic (>T16) and lumbar spondylosis is more difficult to image due to superimposed abdominal contents.
- **Scintigraphy** can demonstrate increased bone activity in the area (Fig. 6.75).
- new bone may be palpable per rectum from L4 caudally.

Management

- no specific treatment is available.
- recommendations are limited to long-term rest and the administration of systemic anti-inflammatory medication if warranted by the signs of pain.
- when back pain is present, the prognosis is usually poor for return to athletic use.

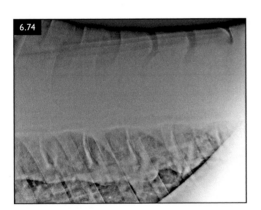 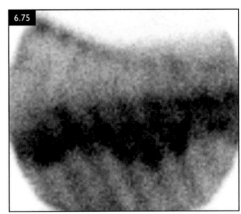

FIGS. 6.74, 6.75 (6.74) Lateral radiograph of the caudal thoracic area of a horse with ventral spondylosis. (6.75) Lateral gamma scintigraphy image of the caudal thoracic area of the same horse showing diffuse uptake along the ventral borders of the vertebral bodies. (Photographs courtesy of Richard Parker)

The Axial Skeleton – Pelvis

Sacroiliac joint and ligament injuries

Definition/overview

- sacroiliac (SI) joints are configured anatomically to transmit forces from the hindlimbs to the vertebral column, via the pelvis:
  - diarthrodial joints with a low range of gliding movement, flexion and extension.
  - joint surfaces undergo significant remodelling in response to changes in body weight.
    - ◆ joint can vary in size and shape with age.
  - joint is associated with:
    - ◆ several muscle groups:
      - – middle and accessory gluteal, internal obturator and iliacus.
    - ◆ sciatic nerve and cranial gluteal nerve, artery and vein:
      - – run ventromedial to the joint through the greater sciatic foramen.
  - joint is strengthened by a fibrous joint capsule and well-developed dorsal, ventral and interosseous ligaments.
- pain from SI joint or region is a common diagnosis in horses with hindlimb lameness:
  - can be difficult to confirm and is frequently a diagnosis of exclusion.

Aetiology/pathophysiology

- true incidence of clinically significant SI pathology is unknown.
- osteoarthritis of the SI joint is common and degenerative changes in the joints have been found in a high proportion of horses in post-mortem studies.
- most common soft tissue injury of the SI joint is desmitis of the SI ligaments:

- some reports suggest may lead to joint instability and subluxation.
- contribute towards the development of painful osteoarthritic changes.
- dorsal portion of the dorsal ligament is the most affected.
- acute damage to soft tissues in the SI region may be caused by trauma (falls).
- commonly an insidious onset.
- cause/pathophysiology is unknown.

Clinical presentation

- horses used for show jumping and dressage are considered most at risk.
- affected horses are significantly older, taller and of greater body weight than the general population.
- reported clinical signs include in order of prevalence:
  - canter quality worse than trot when ridden.
  - poor contact with bit.
  - trunk stiffness during exercise.
  - poor hindlimb impulsion.
  - restricted flexibility of the thoracolumbar region.
  - increased tension in the longissimus dorsi muscles.
  - bucking or kicking out with a hindlimb during ridden canter.
- **many of these clinical signs were more apparent when horses were ridden.**
  - 14% of horses diagnosed with SI pain in isolation **displayed a unilateral hindlimb lameness** when ridden.
- affected horses may have poor or asymmetric muscling.
- asymmetric tubera sacrale are frequently encountered as an incidental finding:
  - 95% of tubera sacrale grossly symmetrical in horses with SI pain in recent study (Fig. 6.76).
- small proportion of horses with SI pain show an exaggerated response to the

DOI: 10.1201/9781003369226-8

FIG. 6.76 Eight-year-old Thoroughbred gelding with right gluteal atrophy and asymmetric tuber sacrale due to a chronic right hindlimb lameness localised to the right stifle.

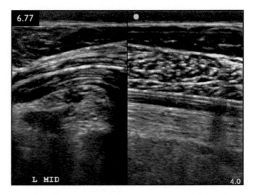

FIG. 6.77 Transcutaneous ultrasound image of a lesion in the short dorsal sacroiliac ligament: transverse view (left) and longitudinal view (right). (Photo courtesy Diane Isbell)

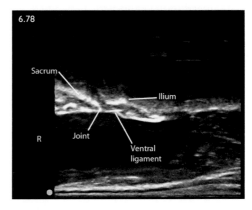

FIG. 6.78 Endorectal reference ultrasound image of a normal sacroiliac joint. (Photo courtesy Diane Isbell)

application of pressure over the tuber sacrale:
- ○ most cases are chronic, and pain is unlikely to be seen on deep palpation.

Differential diagnoses

- hindlimb lameness
- lumbosacral or thoracolumbar pain.
- stress fractures of the ilial wing adjacent to the SI joint.

Diagnosis

- **Ultrasonography** both transcutaneously and rectally (Figs. 6.77–6.79).
  - ○ useful for diagnosis but findings should be interpreted in combination with other diagnostic modalities.

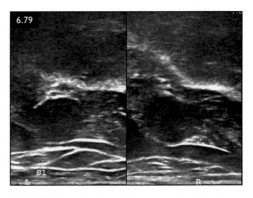

FIG. 6.79 The ultrasound image on the left shows an abnormal ventral sacroiliac joint with significant periarticular bone remodelling. The image on the right is a normal joint. (Photos courtesy Diane Isbell)

- transcutaneously from a dorsoventral direction, on either side of the sacral spinous processes.
  - ◆ dorsal part of the dorsal SI ligament.
  - ◆ bony margins of the sacrum and ilium.
- transrectal approach.
  - ◆ ventral ligament
  - ◆ angle and width of the joint space.
- dorsal ligament appears to be most affected, showing alterations in size, loss of normal echogenicity and parallel fibre pattern.
  - ◆ calcifications in the ligament may indicate a chronic injury.
- **Scintigraphy** of the SI region:
  - wide variation between reports of the association between SI joint pathology and abnormal radiopharmaceutical uptake.
  - in recent study, only 43% of horses that showed a positive response to infiltration of local anaesthesia in the SI region had abnormal radiopharmaceutical uptake on nuclear scintigraphy (Fig. 6.80).
- **Radiography** of the SI region is difficult, requires general anaesthesia and is seldom performed.
- infrared thermographic imaging has been used but its use in SI disease is controversial.
- direct intra-articular injection of the SI joint is not possible in the living horse:
  - periarticular analgesia with a local anaesthetic agent to confirm a diagnosis of pain in this region has been described.
    - ◆ medial approach is favoured by most clinicians and can be used to medicate this area with anti-inflammatories.
    - ◆ recommended that no more than 8 ml of solution containing local anaesthetic is injected per SI region.
    - ◆ diffusion of local anaesthetic into large motor nerves (sciatic and obturator nerves) could produce hindlimb collapse and is a known but uncommon complication.
    - ◆ ultrasound guidance (Fig. 6.81) may be used to improve accuracy and help in preventing inadvertent penetration of vital neurovascular structures in the caudal region.
    - ◆ **caution should always be used when injecting into the SI region.**

Management

- no well-founded reports exist on what treatments are most effective.
- altered work programme focusing on building core strength:
  - poor muscling may make the problem worse and therefore complete (stall) rest is contraindicated.
  - initially, the horse should be worked without a rider, then graduating to

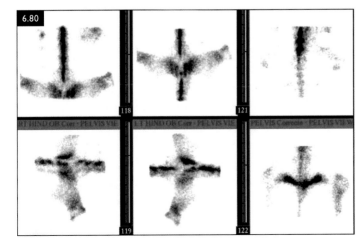

FIG. 6.80 Nuclear scintigraphy of a 10-year-old Belgian Warmblood mare with a 1-month history of stiffness in the trot and resistance to jumping. The images show increased radiopharmaceutical uptake in the right sacroiliac region.

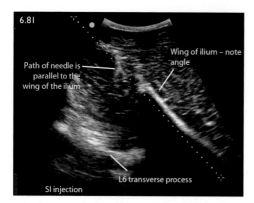

FIG. 6.81 Cranial approach to sacroiliac region. (Photo courtesy Diane Isbell)

work under saddle as the horse's strength and comfort increases.
- Equiband® (Equicore Concepts) can be very beneficial to help engage the abdominal musculature and the hind end.
- critical to address and treat other concomitant sources of pain.
- local and systemic medication:
  - systemic NSAIDs are important in the treatment of SI joint OA.
    - dose tapering as the horse's comfort level improves.
  - periarticular injection of the SI region using both corticosteroids and sarapin, alone or in combination, have been described.
- acupuncture, chiropractic, mesotherapy, physiotherapy and shock-wave therapy:
  - commonly used to manage and treat horses with pelvic and back pain.
  - clinical studies on the efficacy of these treatments specifically for SI pain are lacking, although anecdotal information abounds.
- no specific recommendations for the treatment and rehabilitation of horses with dorsal SI ligament desmitis are available:
  - acute traumatic ligamentous injuries:
    - initial period of complete stall rest to allow healing prior to a gradual reintroduction to work.
    - intralesional injection with platelet-rich plasma and stem cells has been performed with equivocal results.

- chronic injuries.
  - shock-wave and cold laser therapy may also be performed.

Prognosis
- limited literature is available regarding the response to treatment in horses with suspected or confirmed SI pain.
  - prognosis for return to previous athletic activity is fair if adequate rehabilitation is instituted and underlying/concomitant lameness is treated.

Lumbosacral region problems

Definition/overview
- pathology in the lumbosacral region is a source of back pain and poor performance in horses of all disciplines.
- difficult to differentiate from other sources of pain in the thoracolumbar and SI regions.
- lumbosacral joint is the most mobile intervertebral joint of the back:
  - more susceptible to injury.
  - joint is composed of five separate joints.
    - intervertebral disc space between the last lumbar and the first sacral vertebrae.
    - medial and lateral synovial joints between the articular processes.
    - medial and lateral intertransverse joints between the transverse processes of the last lumbar vertebra and the sacrum.
- variation in anatomy of the lumbosacral region has been described including:
  - changes in traditional vertebral formula (number of lumbar and sacral vertebrae).
  - bony fusion of the lumbar vertebra to the sacrum (sacralisation).
  - changes in angulation of the vertebral joints.
  - presence of transitional vertebrae.

Clinical presentation
- **not possible to differentiate lumbosacral pain from other sources of pain in the caudal lumbar, lumbosacral or SI region on clinical signs.**

- local analgesia of the lumbosacral joint is not possible.
- local analgesia of the SI joint can potentially decrease pain in this region and cause false-positive responses.

Diagnosis

- diagnosing the exact problem can be challenging:
  - no definitive clinical signs.
  - radiography, ultrasonography and scintigraphy of the region all have limitations.
    - ◆ superimposition of osseous and soft tissue structures and large size of animal.
      - radiography in adult horses is limited to visualisation of the DSPs.
    - ◆ mild scintigraphic, radiographic and ultrasonographic changes may be present in both painful and asymptomatic horses.
    - ◆ **transrectal ultrasonography** can be used to image the lumbosacral joint from the ventral aspect (Figs. 6.82, 6.83).
      - lumbosacral joint and disc abnormalities have been reported.
      - not been directly correlated with clinical signs.
- clinical examination findings and elimination of other possible differential diagnoses must therefore be the basis of any evaluation of these cases.
- MRI and CT would be the ideal imaging modality for this region but at present it is not available in adult horses.

Management

- no well-founded reports exist on which to base exact recommendations.
- altered work focusing on building core strength:
  - poor muscling may make the problem worse.
  - complete (stall) rest is contraindicated.
- local and systemic medication.
- acupuncture, chiropractic, mesotherapy and shock-wave therapy.
- it is critical to address and treat other concomitant sources of pain that may exist.

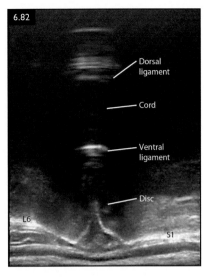

FIG. 6.82 Endorectal ultrasound image of a normal lumbosacral joint space showing the lumbosacral disc. (Photo courtesy Diane Isbell)

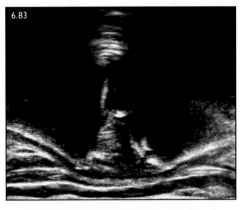

FIG. 6.83 Endorectal examination of a lumbosacral disc with calcification. (Photo courtesy Diane Isbell)

Fractures of the pelvis

Definition/overview

- any of the bony components of the pelvis can fracture:
  - ilial wing, tuber coxae, ilial shaft, pubis, obturator foramen, acetabulum, sacrum and ischium.
- pelvic stress fractures, mainly at the wing of the ilium, are known to be a common cause of hindlimb lameness in the racing Thoroughbred.

Aetiology/pathophysiology

- complete fractures of the pelvis occur as a result of external trauma, falls or as an end stage of stress fractures in horses in training or racing:
  - stress fractures occur as a result of repetitive, high intensity loading of the skeleton, leading to damage accumulation and bone fatigue.
  - pelvic stress fractures in racehorses increase with increasing distance cantered, and both track surface and trainer affect their incidence.
- wing of the ilium most commonly fractured in horses older than 6 years.
- other types of fracture occur at similar rates in younger animals.

Clinical presentation

- usually presented shortly after the time of the suspected incident.
- variable degree of lameness, depending on the location of the fracture and duration.
  - fractures of ilial shaft or acetabulum commonly non-weight-bearing lameness.
  - other types of pelvic fracture vary but can be very lame.
  - commonly evident at the walk.
  - fractures of the sacrum or pubis commonly result in bilateral hindlimb lameness.
  - fractures of the acetabulum have an extremely short limb protraction at the walk.
- pelvic asymmetry is common with fractures of the tuber coxae and complete fractures of the ilial wing or shaft.
- fractures of the tuber coxae frequently present with unilateral lameness and painful swelling over the affected bone, with or without palpable crepitus.
- fractures of the tuber ischia usually show asymmetrical positioning and swelling (Fig. 6.84).
- other clinical signs may include muscle spasm, subcutaneous haematoma and, occasionally, penetration of the skin by the sharp edges of the fractured bone.
  - fatal haemorrhage if a major artery is severed by the sharp edges of displaced bones, particularly in ilial shaft fractures.
- complete fractures of the ilial wing in a racehorse may be present on one side only.
  - signs of subclinical stress fractures in the same site on the contralateral limb are commonly reported (Fig. 6.85).
- mares with fractures of the pubis or ischium may have vaginal or vulvar swelling from oedema and haemorrhage.

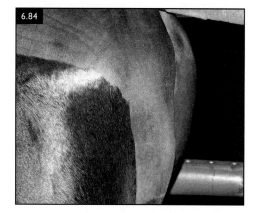

FIG. 6.84 View from the left side of a chronic left ischial fracture showing asymmetry of outline of the left and right tuber ischia. The pull of the thigh muscles displaces the fracture distally.

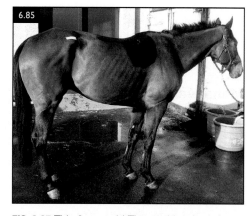

FIG. 6.85 This 6-year-old Thoroughbred racehorse has bilateral ilial wing stress fractures. Note the way the horse is standing with the hindlimbs tucked underneath the body and the pelvis appearing to be more vertical in orientation than normal. There was palpable pain over both ilial wings.

Differential diagnosis

- fracture of the femur
- separation of the femoral head
- coxofemoral luxation.
- exertional rhabdomyolysis.

Diagnosis

- based on history and clinical signs.
- findings on external and rectal palpation.
- imaging.

Imaging

Radiography:

- easily performed in a foal.
- most fractures only imaged in the adult horse under general anaesthesia (Fig. 6.86).
  - risk of further damage or displacement during recovery and may be contraindicated.
  - technique has been described for obtaining images of the caudal ilial shaft, femoral head and greater trochanter, acetabulum and coxofemoral joint in the standing sedated horse (Fig. 6.87).
    - full extent of ischial and pubic involvement cannot be reliably determined using this technique.
    - unlikely to identify iliac wing fractures.
- **Scintigraphy** is very useful for identification of pelvic fractures:
  - increased radiopharmaceutical uptake in region of the fracture in 64% of horses.

- positive correlation between the age of the fracture and the degree of radiopharmaceutical uptake:
  - may fail to identify more recent fractures (Figs. 6.88, Fig 6.90).
- **Ultrasonographic** examination is easy, quick and relatively inexpensive compared with other imaging modalities:
  - useful for fractures of the tuber coxae (Fig. 6.89), iliac wing (Fig. 6.91) and shaft, acetabulum (Fig. 6.92), sacrum and tuber ischii.
  - minimally displaced or incomplete fractures, as well as poor callus formation, may not be identified with ultrasonography.
- CT has been reported in a number of cases with pelvic fractures:
  - may provide additional information regarding the nature of the fracture.
  - may only be feasible in small horses or foals (Fig. 6.93).

Management

- surgical fixation of pelvic fractures in the adult horse is difficult and rarely performed:
  - removal of bony fragments that have formed a sequestrum may be indicated (Fig. 6.94).
- only treatment option is stall confinement for 3–6 months depending on the fracture type and displacement (Fig. 6.95).

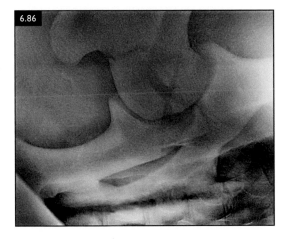

FIG. 6.86 Radiograph of the coxofemoral joint and acetabulum taken under general anaesthesia. Note the comminuted fracture of the acetabulum with a number of displaced bone fragments. (Photo courtesy Henk van der Veen)

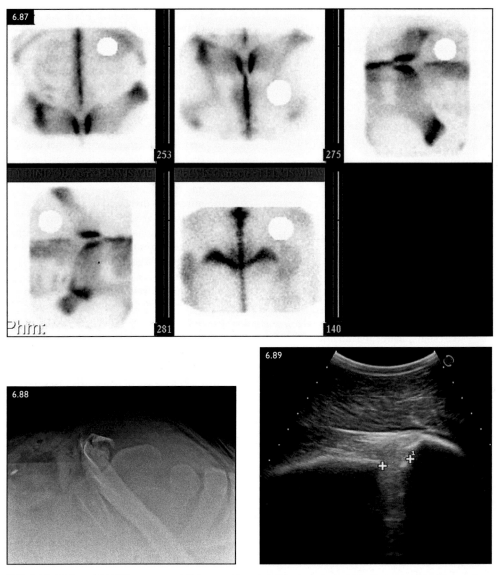

FIGS 6.87, 6.88 and **6.89** Three-year-old Thoroughbred stallion in race training that presented for evaluation of an acute-onset left hindlimb lameness. (6.88) Nuclear scintigraphy was performed and revealed a displaced transverse fracture of left ilium at the level of the tuber coxae. This was confirmed on radiography (6.87) and ultrasound (6.89) (+ marks the fracture gap seen on ultrasound within the left ilium).

Prognosis

- outcome depends on the site and initial displacement of the fracture and the extent of further distraction of the fragments by the subsequent muscle contracture:
  - horses with displaced fractures had fewer race starts than those with non-displaced fractures, but there was no significant difference in race earnings.
  - horses that do not present acutely with severe pain necessitating euthanasia on humane grounds have a good prognosis for return to racing.

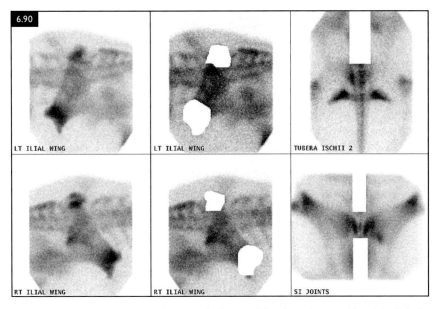

FIG. 6.90 Bone scintigraphy results for a 5-year-old Thoroughbred racehorse with a right ilial wing stress fracture. Note the focal radiopharmaceutical uptake on the caudal aspect of the right ilial wing.

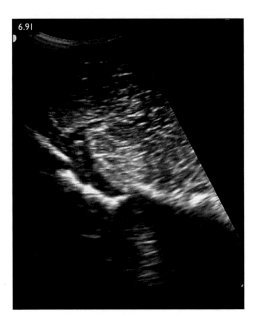

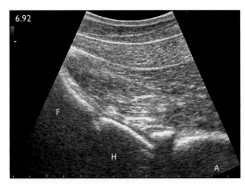

FIG. 6.91 Ultrasound image of a left ilial wing fracture in a 3-year-old Thoroughbred racehorse. (Photo courtesy Diane Isbell)

FIG. 6.92 Ultrasound image of the right hindlimb coxofemoral joint showing a displaced bone fragment off the dorsal margin of the acetabular rim. F = femoral neck; H = femoral head; A = acetabular rim. (Photo courtesy Henk van der Veen)

- ○ other studies have found that 60–93% of horses with pelvic fractures resumed racing at a median of 210 days post injury:
  - ♦ median duration until subsequent retirement was just 10 months.
- horses with partial fractures of the tuber coxae return to function significantly sooner than those with complete fractures (3 months versus 6.5 months).
- horses with fractures of the tuber coxae and ilial wings have a good prognosis for athletic activity:
  - ○ ilial shaft involvement significantly worsens the prognosis.

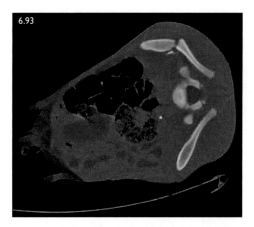

FIG. 6.93 CT scan of a 7-day-old foal found in a ravine with 5/5 left hindlimb lameness. Radiographs revealed a fracture of tuber coxae but it was unclear if the ilial body was involved. CT confirmed a closed, comminuted, displaced tuber coxae fracture without ilial body involvement.

FIG. 6.95 Conservative management of a 3-year-old Thoroughbred gelding with a pelvic fracture. The Anderson Sling is used to prevent the horse from lying down, averting further fracture displacement.

FIG. 6.94 Chronic tuber coxae fracture with sequestration. The sequestered piece was surgically removed, and the draining tract debrided. The horse recovered uneventfully.

- fractures involving the acetabulum carry the worst prognosis for returning to athletic activity due to the development of coxofemoral OA.

Sacral and coccygeal injuries

Definition/overview
- injuries of the sacrum and tail occur rarely in the horse.

Aetiology/pathophysiology
- tail injuries are frequently caused by:
  - inappropriate bandaging (tight tail bandage left in place for 2–3 days will result in ischaemia of the distal tail).
  - trauma.

Clinical presentation
- cranial sacral fractures:
  - lameness/hindlimb weakness.
- caudal sacral and cranial coccygeal vertebral (Co1–Co3) fractures:
  - neuropraxia resulting in tail motor weakness and faecal and urine (mares) staining.
- some injuries may go unnoticed until swelling of the tail or excessive faecal contamination becomes apparent.

- more severe neural injury may result in failure of defaecation or micturition.
- sacrococcygeal fractures or luxations will result in local swelling, abnormal movement, anatomical distortion (Fig. 6.96) and pain on palpation and/or manipulation.
- hindlimb lameness, which may be worse under saddle, particularly if the caudal sacrum is involved.
- ischaemic damage is characterised by necrosis (blackening) of the skin (Fig. 6.97).
- sacral trauma can also cause tail weakness (Fig. 6.98).

Differential diagnosis

- other conditions that can cause similar clinical signs including:
  - Polyneuritis equi
  - Equine protozoal myelitis.
  - Equine herpesvirus myeloencephalopathy.

Diagnosis

- **Radiography** of the sacrum and tail will confirm sacrococcygeal fractures, luxations and osteomyelitis (Figs. 6.98, 6.99).

FIG. 6.97 A tail that was bandaged too tightly and sloughed off the distal end. In addition, it caused varying thickness pressure rubs and ulcerations, which are now healing.

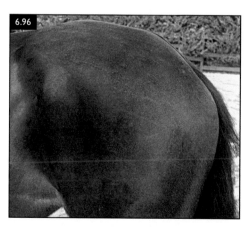

FIG. 6.96 This horse was hit from behind by a vehicle and sustained a fracture of the sacrum. The fracture has healed but left an obvious angulation in the line of the sacrum halfway between the tail head and the tuber sacrale.

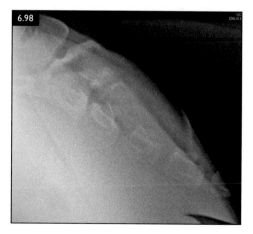

FIG. 6.98 Radiograph of a 10-year-old Friesian gelding with abnormal tail carriage and sensitivity to palpation of the tail head. No lameness was appreciated at the walk. There is a complete, displaced, comminuted fracture of the caudal sacral vertebra. This fracture healed with conservative management.

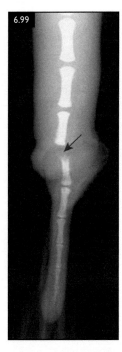

FIG. 6.99 Radiograph of the caudal coccygeal vertebrae of a pony 1 month after a tail bandage was inadvertently left in place for 2 days. Note the subluxation of one intervertebral joint, with lysis of one vertebral endplate (arrow). The distal third of the tail was subsequently amputated and the animal recovered uneventfully.

- **Scintigraphy** may reveal increased radiopharmaceutical uptake on both the soft tissue and bone phase in the region of traumatic injury.
- infiltration of local anaesthetic solution in the region of injury can establish the significance of coccygeal injury, particularly in relation to clinical signs such as hindlimb lameness.

Management

- conservative management options for horses with sacrococcygeal fractures range from reduced exercise alone to epidural injections of corticosteroids.
- fractured or luxated coccygeal vertebrae:
  - can be stabilised with an intramedullary pin if cosmesis is important.
  - surgical amputation will achieve swift pain relief.
- traumatic partial amputation of the tail:
  - removal of the most caudal remaining coccygeal vertebra to allow wound closure.
  - systemic antibiotic and anti-inflammatory therapy.
- extent of ischaemic damage following bandaging injury takes time to become evident:
  - if no signs of healing after 2–3 weeks, amputation is indicated.
- severe displaced sacral fractures warrant euthanasia.

Muscle Disorders of the Horse

Clinical History and Presentation

- clinical history should establish the following facts:
  - signalment of the animal.
  - history of lameness.
  - previous episodes of muscle disease.
  - history of known muscle disease in related animals.
  - discipline and level of work the horse is currently undertaking:
    - has this changed?
    - what is the performance level?
  - management including diet, turnout, training, recent competition etc.
  - changes in body condition, muscle mass and symmetry.
- full physical and lameness examination should include:
  - observation of horse at rest and at exercise on straight line and ridden:
    - observe for gait abnormalities such as stiffness, weakness or lameness.
  - vital parameters.
  - visual examination, palpation and percussion of major muscle groups.
  - flexibility of head, neck, backs and limbs.

Diagnostic techniques

Muscle enzymes

- serum muscle enzyme activity is often measured to screen for muscle disease:
  - usually includes Creatine kinase (CK) and Aspartate aminotransferase (AST).
- CK rises quickly following injury, peaking at 4–6 hours:
  - returns to baseline rapidly with half-life of approximately 12 hours.
  - fairly specific for muscle damage.
- AST rises more slowly, peaking at around 24 hours:

- declines over days to weeks.
- not specific for muscle damage.
  - evaluate liver enzymes to rule out liver disease.
- **Exercise test.**
  - measure muscle enzymes after 10–15 minutes of submaximal exercise:
    - increases sensitivity of detecting an exercise-related myopathy.
    - baseline blood sample collected prior to exercise.
    - second sample 4–6 hours later.

Urinalysis

- extensive muscle damage may lead to myoglobinuria (dark brown urine) (Fig. 7.1).
  - pigmenturia also caused by haemolysis and haematuria.
  - urinalysis reagent strips are unable to differentiate between different causes:
    - high serum muscle enzymes + pigmenturia highly suggestive of myoglobinuria.
- Myoglobin is highly nephrotoxic:
  - full evaluation of renal function is essential.

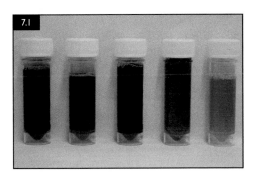

FIG. 7.1 Sequential urine samples from a horse with rhabdomyolysis, showing urine discolouration due to the presence of myoglobin.

DOI: 10.1201/9781003369226-9

Electromyography (EMG)

- used to quantify electrical activity within muscle:
  - specialised technique used to separate neurological and primary myopathic diseases.

Ultrasonography

- used to evaluate:
  - structure of muscle
  - changes in overall muscle volume.
  - focal muscle injuries such as muscle tears and strains.
- overlying skin is usually clipped and cleaned.
- comparing to the contralateral muscle group can be useful (Fig. 7.2).
  - important to ensure horse is equally weight bearing.

Muscle biopsy

- histopathology of biopsy samples provides information on underlying disease process:
  - allows a specific diagnosis and prognosis to be given.

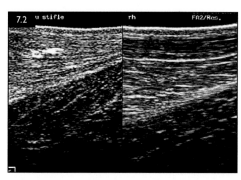

FIG. 7.2 Ultrasonographic images from the left and right hindlimbs of the caudal stifle region muscles. Note the left hindlimb image (left) has an irregular hyperechoic scar within a generally more hyperechoic muscle belly compared to the right hindlimb (right). This is a healed muscle injury in the left hindlimb.

- open incisional biopsies are usually collected from:
  - semimembranosis or gluteal muscles in metabolic muscle disease.
  - focally if specific muscles are affected.

Muscle biopsy procedure:

1. usually sedated prior to biopsy collection.
2. area clipped and aseptically prepared.
3. 5–10 ml of local anaesthetic (mepivicaine) infiltrated subcutaneously along a vertical line at the biopsy site.
4. incise through the skin and subcutaneous tissue down to the surface of the muscle:
   - retractors helpful to improve visualisation.
5. two parallel vertical incisions 1 cm apart and 2 cm long into the underlying muscle.
6. area under the muscle, between the incisions, is undermined using haemostats.
7. top and bottom of the elevated piece of muscle is then sharply incised whilst holding a corner securely with haemostats.
8. specimen should be divided into two equal portions:
   - one placed in formal saline.
   - other into an empty sterile pot and placed on ice packs.
9. subcutaneous tissue closed using continuous suture pattern (absorbable suture material).
10. staples placed in skin.
11. biopsy transported by same day/next day courier service to an experienced laboratory.
12. horses are rested until staple removal at 10–14 days.

- less invasive biopsy techniques are available but yield smaller muscle samples:
  - minimise the need for rest following procedure.

MUSCLE DISORDERS

Exercise induced myopathies

Exertional rhabdomyolysis

Definition/overview

- syndrome of muscle pain that occurs during or following exercise:
  - caused by several different conditions that each affect skeletal muscle function in different ways and result in a common clinical presentation.
- isolated bouts of tying-up, thought to be trigged by acquired environmental factors such as overexertion (Table 7.1).
- recurrent bouts of rhabdomyolysis due to an underlying susceptibility caused by specific diseases, and possibly triggered by environmental factors:
  - some of these diseases have a proven or likely genetic basis to their aetiology.

TABLE 7.1: Causes of exertional rhabdomyolysis in the horse.

ACQUIRED	GENETIC
Overexertion, inadequate training	Recurrent exertional rhabdomyolysis
Electrolyte abnormalities	Type 1 Polysaccharide storage myopathy
Antioxidant deficiencies, e.g. selenium and vitamin E	Type 2 Polysaccharide storage myopathy
Hormones	Myofibrillar myopathy
Infectious agents esp. post viral infections	
Dietary imbalances such as excess energy (grain)	

Recurrent Exertional rhabdomyolysis (RER)

Aetiology/Pathophysiology

- usually Thoroughbred and Standardbred horses, mainly when they are in training:
  - approximately 5–6 % of both breeds.
  - possibly more common in nervous animals and fillies.

- abnormal release of calcium within muscle cells, resulting in abnormal muscle contracture and relaxation.
- pattern of inheritance suggests genetic condition, transmitted as an autosomal dominant trait (not proven).
- recurrent episodes triggered by factors such as:
  - stress
  - variations in exercise intensity or duration
  - diet.

Clinical presentation

- muscle pain and stiffness, during or immediately following exercise.
  - varies from mild gait abnormalities, reluctance to move or, rarely, recumbency.
  - some cases can be particularly distressed.
- severe cases may be dehydrated with congested mucous membranes:
  - tachycardia/tachypnoea
  - colic-like signs.
  - myoglobinuria, and hyperthermia.
- moving the horse may worsen the clinical signs.
- muscles often firm on palpation, with sweating and muscle fasciculations:
  - hindlimb, pelvic and epaxial muscles are most affected.
- subtle signs may include poor performance and lack of power/collection.

Differential diagnosis

- other acquired or genetic causes of exertional rhabdomyolysis (see Table 7.1).
- other causes of sudden-onset lameness including laminitis.
- colic • severe neurological disorders.

Diagnosis

- diagnosis on basis of clinical history and elevated muscle enzymes in a Thoroughbred, Standardbred or related horse.
- histopathology of muscle biopsy samples is non-specific:

- evidence of previous muscle damage without abnormalities associated with PSSM on histopathology is supportive.
- no specific genetic mutation has been identified.
- severe cases, assess blood samples for hydration status, electrolyte levels and renal function, and urine samples for myoglobinuria.

Treatment

- same, regardless of the underlying cause:
  - correct dehydration and maintain diuresis.
  - prevent further muscle damage and provide analgesia.
- fluid therapy – with pigmenturia or marked hypovolaemia:
  - intravenous crystalloids fluids such as compound sodium lactate.
  - less severe cases, where myoglobinuria is absent, use oral fluids.
- NSAIDs such as:
  - meloxicam, flunixin or phenylbutazone are useful analgesics.
  - use with care in hypovolaemic patient, to avoid renal injury.
- Acepromazine (0.04–0.07 mg/kg) and Xylazine (0.4–0.8 mg/kg) beneficial in maintaining muscle perfusion and reducing anxiety in a stressed patient:
  - used with care in dehydrated patients.
- antioxidants such as vitamin E (4000–6000 IU/day p/o):
  - may help scavenge increased free radical production.
- rest horse in a quiet stress-free box to prevent further muscle damage:
  - gentle walking may be beneficial within 1–2 days.
  - when enzyme levels have returned to normal:
    - gradual increase in in-hand walking with small paddock turnout.
    - followed by return to ridden exercise.
- serum muscle enzymes useful in guiding ongoing exercise regimes:
  - sample taken at 24–48 hours after exercise will detect peak AST activity.
  - failure of enzyme levels to fall over time suggests ongoing muscle damage.

Management

- feed a diet low in starch, whilst providing necessary calories in the form of fat.
  - provide high quality forage (1.5–2% body weight) and a balanced vitamin and mineral supplement.
- regular daily exercise with no rest days is essential:
  - avoid any sudden changes in exercise patterns.
  - provide adequate warm-up and cool-down periods.
  - daily turnout is recommended.
- stress and changes in routine can trigger clinical episodes and should be minimised.
- Dantrolene sodium (Dantrium™) prevents the release of calcium from the sarcoplasmic reticulum within muscle:
  - reduces muscle enzyme activity.
  - reduces clinical signs of rhabdomyolysis in Thoroughbreds in training.
  - orally 1–2 hours prior to exercise.
  - not licensed for use in horses in many countries.
    - detection time of 3 days for 500 mg dose given daily for 3 days.

Prognosis

- determined by severity and frequency of clinical bouts, and response to treatment.
- many horses, with appropriate management changes, can fulfil a useful athletic career.
- chronic recurring cases may develop muscle fibrosis or atrophy.
- very severe cases that are recumbent may die or be euthanased.

Polysaccharide Storage Myopathy (PSSM type 1 and 2)

Aetiology/pathophysiology

- first described in Quarter Horse and Appaloosa related breeds:
  - now identified in a variety of different breeds across Europe and North America.
  - high prevalence in continental draught breeds (Table 7.2).

- accumulation of glycogen and abnormal polysaccharide inclusions within muscle fibres and reduction in cellular energy availability.
- autosomal dominant mutation in the glycogen synthase 1 (*GYS1*) gene has been identified as a cause in approximately two-thirds of horses with the disease:
  - PSSM1 for horses with the *GYS1* mutation.
  - PSSM2 for horses that lack this gene mutation.
    - ◆ abnormal polysaccharide within their skeletal muscle from disruption of the myofibrillar proteins rather than a primary glycogen storage disease.

TABLE 7.2 Breeds in which the GYS1 mutation has been identified to date (PSSM1).

PSSM 1	
• Quarter Horses	• Suffolk Punch
• Appaloosa	• Hanoverian
• Paint	• Warmblood
• Belgian Draught horse	• Morgan
• Percheron	• Exmoor pony
• Haflinger	• Cob
• Shire	• Connemara cross
• Polo ponies	• Arab cross

Clinical presentation

- varies from vague signs of poor performance, back pain, loss of muscle, and hindlimb weakness to acute rhabdomyolysis.
- PSSM1 are more likely to present with rhabdomyolysis.
- PSSM 2 are more likely to present with subtle signs of gait abnormality such as an undiagnosed lameness or weakness.

Differential diagnosis

- other causes of exertional rhabdomyolysis listed in Table 7.1.
  - type 1 PSSM has not been identified in a purebred Thoroughbred horse.
- subtle signs of poor performance:
  - other common causes such as orthopaedic conditions, upper and lower respiratory tract disease and cardiac abnormalities should be considered.

Diagnosis

- serum muscle enzymes may be persistently elevated, even at rest, in many cases:
  - 3–4 times rise from normal after exercise test in PSSM1 cases is common.
  - some draught breeds may only reveal mild elevations following exercise.
- genetic test for the GYS1 mutation is available for PSSM1:
  - DNA extracted from blood collected into EDTA, or hair roots.
- PSSM1 can also be diagnosed from a muscle biopsy in animals over 2 years old.
- PSSM2 can only be diagnosed from a muscle biopsy.

Treatment

- cases presenting with acute rhabdomyolysis are treated as described previously (See page 254).
- other presentations, long-term management changes essential in resolving clinical signs.

Management

- regular graduated daily exercise programme plus daily pasture turnout is advised.
- diet low in starch and sugar:
  - <10% digestible energy (DE) as non-structural carbohydrates (NSC).
  - relatively high in fat (13–20% DE).
- continue to receive 1–2% of their bodyweight as good quality forage:
  - low (<12%) NSC content.
- decrease weight in overweight animals.
- grazing may need to be restricted at certain times of the year when NSC content of grass is particularly high.
- specifically formulated commercial diets available:
  - Dodson and Horrell ERS Pellets® or Saracen ReLeve®.
- alternatively, a low starch diet may be supplemented with vegetable oil, up to a maximum of 1 ml/kg bodyweight, to provide sufficient calories.
- diets with slightly lower fat content may be more palatable but still sufficient to control the condition.
- high lipid diets may increase the requirement for antioxidants:

- ○ feed balancer containing vitamin E may be beneficial.
- recommendations are based on studies of horses with PSSM1:
  - ○ presumed that similar recommendations apply to horses with PSSM2.
  - ○ further investigation of this is required.

Prognosis

- favourable in cases where dietary and exercise recommendations are both followed:
  - ○ clinical improvement may be observed within 6 weeks.
  - ○ complete adaptation to these diets often takes several months.

Acquired causes of exertional rhabdomyolysis

- several early studies suggested acquired causes of rhabdomyolysis, such as:
  - ○ variations in female sex hormones, hypothyroidism and electrolyte imbalances.
- most of these suggested risk factors are yet to be corroborated, with different studies yielding conflicting results.
- possible that these risk factors may increase the chances of clinical disease in an already genetically susceptible animal.

Less common causes of myopathy in the horse

Mitochondrial myopathies

Aetiology

- several mitochondrial defects in man that lead to rhabdomyolysis with profound exercise intolerance.
- reports of similar pathology in the horse.

Clinical signs

- horses present with marked exercise intolerance.

Differential diagnosis

- other causes of exercise intolerance such as:
  - ○ systemic infection
  - ○ musculoskeletal pain

- ○ cardiac disease.
- ○ lower and upper airway disease.

Diagnosis

- mitochondrial staining patterns in muscle biopsy samples.
- measuring pyruvate and lactate levels following exercise:
  - ○ reduced aerobic metabolism leads to alterations in their ratio.

Treatment

- no specific treatments available.

Prognosis

- generally poor.

Myofibrillar myopathy

Overview

- recently identified disorder in horses presenting with exercise intolerance, abnormal canter or intermittent exertional rhabdomyolysis.
  - ○ recorded in young Warmblood horses and Arabian endurance horses.
- muscle enzymes are usually normal in Warmbloods.
- defined by specific muscle histopathology.
- possible subset of PSSM2 horses.
- no specific recommendations for treatment and management are presently available:
  - ○ use treatments and managements as for PSSM2.

Muscle tears and strains

Definition/overview

- muscle injuries are a common but poorly understood cause of lameness and poor performance.
- may occur in any muscle group within the fore or hindlimbs, and the axial skeleton, including the lumbar and gluteal regions.

Aetiology/pathophysiology

- most common in equine athletes such as:
  - ○ racehorses, showjumpers, eventers, and Western horses.
- causes include direct trauma, excessive or unusual exercise, secondary to other

sources of lameness (especially hindlimb), or a poorly fitting saddle.

Clinical presentation

- varies with degree of damage and position of the muscle.
- acute muscle damage and subsequent haemorrhage/inflammation may lead to:
  - acute onset moderate to severe lameness.
  - palpable swelling/heat/pain in affected muscles if superficial.
  - deeper situated injuries will not be palpable.
- subsequently:
  - muscle may be firm with fibrosis.
  - atrophied with muscle loss or reveal a defect where there is a complete tear.
  - complete muscle tears may present with an abnormal limb position or gait abnormality.
- **Forelimb:**
  - biceps brachii o brachiocephalicus
  - pectoral muscles.
  - superficial digital flexor muscle/tendon junction.
- **Hindlimb:**
  - biceps femoris, semimembranosus and semitendinosus.
  - adductor, gluteal, quadriceps, and gastrocnemius muscles.
- muscle spasm and pain are common in the thoracolumbar region secondary to a hindlimb lameness or primary thoracolumbar or sacroiliac injury.

Diagnosis

- history and clinical examination including:
  - palpation of affected muscles for pain, swelling, and spasm in acute cases.
  - atrophy/fibrosis in chronic cases.
  - visual appraisal is useful.
- full lameness examination is paramount.
- thermography, nuclear scintigraphy, and ultrasonography can help locate the site of injury:
  - ultrasonography can determine the extent of damage (Figs. 7.3, 7.4) and monitor healing.
- mild increases in plasma muscle enzyme activity may be present:

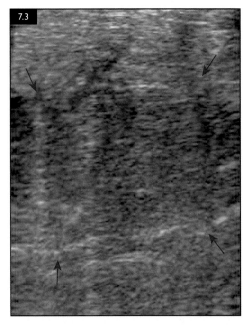

FIG. 7.3 Ultrasonogram of the ascending pectoral muscle on the craniomedial aspect of the forelimb, showing an area of diffuse decrease in echogenicity and loss of normal muscle architecture (arrows). This is indicative of a partial tear and granulation tissue reaction within the torn portion of muscle.

  - depends on severity of muscle damage (often within normal limits in mild tears).
- tears can be classified according to the severity:
  - **First-degree** – mild muscle soreness and low-grade inflammation and swelling with few localising clinical signs.
  - **Second-degree** – partial tear with obvious clinical signs if superficial tear is present.
  - **Third-degree** – complete tear with loss of function.

Management

- type of treatment depends on specific injury and stage at which it is diagnosed.
- acute injuries benefit from cold therapies and NSAIDs to reduce pain and inflammation.
- physiotherapy techniques and machines:
  - laser, therapeutic ultrasound, TENS, and electromagnetic therapy.

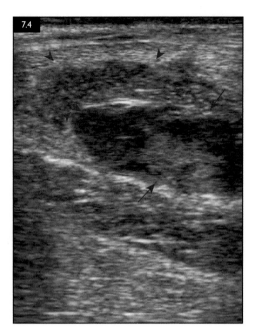

FIG. 7.4 Longitudinal ultrasonogram of the distal part of the triceps muscle in the caudal upper forelimb area of a horse presenting with acute, sudden-onset lameness and pain on flexion of the elbow. There is a large, anechogenic haematoma at the musculotendinous junction (junction between the muscle belly and aponeurosis) (arrows). The distal part of the muscle appears hypoechogenic and heterogeneous because of fibre retraction, cell infiltration, and necrosis (arrowheads). This appearance is typical of a complete muscle tear.

- massage, stretching, and manipulation to improve range of motion.
- box rest (first-degree – 7 days, second-degree – 1–3 weeks, and third-degree – several months).
- controlled graduated exercise programme once the acute stages are resolved.
- treat any primary problem.

Prognosis

- depends on extent, severity, and chronicity of the specific muscle injury.
- varies from good to more guarded:
  - chronic injuries secondary to other problems.
  - subsequent fibrosis or atrophy can lead to permanent mechanical lameness (fibrotic myopathy).

Non-exercise-induced myopathies.

Definition/overview

- number of muscle disorders which affect horses at rest.

Atypical myopathy (AM)

Aetiology/pathophysiology

- severe myopathy with high mortality rate that affects horses at pasture.
- associated with ingestion of toxin hypoglycin A, found in the seeds, leaves and seedlings of trees of the *Acer* genus (maples).
  - United Kingdom, *Acer pseudoplatanus* (Sycamore) is frequently implicated.
  - North America it is *Acer negundo* (Box elder).
- toxin binds to flavin adenine dinucleotide (FAD) thereby inhibiting acetyl-CoA dehydrogenases.
  - blocks energy metabolism, particularly from fat, in type 1 (oxidative) muscle fibres.
  - most prevalent in postural muscles, which are the most severely affected.

Clinical presentation

- usually present following a history of recent turnout.
- signs are often acute with weakness and stiffness and a lowered head carriage.
- horses may appear painful, depressed, trembling and sweating:
  - firm muscles on palpation.
  - may be recumbent or reluctant to move.
- darkened urine is often seen due to myoglobinuria.
- congested mucous membranes and increased respiratory rate and effort.
- rectal examination may reveal a distended bladder.
- some cases may be found dead at pasture.

Differential diagnosis

- colic.
- other causes of weakness and recumbency such as orthopaedic injury or botulism.
- other causes of sudden death.

Diagnosis

- high muscle enzyme activity without history of exercise and above clinical signs.
- *Acer* trees present on the pasture.
- blood samples will reveal evidence of:
  - severe muscle damage
  - hypocalcaemia   ○ hyperglycaemia.
  - variable haematology results.
- analysis of blood acylcarnitines or urinary organic acids can confirm diagnosis:
  - only available at specific laboratories and may take several days.
- muscle biopsy of masseter or intercostal muscles evaluated for abnormal lipid staining.

Treatment

- correct dehydration and other electrolyte abnormalities, provide energy, eliminate the toxin or its metabolites, support mitochondrial function and provide analgesia.
- intravenous crystalloid fluids such as compound sodium lactate are recommended for pigmenturia or marked hypovolaemia.
  - less severe cases and absence of myoglobinuria, oral fluid may suffice.
- glucose can be added to the fluids to provide energy:
  - up to 20 l of a 5% glucose solution can be safely administered to a 500 kg horse:
    - ♦ 500 ml of 50% glucose solution in 5-litre bag of crystalloids.
    - ♦ plasma glucose monitored as frequently become hyperglycaemic.
    - ♦ glucose above 10 mmol/l – insulin therapy may be required:
      - – bolus of 0.4 IU/kg or constant rate infusion of 0.07 IU/kg.
- toxin ingestion within the last 4–6 hours (frequently unknown):
  - laxatives or binding agents may be of value.
- Vitamin E (5000 IU D-α-tocopherol) and selenium (1 mg/kg) as free radical scavengers.
- Riboflavin or vitamin B2 increases FAD availability (multi vitamin preparations).
- NSAIDs such as a meloxicam, flunixin or phenylbutazone are useful analgesics:

  - care in the hypovolaemic patient, to avoid renal injury.
- supportive care is very important:
  - keep warm and, if recumbent, minimise secondary complications such as pressure sores or pneumonia.
  - supporting the head on bales of hay or straw can prevent oedema developing.
  - catheterisation of the bladder may be necessary.

Prognosis

- guarded to poor, with mortality rates of up to 85% reported.
- some horses with an early diagnosis can make a full recovery with appropriate supportive care.

Myotonic disorders

Aetiology/pathophysiology

- abnormal excitability of muscle cell membranes leads to delayed relaxation of muscle following contraction or stimulation.
- most common form is Hyperkalaemic periodic paralysis (HYPP):
  - defect in sodium channel proteins, causing altered sodium/potassium exchange through the skeletal muscle membrane.
  - inherited defect (autosomal dominant) in Quarter Horses descended from the sire 'Impressive'.
  - other breeds such as Appaloosa and American Paint (associated crossbreds) also affected.
- Myotonia congenita is a less common condition:
  - defect in chloride channels, that has been described in New Forest ponies.

Clinical presentation

- episodes of HYPP may be triggered by:
  - fasting, anaesthesia, stress or diets high in potassium.
  - may also occur apparently spontaneously.
  - between episodes horses may appear clinically normal.
- clinical signs vary between individuals and tend to decrease with age.

- affected animals often present with bouts of:
  - sweating, weakness, muscle fasciculations, and prolapse of the third eyelid.
  - may last for minutes to several hours.
- dimpling of the muscles is often detected and may be exacerbated by percussion.
- severe cases:
  - recumbency
  - respiratory distress (laryngeal paralysis)
  - death.

Differential diagnosis

- other causes of severe rhabdomyolysis should be considered.
- renal failure as a cause of hyperkalaemia.
- colic   • tetanus
- viral myeloencephaopathies.

Diagnosis

- history and clinical signs.
- hyperkalaemia occurs in most cases during episodes but not between them.
- EMG has characteristic abnormalities.
- homozygous animals appear to be more severely affected than heterozygous carriers.
- DNA testing can confirm the gene mutation.

Treatment

- mild case – feeding grain or a sugar solution may be helpful.
- other treatments include:
  - epinephrine (3 ml of 1:1000 per 500 kg i/m).
  - acetazolamide (3 mg/kg p/o q12h).
  - intravenous calcium gluconate (0.2–0.4 ml/kg of 23% in 1 l of 5% dextrose) given over 10 minutes helps to decrease muscle hyperexcitability.
  - dextrose (6 ml/kg of 5% solution with/without 1–2 mEq/kg sodium bicarbonate) helps decrease the hyperkalaemia.

Long-term management

- clinical episodes are minimised by:
  - reducing dietary potassium:
    - stop alfalfa, molasses and bran intake.
  - increasing potassium excretion using acetazolamide.
- fasting or sudden changes in diet should be avoided.
- regular graduated exercise is beneficial.

Prognosis

- overall fair with mild episodes often self-resolving.
- severe episodes are emergencies, but death is rare.
- condition can be managed, in most cases, although recurrent bouts may occur.
- **affected horses should not be used for breeding.**

Nutritional myodegeneration ('White muscle disease')

Overview

- rare condition caused by a dietary deficiency of selenium +/- vitamin E, plus unknown factors:
  - cell membrane damage and degeneration in skeletal and cardiac muscles.
- rapidly growing foals (birth to 1 year old), born to dams on selenium deficient pastures (<0.05 ppm), are most frequently affected.
- clinical signs can vary with age but can be split into two forms:
  - **Cardiac form** – sudden death, respiratory difficulty, weakness, recumbency, depression and irregular heart rhythm.
  - **Skeletal form** – slower in onset, difficulty moving, stiffness, stilted gait, muscle fasciculations and weakness.
    - may become recumbent.
    - muscles often firm on palpation.
  - young foals often present with weakness, recumbency and possible dysphagia.
  - older foals often present after periods of brisk exercise with acute or subacute signs.
  - swollen firm muscles may be accompanied by subcutaneous oedema over the rump, neck, and ventral thorax/abdomen.

- adult horses may present as colic, muscle stiffness/soreness, and oedema of the head and neck.
- differential diagnosis includes other causes of heart failure, cerebellar disease, colic, spinal cord compression, tetanus, trauma, meningitis or toxin ingestion and exertional rhabdomyolysis.
- clinical signs, which may include myoglobinuria, are strongly suspicious.
  - increased plasma CK and AST on blood samples.
  - low selenium in whole blood samples.
  - serum vitamin E may also be low.
- post-mortem examination reveals the skeletal muscles as typically pale, with or without white streaks.
- treatment should include:
  - selenium and vitamin E supplementation:
    - rarely effective in acute cases.
  - antibiotics may be required to treat secondary complications such as pneumonia.
  - nursing care to ensure adequate fluid and nutritional intake.
  - systemic NSAIDs to control pain and stiffness.
- daily dietary intake of 10 µg/kg selenium and 2–5 mg/kg vitamin E is recommended for prevention:
  - usually in last trimester pregnant or lactating mares.
  - milder chronic cases for treatment.
- good prognosis in the skeletal form but poor if there is cardiac involvement, particularly in acute cases.

Clostridial myositis

Definition/overview

- acute to per acute condition, characterised by rapid muscle necrosis and associated severe systemic disturbances (toxic shock).

Aetiology/pathophysiology

- Clostridial organisms are ubiquitous in the environment:
  - contamination of an open or puncture wound, necrotic area, or haematoma.
  - such as *C. perfringens*, *C. septicum* and *C. chauvoei*.
- intramuscular or perivascular injections, especially of irritant substances, create a suitable environment for myonecrosis.
- haematogenous contamination may explain abscesses without an apparent wound.
- poorly delineated cellulitis with severe, rapidly developing myonecrosis, abscess formation, and accumulation of gas.
- severe systemic disturbances caused by the release of exotoxins.

Clinical presentation

- rapid onset toxaemia manifesting as depression, fever, tachypnoea, anorexia and lameness.
- progression to tremors, dyspnoea, recumbency and death.
- usually, one affected muscle – hot, swollen and the skin discoloured and thinned.
  - crepitus may be palpable suggesting subcutaneous emphysema.
- previous wounds may have a malodourous discharge.

Differential diagnosis

- colic
- other causes of septic shock
- other causes of abscesses.
- exertional myopathies.

Diagnosis

- clinical signs.
  - exclude other severe toxaemias such as pleuropneumonia, colic or colitis.
- blood samples reflect the severe toxaemia:
  - inflammatory response and mildly increased muscle enzyme activity.
  - hypovolaemia with increased haematocrit and protein concentration.
  - subacute stage, hyperfibrinogenaemia and neutropenia, becoming more obvious.
- ultrasonography will reveal:
  - myonecrosis and presence of exudate and, potentially, of gas (Fig. 7.5).
  - usually very marked oedema in and around the affected muscle(s).
  - ultrasound-guided aspirate/s can be evaluated directly and submitted for anaerobic bacterial culture.

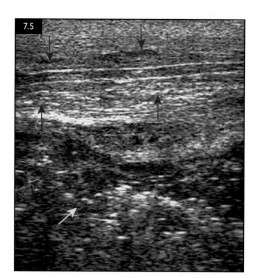

FIG. 7.5 Ultrasonogram of the cranial cervical region of a horse with a clostridial abscess in the neck region. Extensive subcutaneous oedema is shown by dissecting, hypoechogenic lines along the fascial planes (red arrows). Diffuse heterogeneous and hypoechogenic material within the muscle tissue represents exudate from the cellulitis (arrowheads). The exudate contains numerous gas bubbles casting acoustic shadows and comet-tail artefacts (yellow arrow).

Treatment

- aggressive treatment is essential.
    - antibiotic treatment:
        - intravenous penicillin at high doses and very frequently.
        - 44,000 IU/kg i/v every 2–4 hours.
- surgical debridement and fenestration to aerate the tissues is also necessary.
    - daily lavage of the wound with sterile saline to remove debris and improve drainage.
    - wound is left to heal by secondary intention.
- supportive treatment with intravenous fluids and NSAIDs.

Prognosis

- guarded to poor, particularly in acute cases, where the mortality rates are exceedingly high.
- early, aggressive treatment can improve the prognosis and up to a 73% survival rate has been reported.

Immune-mediated myositis

Overview

- inflammatory disease most common in Quarter Horse related breeds:
    - appears to have a hereditary component.
- often a history of respiratory disease or vaccination in the preceding 4–8 weeks:
    - prompts an abnormal autoimmune response to antigens within the host muscle.
- rapid muscle atrophy, usually affecting the gluteal and epaxial muscles:
    - may be associated with muscle stiffness, weakness and depression.
- haematology within normal limits, but muscle enzymes are increased.
- biopsy of affected muscles confirms the diagnosis.
- immunosuppressive doses of corticosteroids are administered:
    - antibiotics may be indicated if any ongoing infection is present.
- full recovery is common but for the muscle mass to fully return may take months:
    - some horses may experience recurrent bouts of muscle atrophy.

Infarctive Purpura Haemorrhagica

Overview

- horses with recent *Strep. equi* infection can develop a type 3 hypersensitivity reaction:
    - leads to leucocytoclastic vasculitis, vascular occlusions and infarction of associated muscle.
- Quarter Horses are most affected.
- early clinical signs include:
    - firm focal muscle swelling, pain and severe lameness and stiffness.
    - muscles in contact with the ground in the recumbent horse are most susceptible.
- other classic signs of purpura haemorrhagica may be observed.
- infarction of the gastrointestinal tract can occur simultaneously leading to signs of colic.

- blood samples usually reveal:
  - leucocytosis with left shift and toxic changes.
  - increased globulins and low albumin.
  - muscle enzymes markedly increased.
- high serum antibody titres to SeM protein (>1:1600) and characteristic histopathology confirm the diagnosis.
- corticosteroids are mainstay of treatment – dexamethasone at 0.04–0.2 mg/kg.
- prognosis is poor and the condition is often fatal.

Post-anaesthetic myoneuropathy

Definition/overview

- identified during the anaesthetic recovery period.
- rapid, progressive degeneration of the muscle fibres.
- often a peripheral neurological component as well:
  - use of the term 'myoneuropathy'.
- two syndromes are recognised:
  - localised myoneuropathy affecting one muscle or muscle group.
  - generalised myopathy causing diffuse myodegeneration.

Aetiology/pathophysiology

- occurs after general anaesthesia.
- intrinsic predisposing factors include:
  - weight of the animal
  - muscular development.
  - intensive training.
  - nervous temperament.
- extrinsic factors relate to the anaesthesia with the risk of problems increasing with:
  - duration of anaesthesia
  - arterial blood pressure ≤70 mmHg.
  - inadequate perfusion of dependent muscles.
  - hard or poorly supporting table padding
  - poor positioning.
  - use of positive pressure ventilation.
- predisposition of certain individuals to generalised myopathy:
  - hypersensitivity to anaesthetic agents.
- decreased perfusion, blood stasis and hypotension in muscles leads to hypoxia and anaerobic metabolism.

- inelastic fascia surrounding the muscle causes the compartment pressure to increase.
- muscle damage may be worsened by reperfusion following recovery and the use of certain anaesthetic agents such as Halothane.

Clinical presentation

- signs apparent during recovery from anaesthesia or soon afterwards:
  - prolonged recovery, with difficulty or inability to stand.
- muscles are swollen, hard and painful to touch, and feel abnormally warm.
- localised to generalised sweating.
- two forms are encountered:
  - **Localised form** – distinct muscle mass is affected:
    - ◆ triceps, biceps and quadriceps femoris, masseter, longissimus dorsi, gluteal, and hindlimb adductor muscles.
    - ◆ mild to severe lameness in the affected limb(s) (Fig. 7.6) or complete paresis.
    - ◆ condition may worsen significantly if the horse is recumbent.
    - ◆ severe cases, the urine is orange to dark chocolate brown (myoglobinuria).

FIG. 7.6 Localised, post-anaesthetic myopathy-induced lameness due to loss of the shoulder muscle mass and triceps functions. The horse is unable to extend its shoulder and elbow. There is focal swelling and patchy sweating over the left shoulder region.

- ◆ usually resolves after a few hours to several days:
  - – muscle atrophy and/or fibrosis may appear in 2–3 weeks and remain.
- ○ Generalised form.
  - ◆ spontaneously or result of prolonged recumbency:
    - – consequence of localised myoneuropathy.
  - ◆ recumbent and show an inability to stand up:
    - – most muscle masses are affected.
    - – increased heart and respiratory rates.
    - – myoglobinuria to anuria.
    - – profuse sweating, distress, shock and death.

Differential diagnosis

- other orthopaedic injuries, such as fractures and post-anaesthetic neuropathy.
- other causes of prolonged recumbency, including myelomalacia of the spinal cord.

Diagnosis

- history, clinical signs and significant increases in muscle enzymes following general anaesthesia confirm the diagnosis.
- hyperkalaemia, hypocalcaemia, and acidosis with early hyperlactacidaemia.
- signs of renal failure (increased urea and creatinine concentrations) and other vascular imbalances:
  - ○ urine analysis reveals myoglobinuria and presence of blood.

Management

- intravenous fluid therapy and NSAIDs:
  - ○ opioids in particularly painful cases.
- clinical signs usually improve rapidly once the animal is standing:
  - ○ hoist and harness may be used if the horse's temperament allows.
- supportive therapies to limit muscle damage from recumbency include:
  - ○ provision of adequate padding.
  - ○ maintenance in sternal recumbency
  - ○ regular turning over.

- Dantrolene has been used in Thoroughbreds with a history of RER that develop myopathy following anaesthesia.

Prognosis

- variable depending on severity and extent of muscle damage, and animal's temperament.
- prognosis is good if able to stand.
- recovery may take a few hours to several days.
- prolonged recumbency leads to a poor prognosis.

Fibrotic/Ossifying myopathy

Definition/overview

- uncommon condition characterised by a severe muscle tear and subsequently:
  - ○ formation of fibrous scar tissue (fibrotic myopathy).
  - ○ may become mineralised (ossifying myopathy).
- characteristic mechanical, non-painful gait abnormality.
- encountered in Quarter Horses and barrel racers and, rarely, in other breeds.
- Semitendinosus muscle and, less commonly, semimembranosus, biceps femoris, and adductor muscles.

Aetiology/pathophysiology

- spontaneous trauma (muscle strain) through repeated overstretching of the muscle.
- external trauma from ropes, kicks, and falls.
- secondary to wounds, intramuscular injection, or surgical trauma.
- chronic strain on the healing muscle leads to:
  - ○ muscle atrophy.
  - ○ formation of exuberant fibrous tissue within the muscle, often at the muscle/tendon junction.
  - ○ recurrent inflammation may lead to fibrocartilaginous or osseous metaplasia.
- congenital form has been described in yearlings, supposedly through perinatal trauma.

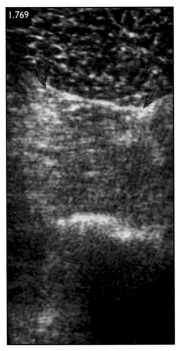

FIG. 7.7 Fibrotic and ossifying myopathy. Transverse ultrasound scan of the distocaudal thigh region showing loss of the normal echo-structure of the semitendinosus muscle, which appears echogenic and amorphous (arrows) because of diffuse fibrosis. It contains hyperechogenic areas casting acoustic shadowing (arrowheads) characteristic of ectopic mineralisation.

Clinical presentation

- acute forms may have sudden onset of lameness with focal swelling over the caudal aspect of the thigh.
- usually presents at the chronic stage:
  - characteristic mechanical gait abnormality most obvious at the walk.
  - slapping of hoof to the ground following protraction of the hindlimb.
  - restriction of cranial phase of stride by functional shortening of caudal thigh muscles.
- usually unilateral:
  - bilateral cases recorded secondary to external trauma to both hindlimbs during transporting.

Differential diagnosis

- other orthopaedic injuries or neuromuscular disorders affecting the pelvic limbs:

  - stringhalt and shivering, and ataxic neurological cases.

Diagnosis

- history, clinical findings, and characteristic features of the gait abnormality.
- abnormal area of damaged muscle is usually palpable.
- no improvement in the gait following regional or intrasynovial analgesia techniques.
- ultrasonography will confirm the presence and extent of an echogenic or hyperechogenic lesion at the muscle/tendon junction (Fig. 7.7).

Management

- acute muscle tears treated conservatively:
  - rest, NSAIDs, local cold application, and physiotherapy.
- surgical treatment is the only option in chronic cases.

Prognosis

- guarded for complete resolution.
- no improvement in chronic cases without surgery.

Corynebacterium pseudotuberculosis abscesses

Overview

- spontaneously occurring intramuscular abscesses and muscle necrosis due to infection by *C. pseudotuberculosis*.
  - Gram-positive, intracellular bacterium present in soil.
  - organism transmitted by flies, particularly *Haematobia irritans*, *Stomoxys calcitrans* and *Musca domestica*.
  - haematogenous or lymphatic spread of the bacteria.
- USA and parts of Canada in horses of all ages and during all months of the year.
- **Three clinical manifestations:**
  - subcutaneous or muscular abscesses, mainly in the pectoral and ventral abdominal/inguinal regions, commonly called Pigeon fever.

- o internal infection causing visceral abscesses.
- o ulcerative lymphangitis – limb inflammation, cellulitis and severe lameness.
- differential diagnosis includes:
  - o abscesses of other origin, including brucellosis.
  - o tumours   o reaction to injections
  - o insect stings or snake bites.
- diagnosis is based on clinical signs, fine needle aspirate for culture and sensitivity, and serology for intra-abdominal or thoracic abscesses but not muscular abscesses.
- antimicrobials are generally not recommended.
- abscesses, under ultrasound guidance, should be:
  - o located, poulticed to stimulate maturation, and lanced and drained thoroughly.
  - o pectoral abscesses are frequently deep (5–8 inches) and difficult to drain.
  - o daily chlorhexidine flushes eliminate purulent exudate and bacteria.
- NSAIDs are used for analgesia.
- controlling the fly population is crucial to prevent and control infection.
- collect and properly dispose of any material contaminated with exudate to minimise exposure to other horses and man.
- vaccine may be available in the future.
- prognosis is generally fair if treatment is initiated rapidly to prevent debilitation and disfigurement.
  - o healing after drainage may be prolonged/delayed and there may be recurrence.
  - o condition is rarely life-threatening.

Hypocalcaemia (see
Endocrine chapter in Concise Textbook of Equine Medicine and Surgery Book 4)

Overview

- decreased blood calcium concentration causing a stiff gait, weakness, and multiple muscle fasciculations (spasms).

- associated with a range of conditions including lactation in the mare.
- decrease in calcium concentration leads to hyperactivity of the sodium channels and muscle fasciculations, tremors, tetany and seizures.
- clinical signs in a tetanus-vaccinated horse with no wounds should lead to a suspicion of this disease.
- serum ionised calcium concentration should be ≥10 mmol/l:
  - o signs become obvious below 8 mmol/l and severe below 5 mmol/l.
  - o other electrolytes (P, K, Cl, Na, and Mg) should also be assessed.
- signs of dehydration and metabolic alkalosis.
- intravenous administration of calcium gluconate solution.
  - o repeated as required to effect.
  - o may not resolve in some horses until they receive magnesium supplementation.
- ionised calcium and magnesium levels should be monitored frequently to assess response to therapy.
- usually good prognosis with prompt treatment, but severe forms may not respond to treatment.

Exhausted horse syndrome

Definition/overview

- very similar to hypocalcaemia but relates to a syndrome of muscle hypercontractility induced by severe electrolyte and acid–base imbalance in horses after a period of hard, sustained exercise.
- most common in endurance horses competing over long distances, and especially in hot weather.

Aetiology/pathophysiology

- endurance exercise leads to the generation of considerable intramuscular heat:
  - o shed to external environment via peripheral vasodilation and increased skin temperature.
  - o increased ambient temperature and humidity make this evaporative gradient less efficient.

- subsequently may develop considerable sweating, muscle fasciculations and 'cramps'.
  - ○ heavy losses of fluid and ions, including sodium, potassium, chloride, calcium.
  - ○ fluid losses cause hypovolaemia and haemoconcentration.
  - ○ poor oxygenation of muscles promotes anaerobic metabolism, with subsequent lactate accumulation and consumption of the stored glycogen.

Clinical presentation

- myositis-like syndrome with a similar presentation to hypocalcaemia or exertional rhabdomyolysis.
- serum calcium concentration is often normal and no myoglobinuria is noted.
- elevated temperature, pulse rate and respiratory rate:
  - ○ body temperature often fails to return to normal on cessation of exercise.
- other signs include depression, anorexia (including for water), a stiff, stilted gait, localised or more diffuse muscle cramping, muscle hardening and pain on palpation.
- dehydration, synchronous diaphragmatic flutter (uncommon), atrial fibrillation, diarrhoea, colic and laminitis.

Differential diagnosis

- Hypocalcaemia
- exertional rhabdomyolysis
- various causes of colic.

Diagnosis

- history, clinical signs and laboratory evidence.
- serum muscle enzymes often within normal limits or only slightly increased.
- PCV and total blood protein are increased.
- blood hyponatraemia, hypokalaemia, and mild hypocalcaemia (ionised fraction).
- metabolic alkalosis, and a partial respiratory alkalosis due to hyperventilation.
- raised kidney enzymes, highly concentrated urine (with dehydration and normal renal function), or poorly concentrated urine (indicative of intrinsic renal disease).

Management

- cease exercise immediately and place horse in a cool shaded area:
  - ○ preferably with fans to cool the immediate environment.
- cooling by continued application of large volumes of cold water over the entire body is particularly effective to reduce hyperthermia.
- oral fluid therapy may be considered if the patient is mildly affected:
  - ○ usually unwilling to drink sufficiently.
  - ○ isotonic mixed electrolyte solutions administered by nasogastric tube.
  - ○ only if horse has normal gut sounds and no evidence of colic or gastric reflux.
- failure to respond to oral fluid therapy or more severely affected:
  - ○ intravenous fluid therapy instigated immediately.
  - ○ large volumes of balanced, polyionic solutions to establish normal circulation.
- subsequently, a more specific careful correction of electrolyte imbalances:
  - ○ including potassium (10 mEq/l).
  - ○ hypocalcaemia is treated as described above.
  - ○ glucose should be added to the fluids (10 g/l).
- NSAIDs if the horse is in pain:
  - ○ **caution if the horse is severely dehydrated.**

Prevention

- provide water regularly during strenuous exercise:
  - ○ use electrolyte supplements in the water (prior to, during, and after a race).
- proper conditioning results in physiological changes that aid in prevention of exhaustion.
- rider management and education are important to ensure prevention.

Prognosis

- usually good with adequate care, but severe cardiovascular shock may lead to organ failure and death.

Diaphragmatic flutter

Overview

- synchronous contraction of the cardiac and diaphragmatic muscles.
- observed in exerted animals.
- severe metabolic and electrolyte imbalance leads to phrenic nerve hyperexcitability:
  - depolarisation of the right atrium stimulates action potentials in the hyperexcitable phrenic nerve as it crosses over the heart.
  - diaphragm submitted to violent contractions at each heartbeat.
- brisk, sudden, rhythmic, and painful contraction of the diaphragm visible at the costal arch and flank synchronously with the heartbeat.
- may occur in isolation or, more commonly, in association with other disturbances as described in hypocalcaemia and exhausted horse syndrome.
- condition disappears if the underlying anomalies are treated and corrected.
- see prevention of hypocalcaemia and exhausted horse syndrome.
  - suggested food rich in calcium (e.g. alfalfa hay) should be avoided in predisposed animals and in endurance horses.
- good prognosis if the primary problem can be addressed.

Soft Tissue Injuries

Types of injury

- certain factors induce enormous loading upon equine tendons and ligaments:
  - horse's weight.
  - anatomical simplification of their distal limb.
  - speed at which they can work.
- most limb soft tissue structures can be involved in either traumatic or degenerative injury and appear to have a similar pathophysiology based on:
  - repeated, cyclic and overuse trauma.
  - less commonly, direct trauma to the limb.
- injuries to the superficial digital flexor tendon (SDFT) and suspensory ligament (SL) are the most common and affect a large number of race and sports horses.
- Tenosynovitis (inflammation of tendon sheaths) is commonly recognised in all types of horses.
  - underlying pathology is often complex, and the prognosis guarded to poor.
- thorough knowledge of soft tissue limb anatomy and the use of specific diagnostic techniques is required for an accurate diagnosis.
- early diagnosis is paramount to improve the chances for recovery.

Diagnostic techniques

- soft tissue injuries can be:
  - subtle and difficult to detect using observation and palpation alone.
  - obvious alterations are not necessarily clinically significant or the cause of the current lameness.
- localisation of the site of pain frequently requires regional or intrasynovial anaesthesia.

Radiography

- rule out concurrent or associated bony/ articular abnormalities.

- locate abnormal bone reaction at tendon or ligament insertion sites:
  - sclerosis   ○ new bone formation
  - entheseopathy.
- positive contrast techniques may be useful:
  - bursography, tenosynoviography or arthrography.
  - **Positive contrast tenosynoviography** has become more popular in differentiating pathology in digital flexor tendon sheath injuries including the deep digital flexor tendon (DDFT) and manica flexoria.
  - injection of an iodinated, positive-contrast material into a given synovial cavity:
    - leakage of the fluid through a wound.
    - opacification of another synovial cavity that does not normally communicate.

Ultrasonography

- examination of all soft tissue structures:
  - not separated from the skin by bone or gas or hoof wall.
- high-definition equipment is required:
  - greater the frequency of the transducer, the better the spatial resolution.
  - 7.5 MHz or higher transducers are generally adequate.
    - 14 MhZ or higher are rapidly attenuated by superficial tissues.
- optimal probe depends on:
  - location, size, and depth of the structures.
  - transmitting properties of the individual patient's skin.
    - linear array transducers most practical in the distal limb.
    - curved array (microconvex/convex) transducers:
      - areas covered by large muscle masses (e.g. ilial wing fractures).

DOI: 10.1201/9781003369226-10

–   overlying irregular bony
    contours.
–   structures that lie at an angle
    to the skin (e.g. intra-articular
    ligaments, distal fetlock).

MRI

- provides accurate means of detecting and
  assessing soft tissue lesions.
- low-field, open MRI magnets allow
  imaging of distal limb in standing horse
  (Fig. 8.1).
- useful for imaging soft tissue
  structures not adequately visualised by
  ultrasonography:
  o injuries in the proximal plantar
    metatarsus or palmar metacarpus
    regions.
  o soft tissue lesions and complex bony
    abnormalities in the foot.
  o flexor tendons and sesamoidean
    ligaments located around the ergot
    area.
- requires excellent knowledge of the
  technology to optimise acquisition
  protocols, and specialist-standard
  experience for adequate interpretation.

FIG. 8.1 Low-field, open MRI magnet machine
used to acquire distal limb images. (Photo courtesy of Alex Font)

CT

- soft tissue windowing of high-definition
  CT provides detailed information
  regarding soft tissue structures.
- poor inherent radiographic contrast
  between soft tissues is a limitation.
- enhancement by iodinated contrast media
  (contrast-enhanced CT) significantly
  improves detection of abnormalities.

CONDITIONS AFFECTING TENDONS AND LIGAMENTS

Superficial digital flexor tendinopathy (tendon disease)

Definition/overview

- 'Tendonitis' is a common condition
  affecting all breeds:
  o particularly racehorses, eventers, and
    horses engaged in high-speed pursuits.
  o 'bowed tendon', 'tendon strain',
    'tendinitis', and 'tendinopathy'.
- progressive, exercise-induced degeneration
  of the collagen matrix leads to acute,
  partial rupture of tendon fibres
  ('tendinopathy').
- healing is slow and characterised by a
  high rate of recurrence.
- major source of economic loss in the
  equine industry.

Aetiology/pathophysiology

- incompletely understood.
- main type of ligament and tendon injury
  is an overstrain (Fig. 8.2):
  o mechanical failure of the collagen
    fibre-based connective tissue matrix.
  o due to recurrent overstretching of
    fibres beyond the tolerance of the
    tissue.
  o cyclic strain during fast stretching–
    release cycles at fast paces gradually
    weakens the tendon.
  o subclinical microtears occur and
    accumulate during training or fast
    exercise.
    ♦ progressively weaken the tendon
      matrix.
- ageing, cell apoptosis and hormonal
  imbalances in older females contribute to

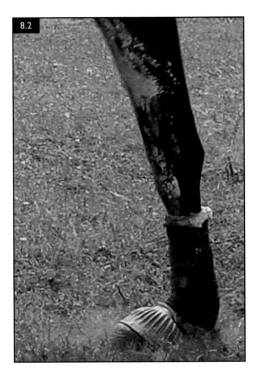

FIG. 8.2 View of the distal limb of a racehorse at full gallop. The fetlock drops at maximum weight bearing during the stance phase of the stride because of stretching of the flexor tendons and interosseous ligament. The SDFT, being most pal-mar, is stretched to a greater degree, nearing its physiological limits (yield point).

progressive weakening following matrix compositional change in older animals.
- normal tendon consists mainly of longitudinally arranged type I collagen fibres.
- cumulative matrix damage results in:
  - increasing proportion of poorly organised collagens (especially type III).
  - altered fibre arrangement.
  - decreased elasticity, increased stiffness and reduced resistance to cyclic strain.
- stretching of the tendon at fast speed (within theoretically normal tolerance levels) can then cause:
  - sudden rupture/strain of a large proportion of collagen fibres (clinical tendinopathy).
  - intralesional haemorrhage from ruptured endotenon capillary vessels.

  - platelet degranulation, inflammatory mediator release, growth factors and collagenases that directly destroy fibres.
  - influx of inflammatory cells.
- core lesions may spread or increase in volume over several days after injury.
- predisposing factors include:
  - work type (racing, athletic activities versus leisure or pleasure activities).
  - genetics   age   career duration.
  - abnormal hoof shape (especially long toe/low heel conformation).
  - ground surface quality (especially soft, deep soil and uneven surfaces).
  - method of training (duration and length of high-speed training).
- direct trauma, especially from limb interference (overreach injuries) or hitting jumps or obstacles, may also cause localised peripheral tendon lesions.
  - different condition to exercise-induced overstrain injury.
  - healing mechanisms are the same.
  - concurrent paratendonitis may exist (inflammatory thickening of the paratenon).
- healing occurs through:
  - removal of the haematoma by white blood cells.
  - fibroblasts produce a temporary, immature matrix (lacks resistance to strain) with haphazard ingrowth of vascular buds (fibroplasia).
    - substrate for subsequent scar tissue formation.
  - slow maturation and remodelling of the strong, inelastic fibrous scar tissue.
    - relative increase of more longitudinally arranged type I fibres.
    - decrease of type III fibres (over 18 months).
    - resulting tendon scar remains less elastic and resistant to cyclic strain than normal tendon.
  - tendon healing is usually split into three stages:
    - acute (inflammatory) phase lasting 7–10 days.
    - subacute or repair phase lasting 2–4 months.
    - remodelling phase (4–18 months).
- recurrence with worsening of the lesion is common:

- o premature return to exercise may cause more severe damage with significant worsening of the prognosis.
- o 40–80% of injured racehorses suffer recurrence in various studies.
- o lesions often occur at the extremities or along the periphery of the tendon scar:
  - ♦ area of transition between inelastic scar tissue and normal elastic matrix.
  - ♦ mechanical stress is concentrated.

Clinical presentation

- swelling, heat, and pain over the palmar metacarpal or plantar metatarsal area.
- marked pain on palpation (compare with contralateral limb):
  - o gradually recedes over 4–10 days.
  - o leads to firm, bow-shaped swelling over the 'tendon' area ('bowed tendon') (Fig. 8.3).
- often sudden-onset lameness, during or immediately after exercise:
  - o variable, with up to 40% horses not showing overt lameness.
  - o usually recedes within 2–10 days.

FIG. 8.3 Typical swelling ('bowed tendon') over the mid-metacarpal region of a horse suffering from subacute tendonitis of the SDFT.

- o poor correlation between severity of clinical signs and severity of tendon injury.
- may be associated distension of the carpal and/or digital sheath(s).

Differential diagnosis

- subcutaneous or paratendonous trauma (paratendonitis).
- swelling/oedema from inflammation or infection elsewhere in the limb.
- tenosynovitis of the digital sheath
- inferior check ligament desmitis.
- SL desmitis.
- Lymphangitis.

Diagnosis

- clinical examination.

Ultrasonography

- diagnostic method of choice.
- **initial examination during the first week after injury:**
  - o confirm the presence of injury.
  - o may underestimate the severity and extent of the damage.
    - ♦ early haematoma may have a similar echogenicity to normal parenchyma.
    - ♦ lesion often enlarges over several days (Figs. 8.4, 8.5).
- **definitive ultrasonographic examination around 10–15 days after injury:**
  - o determine the base-line localisation, size, extent, and severity of the lesion.
  - o lesion severity is based on:
    - ♦ increased cross-sectional area of the tendon.
    - ♦ decrease of echogenicity.
    - ♦ cross-sectional surface area of the lesion (relative to that of the tendon).
    - ♦ loss of fibre alignment.
    - ♦ proximodistal length of the lesion.
- **systematic approach must be used and comparison with opposite limb is essential.**
  - o clipping, cleaning and degreasing of the skin is essential.
  - o standardisation of the level of the image planes is needed for contralateral and follow-up comparison:

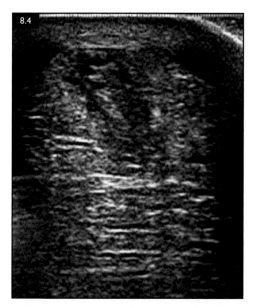

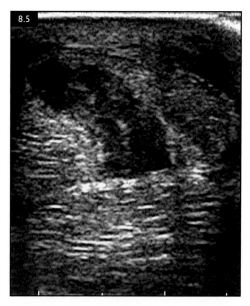

FIGS. 8.4, 8.5 Transverse ultrasound scans obtained from the same horse 3 days after injury (8.4) and 16 days after injury (8.5). In (8.4) the lesion is ill defined, heterogeneous and echogenic. In (8.5), the lesion now appears hypoechogenic and better defined and has moderately enlarged.

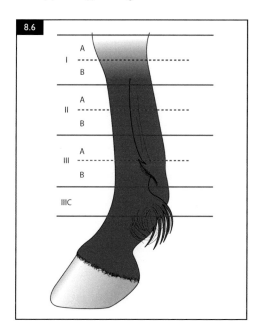

FIG. 8.6 Zones I to III and subdivision into zones A and B for each area of the palmar metacarpus.

♦ distance in centimetres from the distal border of the accessory carpal bone/point of the calcaneus.
♦ divide metacarpal region from the carpometacarpal joint to the ergot into seven equal areas (Fig. 8.6).

♦ divide metatarsal region from the tuber calcanei to the ergot into nine zones, with zone I A and I B spanning the plantar aspect of the tarsus, through the plantar aspect of the fetlock at IV C.
♦ divide palmar or plantar pastern area into three equal zones (P I to P III) from the ergot to the deepest part of the hollow of the pastern.
○ assessment in transverse planes, then in longitudinal sections (sagittal and parasagittal planes).
• **ultrasonographic appearance of normal tendon tissue is:**
  ○ homogeneously granular on cross-sectional images (Fig. 8.7).
    ♦ regularly arranged fibre bundles.
  ○ longitudinal (long-axis) images (Fig. 8.8).
    ♦ fibre bundles produce parallel, linear echoes throughout the section.
• **cross-sectional surface area (CSA) of the tendon:**
  ○ serves to evaluate tendon thickening (Fig. 8.9).
  ○ huge variation in size of normal SDFT CSAs.

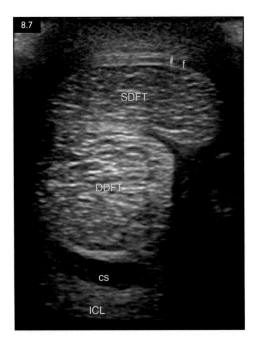

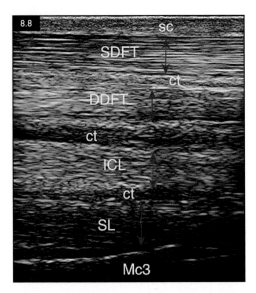

FIG. 8.7 Transverse ultrasonographic image of the palmar metacarpus using a high-frequency trans-ducer (18MHz), showing the appearance of the normal flexor tendons. The larger bundles in the SDFT produce a coarser pattern than in the DDFT. f = deep fascia (the paratenon is indistinguishable from the fascia); cs = carpal sheath (distal recess); ICL = inferior check ligament.

FIG. 8.8 Longitudinal (sagittal) ultrasonographic image of the palmar metacarpus (13 MHz). Normal fibre bundles in the SDFT form a hyper-echogenic, longitudinally arranged striation. Fibres in the the other tendons (DDFT, ICL and SL) are also aligned but with a coarser pattern. sc = subcutaneous connective tissue; ct = connective tissue; ICL inferior check ligament; SL = suspensory ligament; Mc3 = third metacarpal bone.

- ♦ mean CSA depends on anatomical zone, age, exercise level, breed, and size.
- ♦ both forelimb tendons compared.
- • greater than 20% difference is significant for an injury.
- • greater than 10% difference suspicious for predisposition to injury.
  - ○ Havemeyer consensus for grading severity of injury based on tendon CSA at the maximum injury zone (Thoroughbred forelimb):
    - ♦ mild (grade 1) if CSA < 2cm².
    - ♦ moderate (grade 2) if CSA = 2–5cm².
    - ♦ severe (grade 3) if CSA > 5cm².
  - ○ CSA lesion ratio:
    - ♦ simplest technique of lesion grading.
    - ♦ calculate the ratio of the lesion CSA (LCA) at the 'maximum injury zone' (MIZ - largest point) to the tendon CSA (TCA) at the same level (LCA × 100/TCA) (Fig. 8.10).

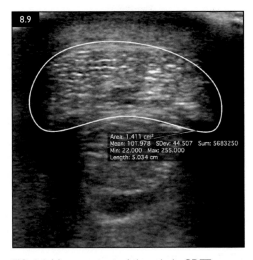

FIG. 8.9 Measurement of the whole SDFT cross-sectional area may be performed using specific software by tracing the contours of the tendon.

  - ○ Havemeyer consensus for grading severity of injury based on the CSA ratio at MIZ (Thoroughbred forelimb):
    - ♦ mild (grade 1) if ratio < 10%.

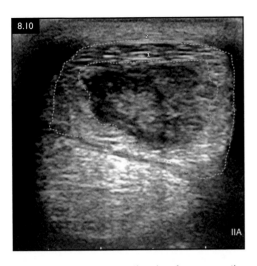

FIG. 8.10 Lesion cross-sectional surface area ratio. The lesion CSA (LCA) (1) at the 'maximum injury zone' is divided by the total tendon CSA (TCA) (2) to obtain a ratio (LCA × 100/TCA). In this case, 51% is indicative of a severe lesion. The ratio may be measured at each standard point in the meta-carpus to obtain a severity score.

- ◆ moderate (grade 2) if ratio = 10–40%.
  - ◆ severe (grade 3) if ratio > 40%.
- **Lesion echogenicity grade/score:**
  - ○ semi-quantitive assessment of lesion echogenicity from good quality images.
  - ○ commonly used grading scale from 0 to 4:
    - ◆ 0 = normal; 1 = mostly echogenic; 2 = half echogenic/half anechogenic
    - ◆ 3 = mostly hypoechogenic; 4 = anechogenic (Figs. 8.11–8.14).
- **Fibre alignment score (FAS):**
  - ○ loss of the normal collagen fibre alignment is a major factor when assessing the severity of the injury (Figs. 8.15–8.17).
  - ○ non-aligned fibres (immature or poor-quality scar tissue) may not be distinguishable from total absence of fibres (haematomas or early granulation tissue).
  - ○ semi-quantitative grading scale from 0 to 3 has been proposed:
    - ◆ FAS = 0 (normal) if fibres normally aligned

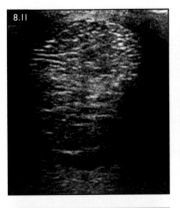

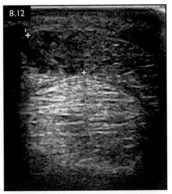

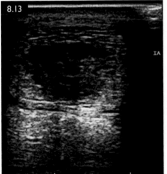

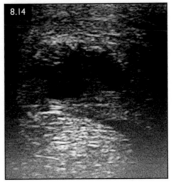

FIGS. 8.11–8.14 Echogenicity may be subjectively assessed by comparing the lesion echogenicity to that of the normal surrounding parenchyma. This is very subjective as it will depend on the timing of the examination (acute vs. subacute), and machine settings (gain, contrast etc.). Examples on a grade of 0–4: (8.11) grade 1; (8.12) grade 2; (8.13) grade 3 and (8.14) grade 4.

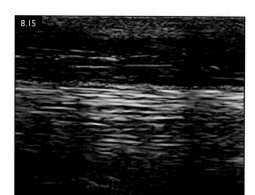

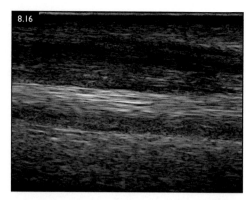

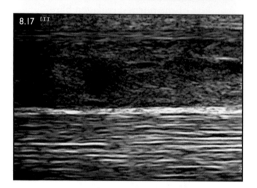

FIGS. 8.15–8.17 Fibre alignment score is assessed on longitudinal sonograms and may be used both at the acute/subacute stage or for follow-up. (8.15) Grade 1 lesion with persistance of more than half the fibres in the lesion; (8.16) grade 2 with marked loss of fibre; (8.17) grade 3 with near total loss of fibre (acute lesion filled by an echogenic haematoma).

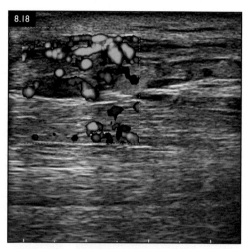

♦ FAS = 1 (mild injury) with slight loss of fibre pattern
♦ FAS = 2 (moderate) with moderate loss of striation
♦ FAS = 3 with marked loss of fibre alignment (little to no striation).

- **Doppler ultrasound:**
  ○ assess vascularisation of tendon during the healing process (Fig. 8.18).
  ○ performed with limb held in partial flexion.
  ○ incorrect settings and artefacts make interpretation difficult.
  ○ abnormal vascular flow (numerous, large calibre vessels) may be present in:
    ♦ immature scar tissue.
    ♦ chronic, poorly organised, still active fibrous tissue.

Acute/subacute injuries.

- most commonly discrete, well-defined hypo- (dark grey) to anechogenic (black) areas:
  ○ near the centre of the tendon (core lesion) (Fig. 8.19).
  ○ periphery (marginal lesion) (Fig. 8.20).
  ○ longitudinal scans show extent of lesion and loss of fibre pattern (Fig. 8.21).

FIG. 8.18 Colour flow or power Doppler imaging can be helpful to assess lesion vacularisation. Flow in small diameter vessels of normal tendon or early scar tissue cannot be detected with Doppler imaging. Large calibre vessels are detected in chronic, active disease and are associated with a poorer prognosis.

- sometimes diffuse and less well-defined lesions (Fig. 8.22)
- loss of echogenicity due to the loss of large enough fibre bundle interfaces:
  ○ may represent fluid, necrotic material, or poorly organised, immature tissue.

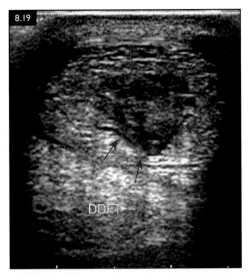

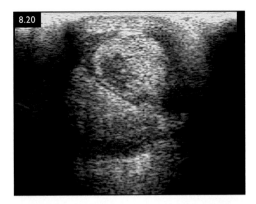

FIG. 8.20 A subacute, peripheral hypoechogenic lesion on a cross-sectional (transverse) ultrasonographic scan of the palmar aspect of the metacarpus.

FIG. 8.19 A subacute, hypoechogenic central core lesion (arrows) on a cross-sectional (transverse) ultrasonographic scan of the palmar aspect of the metacarpus.

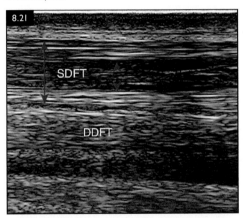

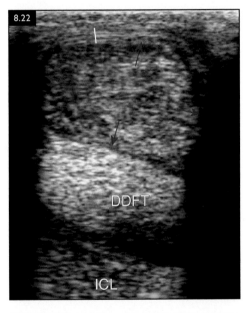

FIG. 8.21 Longitudinal, sagittal ultrasonographic section of the SDFT of a horse with subacute tendonitis. There is loss of the normal fibre alignment within the SDFT (arrowheads), which also appears significantly enlarged and hypoechogenic. The underlying DDFT shows a normal echostructure.

FIG. 8.22 Diffuse lesion characterised by enlargement of the SDFT (arrows) and generalised, heterogeneous decrease of echogenicity. Line shows thickening of the paratendon fascia.

- early haemorrhage = echogenic because of aggregating cells in the thrombus. (Figs. 8.23, 8.24).
- organising haematoma = hypo to anechogenic as the cells are lysed (Fig. 8.25).
- early granulation tissue = hypo to anechogenic (Fig. 8.19).
  - very small fibres, gel-like water-rich matrix.

  - microscopic vascular buds and sparse cells.
- maturing repair tissue = increasing echogenicity (Figs. 8.26, 8.27).
  - vessels and fibres organise into large enough structures to cast echoes.
  - linear organisation is poor in early scar tissue – granular pattern.

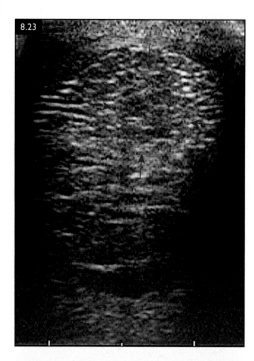

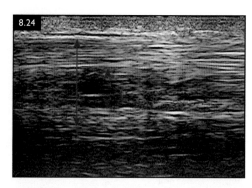

FIGS. 8.23, 8.24 (8.23) Transverse scan image of the palmar metacarpus through zone IIA, showing a typical acute lesion (a few hours after a harness race) with minimal increase in cross-sectional area. The lesion (arrows) is heterogeneous and nearly isoechogenic to the normal parenchyma, representing focal haematoma. (8.24) Longitudinal scan in the same horse showing irregular loss of fibre alignment in the lesion (arrowheads). Repeat examination several days later showed marked increase in both tendon and relative lesion cross-sectional areas.

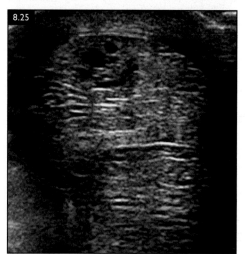

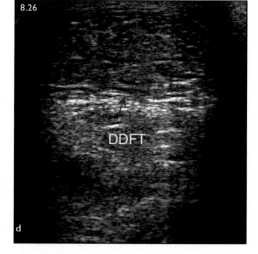

FIG. 8.25 Subacute lesion (arrows) 4 days after injury in a French Trotter racehorse (zone IIA). The early haematoma has organised with anechogenic fluid-filled cavities separated by echogenic material (fibrin).

FIG. 8.26 Transverse image of the SDFT 5 months after injury showing increased echogenicity with a finely granular pattern (arrows). This is created by fibrous scar tissue in which the collagen fibres form thick enough bundles to produce echoes but with poor organisation.

- severe tendon sprains are characterised by:
  - marked, diffuse decrease in echogenicity.
  - significant enlargement (2–3-fold) (Fig. 8.28).

- completely ruptured tendons characterised by a heterogeneous haematoma (Fig. 8.29):
  - gradually replaced by granular, hypoechogenic granulation tissue (Fig. 8.30).

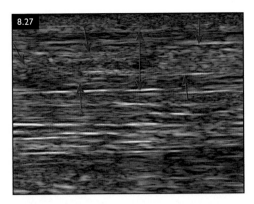

FIG. 8.27 Longitudinal image in the same horse as in 8.26. Poor linear arrangement of fibres produces a granular rather than linear pattern on longitudinal scans.

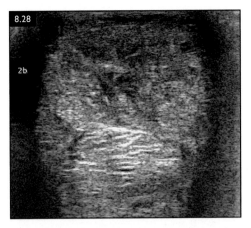

FIG. 8.28 Severe SDFT sprain with marked enlargement of the tendon on transverse images and diffuse, heterogeneous loss of echogenicity.

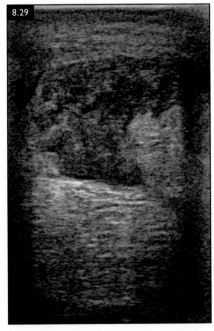

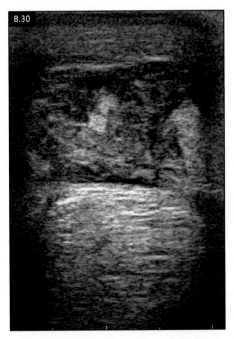

FIGS. 8.29, 8.30 (8.29) Acute SDFT rupture (transverse ultrasonographic image), zone IIB. The tendon parenchyma is very hypoechogenic and heterogeneous with echogenic tissue aggregates surrounded by hypoechogenic fluid and haemorrhage. The tendon and surrounding soft tissues are very enlarged and the tendon margins are ill defined. (8.30) After 2 weeks, the lesion has organised to a diffuse granulation tissue, with a more homogeneous, granular pattern and better defined contours.

- ♦ without a normal fibre pattern.
- ○ frayed ruptured ends are enlarged, hypoechogenic and heterogeneous (Fig. 8.31).
- • lesion location:
- ○ usually metacarpal, unsheathed area:

- ♦ sometimes extend into the digital sheath (Fig. 8.32).
- ♦ differentiate from longitudinal tears in sheath in the fetlock or pastern region.

- ○ rarely primary metacarpal lesion extends proximally into carpal sheath (Fig. 8.33).
- ○ rarely primary location within the carpal canal:

- ♦ severe, diffuse with possible spontaneous rupture (Fig. 8.34).
- ♦ most common in older animals.
- ○ injuries to the distal branches of the SDFT near their insertion on the middle phalanx.

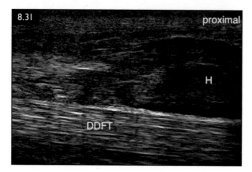

FIG. 8.31 Acute SDFT rupture (Longitudinal ultrasonographic image), zone IIB/IIIA. The ruptured tendon distal end has retracted distally, with frayed fibres ending abruptly (arrowheads). Heterogeneous haemorrhage (H) accumulates between the ruptured ends (right of image). Peritendinous oedematous tissue is markedly thickened and hypoechogenic (arrow).

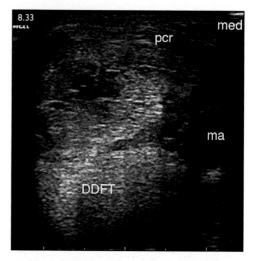

FIG. 8.33 Subacute SDFT tear immediately distal to the accessory carpal bone. Although a discrete 'core' lesion is visible (arrowheads), diffuse damage to the tendon is present. Its contours are ill defined and blend into the thickened synovial membrane (arrows). pcr = palmar carpal retinaculum; ma = medial palmar artery.

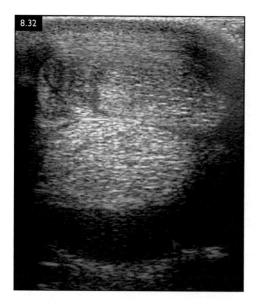

FIG. 8.32 Transverse ultrasonogram of the SDFT in the distal metacarpal area, within the digital flexor tendon sheath. The SDFT is enlarged and there is an irregular, peripheral hypoechogenic lesion in the lateral part of the tendon (arrow). The sheath is distended and a hypoechogenic halo around the tendons suggests synovial inflammation (tenosynovitis) (arrowhead).

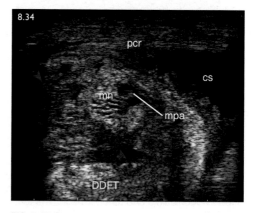

FIG. 8.34 Spontaneous rupture of the SDFT in the carpal canal area. The DDFT is intact. The SDFT is no longer visible; the space is filled by mixed tissue consisting of blood, fibrin and tendon debris (arrowheads). The synovial membrane is markedly thickened. Some fluid accumulates medially in the carpal sheath recess (cs). pcr = palmar carpal retinaculum; mpl = medial palmar artery containing a thrombus, a common occurrence in these cases; mn = median nerve (compressed).

- one or, more rarely, both branches affected.
- enlarged, diffusely hypoechogenic, and heterogeneous branch (Fig. 8.35).
- or well-defined core lesion (Fig. 8.36).
- associated digital sheath effusion, particularly if extends as superficial SDFT tear.
- rarely, complete rupture of both branches or avulsion of their insertions on the palmar processes of P2, with subluxation of the proximal interphalangeal (PIP) joint (Fig. 8.37).

Healing and chronic injuries

- increasing echogenicity:
  - decrease of oedema, matrix water content and diffuse cellular infiltrate.
  - increase in collagen deposition.
- immature and early fibrous scar tissue:
  - fairly homogeneous and hypoechogenic.
  - small vessels and poorly arranged collagen fibres accumulate in a diffuse pattern (Fig. 8.38).

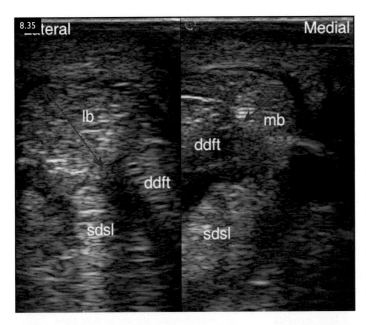

FIG. 8.35 Lateral and medial branches of the SDFT (split screen showing palmarolateral and palmaromedial transverse scans). The lateral branch (lb) is markedly enlarged compared with the medial branch (mb). The DDFT is displaced medially. sdsl = straight distal sesamoidean ligament

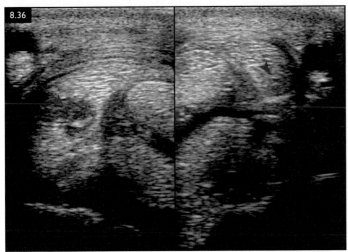

FIG. 8.36 Transverse ultrasonogram of the SDFT in the proximal digital (pastern) area. The left image shows an enlarged lateral SDFT branch. It contains a well-defined hypoechogenic lesion (arrow), although the remainder of this branch also appears heterogeneous. The right image shows the normal medial branch (arrowhead). The latter is smaller but is surrounded by hypoechogenic material representing inflamed synovial tissue.

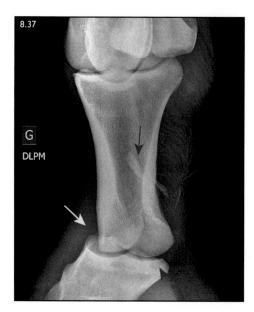

FIG. 8.37 Dorsomedial/palmarolateral radiograph of the digit of a horse with avulsion of both branches of the SDFT. Two bony fragments, avulsed from the P2 palmar processes (arrowhead), are displaced proximally (red arrows). Subluxation and hyperextension of the PIP joint is obvious dorsally (yellow arrow).

- poor linear organisation producing a granular pattern on longitudinal scans (Fig. 8.27).
- maturing scar tissue:
  - iso- to hyperechogenic becoming more brightly echogenic but heterogeneous.
  - lacking normal longitudinal fibrillar pattern on long-axis scans.
  - growing number of larger fibre bundles in a poorly organised pattern (Fig. 8.39).

- healed tendon:
  - coarse, heterogeneous but iso- to hyperechogenic images on transverse scans.
  - short, coarse linear echoes on long-axis images (Figs. 8.40, 8.41).
  - fibre bundles rearranging into a more organised linear pattern, yet never normal.
- chronic tendinopathy due to poor healing and repeat microtearing:
  - active lesion with a heterogenous mixture of mature scar tissue and subacute lesion.
  - persistent increase in tendon CSA.
  - very heterogeneous tissue, often very diffuse:
    - ♦ containing hyper and hypoechogenic areas (Figs. 8.42, 8.43).
  - mineralisation possible in chronic cases, causing hyperechogenic interfaces casting an acoustic shadow.

Follow-up of injuries.
- ultrasonography every 8–12 weeks to monitor progress and quality of healing, through assessment of:
  - CSA of the lesion and tendon.
  - echogenicity score increase (fibrous tissue formation).
  - improvement of fibre alignment score (Figs. 8.44–8.47).
- adequate healing is characterised by:
  - stabilisation, or even reduction, of tendon CSA.
  - isoechogenicity of the damaged portion with normal tendon tissue.

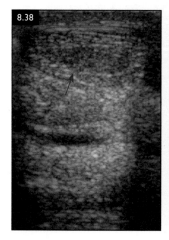

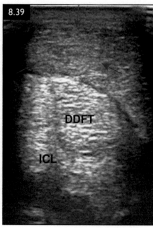

FIG. 8.38 Immature and early fibrous scar tissue is fairly homogeneous and still fairly hypoechogenic as small vessels and poorly arranged collagen fibres accumulate in a diffuse pattern. (Arrows indicate lesion margins).

FIG. 8.39 More mature scar tissue (4–9 months) contains larger fibre bundles in a poorly organised pattern that creates a more brightly echogenic but heterogeneous pattern. (Arrows indicate lesion margins).

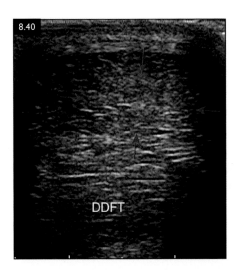

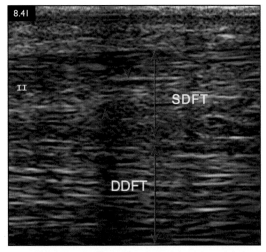

FIGS. 8.40, 8.41 Healed tendon injury (11 months after initial onset). The lesion (arrows) is ill defined and iso to hyperechogenic to the rest of the tendon, with a slightly more densely packed, less regular pattern both in transverse (8.40) and long-axis (8.41) scans. Partial but irregular restoration of fibre alignment is visible on the longitudinal image (arrowheads).

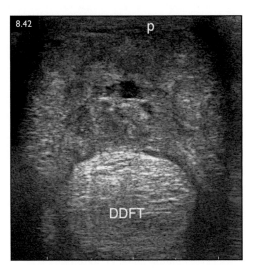

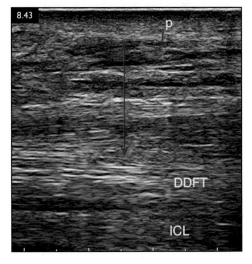

FIGS. 8.42, 8.43 Chronic tendinopathy leads to a mixture of subacute tears interspersed with scar tissue. (8.42) On cross-section, the tendon is severely enlarged (double arrow in 8.43), heterogeneous with hyperechogenic areas and variably hypoechogenic foci (arrow). On longitudinal images (8.43), hypoechogenic lesions, devoid of striation, dissect between hyperechogenic scar tissue. Active peritendonitis causes thickening of the paratenon (p).

- ○ fair longitudinal alignment of the replacement fibres.
- • important to evaluate tissue change in relation to increased mechanical loading:
  - ○ activity level during rehabilitation should be kept the same or reduced if tendon or lesion CSA increase.
    - ♦ activity level is too intense if severity score increases >10% or TCA increases >20% at any level.
  - ○ return to fast work (canter or galloping) only if:
    - ♦ echogenicity score near-normal.
    - ♦ fibre alignment score improved dramatically (0 to 1).

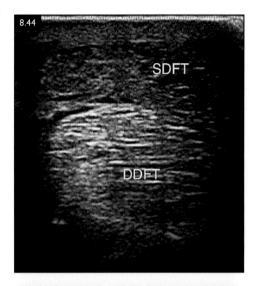

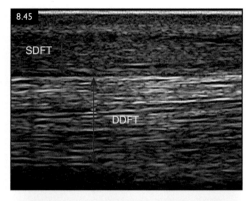

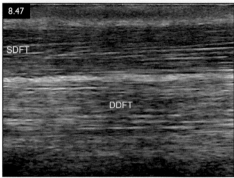

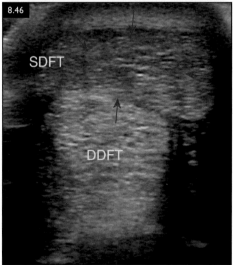

FIGS. 8.44–8.47 Although fibre alignment is only visible on longitudinal views, transverse and longitudinal scans should be used to interpret the image. (8.44, 8.45) Follow-up scan 4 months after injury. The lesion (arrows) has filled in with hyperechogenic tissue poor on transverse scans and fibre alignment is poor on the longitudinal image. (8.46, 8.47) Eight months post injury in another animal, the lesion (arrows) is iso- to slightly hyperechogenic to the normal tissue, the margins are indistinct and there is good restoration of fibre pattern on the longitudinal image.

MRI

- better correlation between histological tissue quality and MRI than with ultrasound.
- ultrasonography tendency to underestimate lesion extent in chronic injuries.
- costly technique with limited availability, and less convenient than ultrasound.

Management

- Three stages of tendon healing:
  - acute (inflammatory) phase lasting 7–10 days.
  - subacute or repair phase lasting 2–4 months.
  - remodelling phase (4–18 months).
- therapeutic steps vary according to the healing phase that is targeted.

Acute phase

- aggressive anti-inflammatory treatment to reduce release and activity of deleterious pro-inflammatory mediators, growth factors and enzymes:
  - steroidal anti-inflammatory drugs interfere with fibroplasia and may delay the early repair.
  - systemic NSAIDs have little effect on local inflammation at current dosages.
  - topical NSAIDs may not reach the tendon tissue.

TABLE 8.1	
GUIDELINES	POSSIBLE CONTROLLED EXERCISE PROGRAMME
Weeks 1–4	Walking for 30 minutes daily (in-hand or on walker or treadmill)
Weeks 5–8	Walking (in-hand, walker, or under saddle) 45 minutes daily
After 8 weeks	Ultrasound exam
Weeks 9-12	Walking 45 minutes + trotting 5 minutes daily
Weeks 13-16	Walking 45 minutes + trotting 10 minutes daily
After 16 weeks	Ultrasound exam
From week 16–28	Gradually add slow canter (15–20 minutes daily) depending on severity of injury and sonographic appearance
After 6 months	Canter one mile daily
After 7 months	Canter two miles daily
After 8–12 months	Reintroduce normal training gradually

The protocol should be altered depending on the severity of the injury (increase exercise regimen earlier if the lesion is mild and sonographic appearance indicates adequate repair; delay if severe and sonographic appearance shows poor healing).

○ DMSO is a potent free-radical scavenger but of questionable usefulness for tendon injuries.
○ cryotherapy (cold water hosing, bathing or ice packs for 20 minutes 3–6 times daily):
  ♦ probably the most efficient local anti-inflammatory treatment.
  ♦ induces vasoconstriction, decreasing haemorrhage and oedema.
  ♦ partially inhibits enzyme activity.
• strict stall rest to limit mechanical strain and avoid exacerbation of injury.
• counter pressure (pressure bandages) decreases haemorrhage, oedema and tendon swelling in early stages.

Subacute phase
• local signs of inflammation have receded (oedema and local heat).
• goals of therapy:
  ○ improve tissue organisation and fibre alignment in the early stages.
  ○ reduce adhesions with surrounding structures.
  ○ reduce paratenon fibrosis.
  ○ prevent reinjury or chronic reactivation of the lesion.
• controlled graduated exercise programme over 6–8 weeks:

○ in-hand walking or use of walkers.
○ treadmills or underwater treadmills, according to a graduated programme.
○ ridden walk from 4 weeks post injury onwards.
○ no galloping before 6 months after injury, as re-sprain is likely.
○ pasture turnout is not recommended.
  ♦ turn out in very small paddocks and only if calm and sensible.
○ controlled exercise (Table 8.1) is superior to stall or field rest or other treatments alone.
○ close monitoring with regular ultrasound examinations to optimise this regimen.
• physiotherapy techniques:
  ○ little scientific evidence to determine their impact on the recovery rate or rate of recurrence of the injury.
  ○ laser, shock-wave, and therapeutic ultrasound may reduce pain and oedema.
  ○ cold water hosing should be used after exercise throughout the recovery period.

Intralesional treatments.
• goals of these treatments are:
  ○ improve healing tissue quality by:

- altering local cell synthetic metabolism.
- supplying cells capable of producing near-normal tendon matrix (i.e. stem cells).
  - best used in early stages of healing.
  - little consensus on efficacy.
- Polysulfated glycosaminoglycans: disappointing.
- Hyaluronic acid: disappointing.
- corticosteroids: deleterious because of inhibition of repair mechanisms.
- Insulin-like growth factor-1 (IGF-1): some gain in speed of repair.
- Autologous Platelet rich plasma (PRP):
  - prepared from venous whole blood.
  - growth factor concentration is variable.
  - contamination with undesirable red or white blood cells is variable.
- Autologous bone marrow:
  - source of growth factors and/or stem cells.
  - little published evidence regarding its benefits.
  - potential adverse effects:
    - large volume required.
    - presence of blood cells, differentiated stromal cells, large amounts of fat and bone material.
    - small or even negligible percentage of stem cells.
- Mesenchymal stems cells (MSC):
  - capable of producing chondro-, osteo-, fibro- or tenoblasts.
  - isolated from bone marrow or adipose tissue aspirates.
  - cultured *in vitro* before being reinjected into the lesion.
  - major improvement in healing tissue quality in early stages of healing.

Surgical treatments

- cautery and blistering have no beneficial effects and may be deleterious:
  - promote adhesions and paratenon fibrosis.
  - raise concerns regarding the animal's welfare.
- Proximal check ligament (accessory ligament of the SDFT) desmotomy:
  - subacute phase
  - increase speed of recovery
  - no change in recurrence rate.
  - reserved for severe or recurrent cases.

- may predispose to contralateral SDFT tendonitis or SL injury in either limb.
- evidence of efficacy is not strong.
- Percutaneous longitudinal tenotomy ('splitting'):
  - subacute phase
  - no change in recurrence rate.
  - reserved for severe or recurrent cases.
  - evidence of efficacy is not strong.

Prevention

- may be more effective than treatment as recurrence of injury is high.
- early training (before age of 2.5 years) may help to increase tendon resistance to strains.
- improved training techniques, adequate care of training surfaces, proper farriery.
- identification of early degenerative changes in the tendon:
  - ultrasonography proved disappointing.
  - MRI may prove more useful as detects alterations in matrix water content.
- ultrasonography useful to monitor adaptation to exercise level and to predict reinjury, especially when monitoring variations of total CSA.

Prognosis

- always good for soundness as even severe tears always heal.
- for return to the same level of activity and performance is guarded:
  - depends on the severity of the lesion.
  - up to 50% of affected racehorses never return to racing.
  - recurrence in up to 80% of cases.
- poorer for SDFT injuries within the confines of the digital or carpal tendon sheaths.
- SDFT branch injuries is fair to good:
  - provided the lesion does not extend to the undivided part of the tendon.
- very poor for avulsions.

Superior check ligament (accessory ligament of the SDFT) desmopathy

Definition/overview

- AL-SDFT (superior check ligament) is strong ligament in the medial wall of the carpal sheath:

- o arises from the caudomedial aspect of the distal radius.
- o joins the SDFT near its myotendinous junction.
- o proximal to the accessory carpal bone.
- injury is rare and always associated with secondary tenosynovitis of the carpal sheath.

Aetiology

- unclear:
  - o overextension of the carpus while the SDFT is under load.
  - o secondary to proximal extension of an SDFT tear.

Clinical presentation

- not specific but include acute onset lameness and swelling of the carpal sheath.
- pain and swelling on the distocaudomedial aspect of the antebrachium.

Diagnosis

- ultrasonography may reveal:
  - o enlarged, hypoechogenic and heterogeneous appearance of AL-SDFT.
  - o complete rupture with haematoma formation between ends.
  - o acute cases may have haemorrhage in the sheath or marked oedema and sheath distension.
- tenoscopy of the carpal sheath.

Treatment

- rest.
- carpal sheath medications.
- tenoscopy of the carpal sheath to remove extruded damaged ligament material.
- injection of PRP and/or cultured stem cells into the residual deeper disrupted region of the ligament under tenoscopic or ultrasound guidance.

Prognosis

- good for isolated injuries.
- poorer for injuries extending into the SDFT.

Suspensory ligament desmopathy ('desmitis')

Definition/overview

- SL is a muscle (interosseous III) with two small muscle bundles running through the origin and body of the ligament, one in each lobe.
- suspensory desmitis is second most common orthopaedic soft tissue injury in the horse.
- encountered in all breeds but most common in:
  - o forelimbs of racing Thoroughbreds.
  - o both fore- and hindlimbs in Standardbreds.
  - o hindlimbs of dressage horses.
- different injuries according to lesion location:
  - o proximal desmopathy (the origin or proximal 3–5 cm of the SL).
  - ó desmopathy of the body (middle third).
  - o desmopathy of the branches.

Aetiology/pathophysiology

- repetitive cyclic strain leads to overuse injury:
  - o reason for specific locations (origin, body and branch) is unclear.
- branch injuries sometimes associated with:
  - o single excessive or asymmetrical weight-bearing event.
  - o proximal sesamoid bone injuries.
  - o fracture of the distal 1/3 of the splint bones.
- mechanical interference by periosteitis ('splints') or fractures of splint bones:
  - o rare cause of focal body or branch desmitis.
- fibrous adhesions between the SL and surrounding structures:
  - o rare cause of recurring desmopathy or failure of treatment of all three lesion locations.
- proximal palmar metacarpal or plantar metatarsal cortex (avulsion or fatigue hairline) fractures are distinct and separate conditions.

Proximal suspensory desmitis (PSLD)

Aetiology/pathophysiology

- pathogenesis unclear but probably gradual cumulative fibre damage from chronic, cyclic strain.
- fore and hindlimbs are different:
  - forelimb PSLD:
    - ♦ may be encountered at any age and in any discipline.
    - ♦ common in racehorses, including Standardbreds and flat racehorses.
  - hindlimb PSLD:
    - ♦ most common in middle-aged, dressage or eventing horses.
    - ♦ straight hock and low fetlock conformation are predisposing factors.
- Compartment syndrome:
  - origin of the SL and periligament neurovascular structures are encased in a non-expandable canal formed by the metatarsal bones dorsally and abaxially, and the deep plantar fascia plantarly.
  - increased pressure caused by ligament enlargement may cause impingement and necrosis of the neurovascular structures.
    - ♦ compressive damage of the deep branch of the lateral plantar nerve has been identified in horses with PSD.
- Proximal entheseopathy of the proximal palmar/plantar metacarpus or metatarsus:
  - proximal attachment of the SL.
  - commonly diagnosed on MR images of horses with proximal metacarpal/metatarsal pain with or without evidence of desmopathy.

Clinical presentation

- lameness that varies from mild to severe, often worsening with exercise:
  - may be intermittent.
  - acute onset more common in forelimbs.
  - chronic, intermittent lameness with insidious onset common in hindlimbs.
  - often increased when lunging in sand:
    - ♦ especially with affected limb on the outside.
  - often worse when ridden, especially when rider is sitting on the affected diagonal.
- exercise intolerance or poor performance, without overt lameness, is common.
- few localising signs:
  - occasionally, swelling between the plantar margin of the fourth metatarsal bone and the lateral margin of the SDFT in the proximal palmar/plantar metacarpus/tarsus region.
  - sometimes digital pressure in proximal palmar/plantar metacarpus/tarsus region resented more in the lame limb:
    - ♦ comparison with contralateral limb is always necessary.

Differential diagnosis

- proximal SDFT tendinopathy
- Plantar ligament desmopathy.
- lateral digital flexor tendonitis within the tarsal sheath.
- proximal metacarpal/metatarsal stress fractures.
- avulsion fractures of the proximal palmar/plantar metacarpal/tarsal cortex.
- osteoarthritis (OA) of the distal tarsal joints.

Diagnosis

- **Clinical examination**
  - response to flexion tests is generally non-specific.
- **Diagnostic analgesia**
  - some improvement following distal metacarpal or metatarsal analgesia (low 6-point nerve block) possible due to proximal diffusion.
  - lameness abolished by:
    - ♦ anaesthesia of the palmar metacarpal or plantar metatarsal nerves 2–3 cm distal to the carpometacarpal/tarsometatarsal joint.
    - ♦ direct infiltration of local anaesthetic 'around' the origin of the ligament.
    - ♦ lateral palmar nerve block inside the accessory carpal bone ('Castro block') or ulnar nerve block in forelimb.

- ◆ anaesthesia of the deep branch of the lateral plantar nerve 2–3 cm distal to the head of the fourth metatarsal bone in the hindlimb.
  - – false-positive results due to analgesia of tarsometatarsal joints are common.
- **Radiography**
  - ○ almost always within normal limits.
  - ○ rarely sclerosis, with loss of the trabecular pattern and thickening of the palmar/plantar cortex of the third metacarpal or metatarsal bone (Fig. 8.48).
  - ○ cortical fissures or fragments near the origin of the ligament if avulsion or stress fracture (Figs. 8.49, 8.50).
  - ○ useful to rule out distal tarsal OA if positive nerve block of the deep branch of the lateral plantar nerve.
- **Ultrasonography**
  - ○ most lesions within the proximal 5 cm of the ligament.
  - ○ always compare with the contralateral limb.
  - ○ ultrasonographic signs can be difficult to identify especially in hindlimbs.
    - ◆ artefacts are commonly overinterpreted.

- ○ ultrasonography signs may include:
  - ◆ enlargement of ligament (Fig. 8.51) seen as decreased space between:
    - – ligament and the palmar/plantar metacarpal/metatarsal cortex.
    - – ligament and the check ligament (accessory ligament of the DDFT).
  - ◆ increased width of the ligament.
  - ◆ convex palmar/plantar outline on longitudinal scans with palmar/plantar displacement of the flexor tendons (Fig. 8.52).

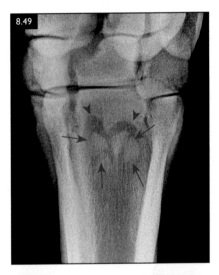

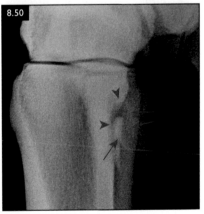

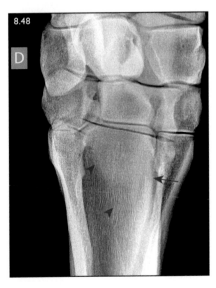

FIG. 8.48 Dorsopalmar radiograph of the proximal metacarpus of a horse presenting with proximal SL desmitis. There is a diffuse increase of mineral opacity and loss of the normal, coarse trabecular pattern in the proximal metacarpus, indicative of sclerosis (delineated by arrowheads and arrow).

FIGS. 8.49, 8.50 Dorsopalmar (8.49) and flexed lateromedial (8.50) radiographs of a horse with avulsion of both heads of the SL. Note the two slightly displaced bony fragments (arrows), leaving a fracture bed visible as a lucent halo with a sclerotic rim (arrowheads).

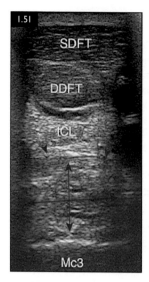

FIG. 8.51 Proximal suspensory desmitis in the forelimb of a horse. The ligament is diffusely enlarged (double arrows) and fills up the entire space between the accessory ligament of the DDFT (ICL = inferior check ligament) and the third metacarpal bone (Mc3). The arrowheads indicate the deep fascia being displaced palmarly against the ICL.

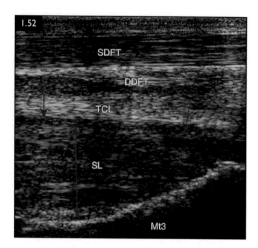

FIG 8.52 On longitudinal views (here in a hindlimb), the enlarged SL (double arrow) fills up the connective tissue space and displaces the deep fascia (short arrows). In severe cases, the plantar ligament margin becomes convex. TCL = tarsal check ligament.

- poor definition of the ligament margins (Fig. 8.53) and loss of the hypoechogenic space (loose connective tissue) around the ligament.
- one or several, diffuse or well-defined hypoechogenic areas within the substance of the ligament (core lesions) (Fig. 8.54).
- diffuse hypoechogenicity of whole ligament (Fig. 8.55), or a varying portion (Fig. 8.56).
- diffuse tears that only affect the dorsal one-third to one-half of the ligament (Fig. 8.57).
- hyperechogenic areas in chronic cases, particularly in the hindlimbs (Figs. 8.58–8.60):
  - occasionally acoustic shadowing due to mineralisation.
- irregularity of the palmar (plantar) surface of the third metacarpal (metatarsal) bone (Fig. 8.61):
  - entheseophyte production (may occur in non-lame horses).
- loss of the normal fibre pattern in longitudinal scans in acute/subacute cases:
  - discrete, longitudinal core lesion (Fig. 8.62).
  - more diffusely across ligament (Fig. 8.63).

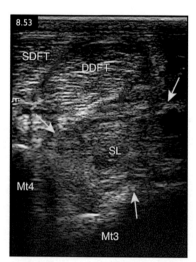

FIG. 8.53 Severe, diffuse proximal suspensory desmitis in the hindlimb of a French warmblood. The suspensory ligament (SL) is markedly enlarged, displacing the DDFT and surrounding tissues. The plantar and dorsal boundaries of the ligament are ill defined (yellow arrows) and merge with the surrounding connective tissue, the deep plantar fascia and the tarsal check ligament (red arrow).

- poor longitudinal fibre alignment in healed or chronic cases (Fig. 8.64).
- avulsion fragments usually located (Figs. 8.65, 8.66):

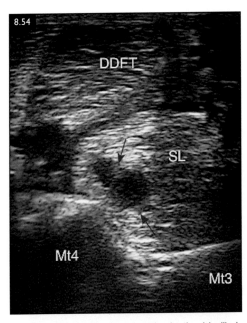

FIG. 8.54 Acute SL desmopathy in the hindlimb of a French Trotter. A well-defined, hypoechogenic lesion is visible in the lateral part of the SL (arrows). The SL is moderately enlarged. Mt3 and 4 = third and fourth metatarsus bone.

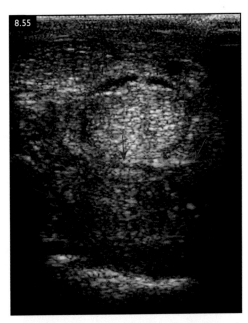

FIG. 8.55 Acute, severe and diffuse hindlimb proximal suspensory desmopathy. The SL is severely enlarged, with ill-defined margins (marked by arrows). The whole cross-section of the SL origin is very hypoechogenic with loss of the normal granular pattern.

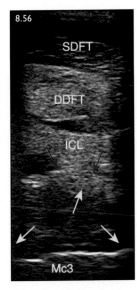

FIG. 8.56 Proximal suspensory desmitis in the forelimb of a French Trotter. The SL is moderately enlarged, displacing the fascia (red arrows). There is diffuse loss of echogenicity over most of the SL's cross-section (yellow arrows). The palmar one-third of the ligament remains intact (between the red and yellow arrows).

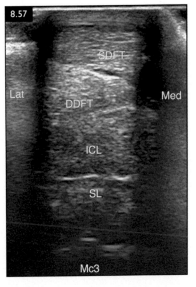

FIG. 8.57 Focal lesion involving the dorsal aspect of the ligament (arrowheads). Note the prominent sagittal ridge of the palmar third metacarpus (arrow). This is a normal variant and should not be confused for an entheseophyte or avulsed fragment.

- several centimetres distal to the ligament origin, dorsal to the SL.
♦ chronic proximal SL desmopathy in hindlimbs often only subtle changes:

- mild enlargement.
- heterogeneous and poorly organised structure and echogenicity in both transverse and longitudinal planes (Figs. 8.67, 8.68).

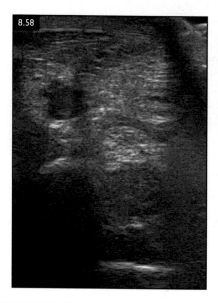

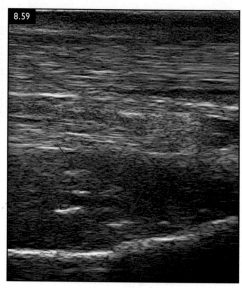

FIGS. 8.58, 8.59 Transverse (8.58) and longitudinal (8.59) scan of the proximal metatarsus. The SL is diffusely enlarged and hypoechogenic with complete loss of its normal architecture. Focal hyperechogenic foci are visible (arrows), these are not casting shadows and probably represent fibrous tissue or partially mineralised material.

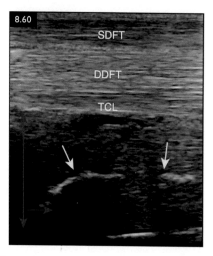

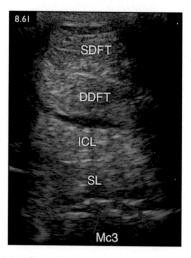

FIG. 8.60 Longitudinal scan of the proximal metatarsus in a horse with chronic proximal desmopathy. Mineralised foci induce discrete hyperechogenic interfaces (yellow arrows), casting acoustic shadows (red arrows). Double arrow = SL.

FIG. 8.61 Chronic, proximal SL desmopathy in the forelimb. Entheseophytes are represented by irregular, hyerchogenic interfaces protruding from the third metacarpal surface (Mc3) into the SL origin (arrows).

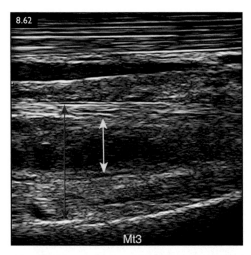

FIG. 8.62 Acute lesion affecting the proximal portion (origin) and extending into the body of the SL in the hindlimb of a French Trotter. Note the enlarged ligament (red arrow) and discrete, hypoechogenic lesion (yellow arrow) with complete loss of the normal fibre pattern.

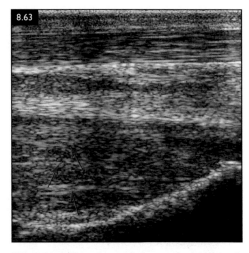

FIG. 8.63 Diffuse loss of the normal longitudinal fibre pattern. This is an acute episode in a chronically evolving desmopathy. Hypoechogenic areas (haemorrhage and oedema) are spreading through a heterogeneous ligament with abnormal clusters of echogenic areas (arrows).

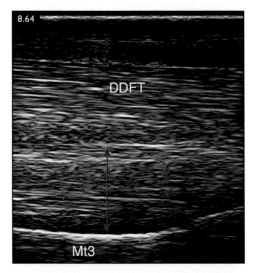

FIG. 8.64 Healed focal lesion. The lesion (dashed arrow) is isoechogenic to the rest of the SL (plain arrow) but the interfaces are short and irregular, creating a granular pattern. This represents immature, or poorly organised fibrous tissue.

- Scintigraphy
  - increased radionucleotide uptake in the area of the origin of the ligament in:
    - approximately 10% of cases:
      - useful to rule out tarsometatarsal joint disease.
      - high specificity but low sensitivity for PSD.
- CT
  - thickening of the ligament and adhesions may be identified on soft tissue windowing.
  - very sensitive for bone abnormalities associated with an injured enthesis.
- MRI
  - most sensitive and accurate technique to diagnose PSLD is high-field MRI.
  - significantly more accurate, sensitive and better specificity than ultrasonography:
  - Low-field MRI more affected by motion artefacts, especially in hindlimbs.
    - may preclude soft tissue assessment.
    - particularly useful for detection of bone pathology:

Management

Forelimb proximal suspensory ligament desmopathy

- conservative treatment as for SDFT tendinopathy:
  - initial box rest with in-hand walking for 5–10 minutes 2–4 times daily for 4 weeks, then gradually increasing to

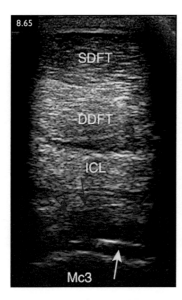

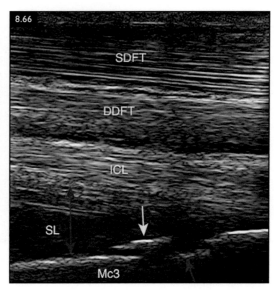

FIGS. 8.65, 8.66 (8.65) Avulsed fragment, forming a sharp interface in the dorsal part of the SL (yellow arrow). A diffuse, hypoechogenic lesion is present in the parenchyma surrounding the fragment (red arrows). (8.66) Longitudinal scan of 8.65 (medial parasagittal section). The avulsed fragment (yellow arrow) is displaced palmarodistally. The SL (double arrow) is hypoechogenic with diffuse loss of fibre alignment. An irregular defect in the palmar aspect of the third metacarpus (Mc3), proximal to the fragment, represents the fracture bed (red arrows).

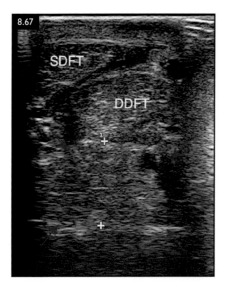

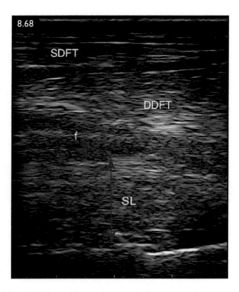

FIG. 8.67 Transverse scan of the proximal metatarsus. In this horse with chronic, proximal SL desmopathy, the SL is markedly enlarged (arrows). The parenchyma is nearly normal in echogenicity but is heterogeneous, with poorly demarcated hypo- and hyperechogenic areas. The connective tissue space between the SL and DDFT is obliterated.

FIG. 8.68 Longitudinal scan in the same horse as 8.67, showing diffuse loss of fibre alignment, with hyperechogenic interfaces mixed with hypoechogenic, finely granular areas. The SL (long double arrow) is enlarged and convex on the plantar aspect. The fascia (f) is displaced and a diffuse, moderately echogenic fibrous tissue fills the space (short double arrow).

20 minutes in-hand walking or in a mechanical walker over a 4–8-week period.
- local hydro- or cryotherapy.
- foot balance should be addressed.
- shoeing is controversial as a treatment aid:
  - ◆ Egg bar shoes (long and wide at the heels) to prevent fetlock sinking in horses with long, sloping pastern.
  - ◆ 'Suspensory shoes' (wide toe, narrow branches) to straighten fetlock in horses with short pastern.
- follow-up ultrasonography at 12 weeks.
- return to work:
  - ◆ between 3- and 6-months post injury for acute injuries in the forelimbs.
  - ◆ absence of evolution of lesion on two consecutive 3-monthly ultrasonographic examinations for chronic cases.
- peri- or intraligamentous injections:
  - IRAP, PRP, MSCs, sodium hyaluronate, or glycosaminoglycans have been used with limited evidence base.
  - intraligament PRP most popular to improve ultrasonographic appearance.
  - periligament short-acting corticosteroids to eliminate early swelling, may avoid compartment syndrome.
- surgical options for rare refractory cases:
  - osteostixis of the proximopalmar cortex:
    - ◆ horses with persistent lameness and extensive osseous pathology.
  - neurectomy of the deep branch of the lateral palmar nerve:
    - ◆ cases with a positive response to diagnostic analgesia.
    - ◆ successful outcome described in a small series.

Hind limb proximal suspensory desmopathy

- conservative treatment as described as above but lameness persists in 80% of cases:
  - worse with increasing severity of the lesion.
  - osseous lesions decrease the chances of full recovery.

- shock-wave therapy:
  - improves lameness with 41% of cases returning to work after 6 months in one study.
- corrective foot trimming and shoeing:
  - improve poor foot balance.
  - shoeing controversial: egg bar shoes with long heels or 'suspensory shoes' with wide toe and narrow branches?
- periligament injections of short-acting corticosteroids:
  - decrease swelling and compartment syndrome.
  - may be detrimental as inhibits healing.
- intralesional orthobiologics:
  - few published reports to support their efficacy.
  - MSCs.
  - PRP.
- surgically based treatments are more commonly used for hindlimb PSD:
  - fasciotomy of the deep plantar fascia to relieve compartment syndrome.
  - desmoplasty (longitudinal tenotomy).
  - osteostixis.
  - neurectomy of the deep branch of the lateral plantar nerve (DBLPN) alone or combined with fasciotomy:
    - ◆ return to full work for 70–80% of selected cases.
    - ◆ **avoid neurectomy in horses with:**
      - – large core lesions.
      - – poor predisposing conformation (straight hocks and low fetlocks).
        - ◇ condition may worsen and lead to suspensory breakdown.
- summary:
  - initial treatment with rest and shock-wave therapy
  - surgery recommended for appropriate cases if medical therapy has failed.

Prognosis

- good for return to exercise in the forelimb (90%).
- guarded to poor for return to exercise in the hindlimbs with conservative treatment alone (22%).
- good for return to exercise in hindlimbs with neurectomy/fasciotomy for selected cases (70–80%).

Avulsion fracture of the origin of the suspensory ligament

Overview

- aetiology similar to proximal desmopathy but separate condition.
- affects all types of horses but most common in Standardbred racehorses.

Clinical presentation

- sudden-onset, mild to severe lameness, often during or shortly after exercise or racing.
  - usually unilateral, occasionally bilateral.
- most common in forelimbs.
- no obvious localising signs as for proximal desmopathy.

Differential diagnosis

- as per proximal suspensory desmopathy and entheseopathy.
- Palmar cortical stress fractures of the third metacarpal bone.

Diagnosis

- diagnostic analgesia of the origin of the SL.
- **Radiography** (Figs. 8.49, 8.50):
  - fragment visible on lateromedial radiographs:
    - ♦ displaced mineralised body immediately palmar/plantar to cortex.
    - ♦ 2–4 cm distal to carpometacarpal/ tarsometatarsal joint.
  - curvilinear to circular radiolucent line may be visible on dorsopalmar/plantar projections.
  - marked sclerosis in proximopalmar/ plantar aspect of third metacarpus/ metatarsus.
- **Ultrasonography** (Figs. 8.65, 8.66)
  - always concurrent damage to dorsal aspect of SL at its origin.
  - displaced fragment – hyperechogenic interface within the dorsal part of the ligament.
  - irregular margin (callus formation or periosteal new bone):

  - ♦ delineates bony interface proud of palmar/plantar metacarpal/ metatarsal bone surface.
- Scintigraphy
  - sensitive technique to detect IRU suggestive of avulsion fragments.
- MRI / CT
  - cross-sectional imaging techniques define the fragment most accurately.
  - MRI shows concurrent bone oedema.
  - MRI and CT are sensitive indicators of concurrent sclerosis.

Management

- conservative management frequently effective:
  - strict stall rest for 6 weeks.
  - stall rest with walking exercise for a further 6 weeks.
  - controlled exercise programme started at 3–6 months, when lameness is absent.
  - suggested ancillary treatments:
    - ♦ shock-wave ♦ perilesional injections.
  - intralesional injections of orthobiologics if concurrent severe desmopathy.
- surgical treatment suggested without much evidence, especially for hindlimb lesions:
  - osteostixis.
  - lag screw fixation in selected cases.

Prognosis

- good for simple avulsion.
- guarded for avulsion with severe desmopathy, especially in hindlimbs.

Desmopathy of the suspensory ligament body

Overview

- most common in racehorses:
  - forelimbs in Thoroughbreds.
  - both fore and hindlimb in Standardbreds and European Trotters.
- uncommon in sports horses:
  - very occasionally in association with exostosis of a splint bone.

Clinical presentation

- severe lameness at outset which decreases gradually within 3–6 weeks.

- obvious swelling and oedema of the SL in the mid-metacarpal region.
- pain on focal pressure of the SL margins.
- may eventually lead to complete breakdown, particularly in Standardbreds.

Differential diagnosis

- superficial digital flexor tendonitis
- inferior check ligament desmitis.
- fracture of a splint bone.

Diagnosis

- **Clinical eaxmination**
  - palpation.
  - diagnostic analgesia rarely required.
- **Radiography**
  - identify concurrent fracture or periostitis of the small metacarpal or metatarsal bones.
- **Ultrasonography**
  - increased CSA of the SL.
  - discrete anechogenic core lesions to diffuse hypoechogenicity of the entire ligament (Figs. 8.69–8.71).
  - chronic injuries may have acoustic shadowing (Figs. 8.72, 8.73):

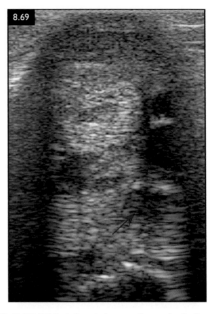

FIG. 8.69 Well-defined, hypoechogenic lesion in the medial part of the SL body (arrow). The ligament is only moderately increased in size around the lesion.

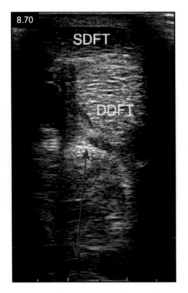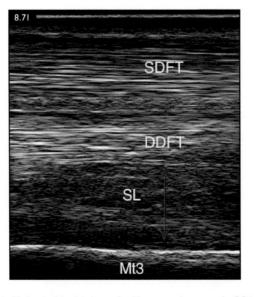

FIGS. 8.70, 8.71 (8.70) Severe tear in a hindlimb SL body (double arrow) with severe increase in CSA displacing the DDFT and tibial collateral ligament and diffusely mottled hypoechogenicity. (8.71) Longitudinal (sagittal) sonogram from a plantar approach in the same horse as 8.70, showing complete loss of fibre alignment and heterogeneous loss of echogenicity. The enlarged SL (double red arrow).fills up the entire space between the third metatarsal bone (Mt3) and the DDFT.

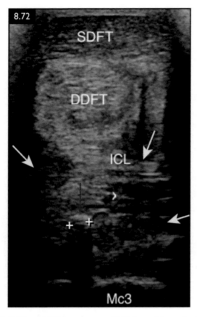
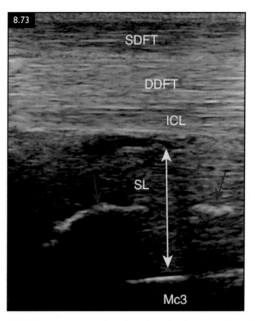

FIGS. 8.72, 8.73 Transverse (8.72) and longitudinal (8.73) sonograms showing chronic desmopathy of the SL body (level IIA). The SL body is markedly enlarged (yellow arrows), and heterogeneous, with a hyperechogenic interface casting acousting shadowing (red arrows).

- ◆ small hyperechogenic areas compatible with focal mineralisation.
- ○ marked periligamentous thickening may exist without signs of desmitis (Fig. 8.74).

- ○ encroachment between an exostosis of a splint bone and the SL may occur:
  - ◆ may be difficult to identify due to presence of marked periligament fibrosis (Figs. 8.75, 8.76).
- ○ dynamic examination may detect adhesions:
  - ◆ examine the limb non-weight bearing.

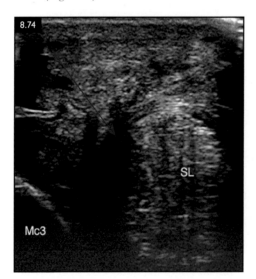

FIG. 8.74 Comminuted fracture of the 2nd metacarpal bone associated with marked periligamentous thickening (arrow) without evidence of damage to the SL.

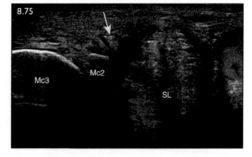

FIG. 8.75 Interference between the splint bone (here Mc2) and the SL body is characterised by soft tissue thickening in the space between the ligament and bone. Here the whole ligament (SL) is enlarged and heterogeneous. There is marked periosteal thickening over the second metacarpal bone (yellow arrow) and diffuse, focal hypoechogenicity representing oedema and/or haemorrhage (red arrows).

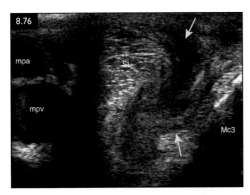

FIG. 8.76 A subacute, displaced fracture of the fourth metacarpal bone with traumatic SL body desmopathy. The SL contains a large, diffuse lesion on its dorsolateral aspect, extending over the dorsal aspect (red arrows). Hypoechogenic tissue (haemorrhage and oedema) fills the triangle left by the displaced fragments (yellow arrows). mpa = medial palmar artery; mpv = medial palmar vein; Mc3 = metacarpal 3.

♦ assess relative movement of the SL in relation to the surrounding tissues.
- **MRI**
  ○ more accurate detection of:
    ♦ focal fibre damage associated with exostosis encroachment on the SL margin.
    ♦ adhesions between the small metacarpal/metatarsal bones and the SL body.

Management
- conservative and medical management:
  ○ rest and controlled exercise:
    ♦ 4–8 weeks of rest with in-hand exercise.
    ♦ followed by slowly increasing amounts of trotting depending on serial follow-up ultrasonographic evaluation.
  ○ faster healing than for the SDFT and most horses can resume work after 4 months.
  ○ biological treatments:
    ♦ intralesional PRP may be effective.
- surgical management:
  ○ longitudinal tenotomy ('splitting') is not usually recommended.
  ○ cautery and blistering are painful, without any benefit and should not be used.

- encroachment of an exostosis:
  ○ local corticosteroid injection coupled with rest may be useful early on.
  ○ surgical adhesiolysis and removal of exostosis.
  ○ partial or segmental amputation of the affected splint bone.

Prognosis
- common for the lesion to persist despite clinical improvement.
- recurrence of injury is common.
- encroachment of exostosis: good prognosis.

Suspensory branch desmopathy

Clinical presentation
- most common injury of the SL:
  ○ all types of horses from racehorses to pleasure riding animals.
- variable lameness depending on the severity of the lesion:
  ○ worse in acute and bilateral cases.
- local signs are usually obvious:
  ○ marked thickening +/- palpable pain of affected branch(es).
  ○ marked oedema extending proximally and distally in acute cases.
  ○ biaxial lesions more common in the forelimb, and lateral lesions in the hindlimb.
  ○ distension of the fetlock and digital sheath synovial pouches may be present in acute stage.
  ○ severe chronic lesions are common in Standardbreds.
- discrete lesions may be undetectable on clinical examination.

Differential diagnosis
- superficial or deep digital flexor tendonitis in the digital sheath.
- tenosynovitis of the digital sheath.
- synovitis of the fetlock joint
- sesamoid bone fractures.
- splint bone fractures.

Diagnosis
- clinical examination:
  ○ localising clinical signs.

- ○ positive fetlock flexion.
- ○ diagnostic analgesia only necessary in discrete focal lesions.
- **Radiography**
  - ○ detect distal entheseopathy at insertion of SL branches on the proximal sesamoid bones ('sesamoiditis').
  - ○ detect distal splint bone fractures that occur frequently concurrently with suspensory branch desmitis.
- **Ultrasonography**
  - ○ increased cross-sectional area of the branch (Fig. 8.77).

- ○ well-defined (Fig. 8.78) or diffuse (Figs. 8.79, 8.80) hypo- to anechogenic lesions with loss of fibre alignment.
- ○ lesions are often axially situated (Fig. 8.81) and may or may not extend to the distal insertion on the sesamoid bone.
- ○ abaxial, focal lesions may be associated with overreach injuries (Figs. 8.82, 8.83).
- ○ marked periligamentous soft tissue thickening:

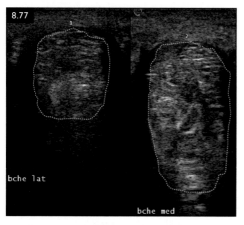

FIG. 8.77 Increased cross-sectional area of the medial SL branch as compared to the lateral branch. This is an acute on chronic injury (note the heterogeneous lateral branch).

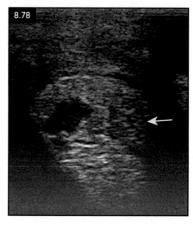

FIG. 8.78 Well-defined, hypoechogenic lesion (red arrows) within the medial branch of the SL due to an acute tear. Note the presence of a more diffusely hypoechogenic area, suggesting an ongoing, chronic injury (yellow arrow).

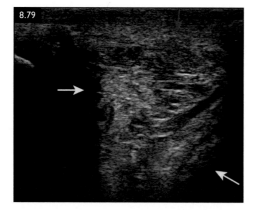

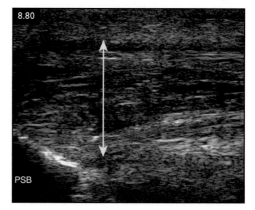

FIGS. 8.79, 8.80 Severe, acute desmopathy of the medial branch of the SL. The lesion is very diffuse with a coarsely heterogeneous appearance on the transverse view (8.79, obtained from a medial approach) and amorphous loss of fibre organisation throughout the thickness of the branch on the longitudinal (frontal) view (8.80). The ligament is severely enlarged (yellow arrows). Note the marked thickening of the paratenon and periligamentous tissues (red arrows).

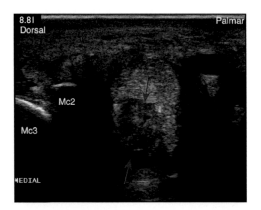

FIG. 8.81 Acute injury in the medial branch of the SL (transverse medial approach). A hypoecho-genic lesion is visible in the axial portion of the branch (arrows). Note the marked periligamentous thickening due to oedema.

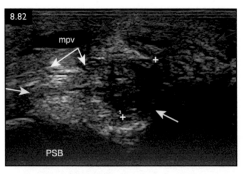

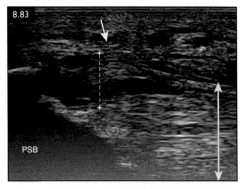

FIGS. 8.82, 8.83 Transverse (8.82) and longitudinal (8.83) unasonograms of a traumatic overreach injury over the medial SL branch. with subcutaneous tissue contusion and haemorrhage (red arrows), transection of the fascia (white arrows) and focal, hypoechogenic lesion on the superficial part of the SL (calipers). The yellow arrows point to the limits of the SL branch. mpv = medial palmar vein

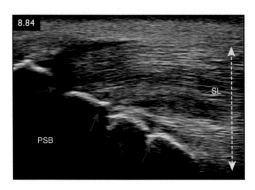

FIG. 8.84 Chronic entheseopathy of the SL branch insertion with a combination of new bone production and defects (widened vascular channels) (red arrows) leding to the typical 'sesamoiditis' appearance on radiographs. Note the loss of fibre alignment at the SL insertion.

- ♦ difficult to locate the actual contours of the thickened ligament branch.
- ○ irregularity or bone formation (entheseopathy) at the proximal margin of the sesamoid bone may be present (Figs. 8.84, 8.85).
- ○ marked ectopic mineralisation around the edges of, or within the ligament:

- ♦ possible in chronic or recurrent cases (Fig. 8.86).
- ○ avulsion fragments produce a sharp hyperechogenic interface within the branch several millimetres or centimetres from the ligament's insertion (Figs. 8.87, 8.88):
  - ♦ hypoechogenic material between the fragment and parent bone.

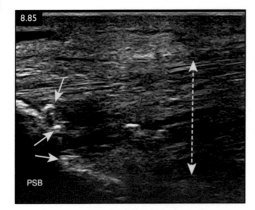

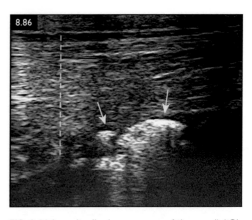

FIG. 8.85 Longitudinal frontal plane sonogram of the medial branch of the SL (dotted yellow line). Chronic desmopathy associated with heterogeneous parenchyma, loss of fibre alignment and periligamentous thickening. New bone spurs at the ligament–bone interface (yellow arrows) represent entheseopathy. Discrete ectopic mineralisation is also present (red arrow) with a hypoechogenic lesion.

FIG. 8.86 Longitudinal sonogram of the medial SL branch with chronic injury (note the loss of fibre alignment). Ectopic mineralisation forms sharp, hyperechogenic interfaces (arrows) casting shadowing artefacts at the axial border of the branch, most likely in the paratenon. Dotted yellow line = extent of SL branch.

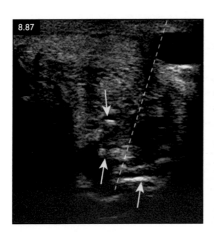

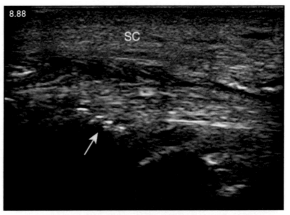

FIGS. 8.87, 8.88 Transverse (8.87) and longitudinal (8.88) (centred on the lesion as indicated by the dotted line) ultrasonograms of a medial SL branch avulsion. The SL ligament is very heterogeneous, irregularly thickened with diffuse hypoechogenic tissue in the deep (axial) part of the ligament. Small, hyperechogenic interfaces casting shadow artefacts represent avulsion fragments (yellow arrows). Red arrows = periligamentous fluid).

- ○ fractures of the distal 1/3 of the splint bones produce:
  - ♦ interruption of the bone interface.
  - ♦ very rarely fibrous tissue adhesions to the SL.
- ○ distal chronic entheseopathy of SL branches specifically in Standardbreds:
  - ♦ symmetrically bilateral with marked thickening of distal portion of the SL branches.

- ♦ diffuse areas with heterogeneous echogenicity and loss of striation of the affected branches (Figs. 8.89, 8.90).
- ♦ hypoechogenic areas in the branch immediately proximal to the insertion.
- ♦ very irregular proximal margin of the sesamoid bones (entheseopathy).

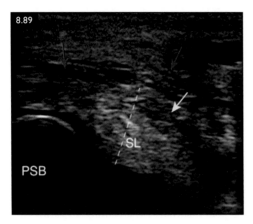 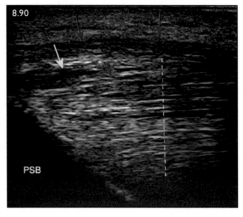

FIGS. 8.89, 8.90 Transverse (8.89) and longitudinal (8.90) sonograms of the medial SL branch (dotted lines) insertion in a French Trotter with non-painful thickening around the proximal sesamoid bones (PSB). A diffuse heterogeneous lesion is noted over the abaxial portion of the SL at its insertion (yellow arrows). Periligamentous thickening extends over the abaxial aspect of the bone (red arrows). This lesion affected both lateral and medial branches in both forelimbs.

Management

- conservative and medical management as for SDFT tendonitis.
  - return to exercise more quickly (3–4 months) with adequate controlled exercise.
  - incomplete ultrasonographic healing tends to evolve as chronic, active lesions.
  - recurrence is extremely common.
- biological treatments:
  - injection is difficult without core lesions.
  - no controlled studies are available.
  - controlled study of use of PRP for chronic entheseopathy in young racehorses failed to show efficacy.
- Extracorporeal shock-wave therapy.
- tendon splitting (longitudinal tenotomy) may be useful in the subacute stage or in the presence of chronic entheseopathy.
- Avulsion fragments:
  - as for desmitis.
  - surgical removal not recommended as invasive.
  - splitting and/or injection of PRP or stem cells.
- distal splint bone fractures:
  - conservative treatment with rest and pressure bandages for 6–8 weeks.
  - most minimally displaced fractures heal satisfactorily.

- surgical removal may be considered if:
  - adhesions (very rare).
  - fragment is severely displaced (yet fibrous union may be sufficient).
  - desmitis of the branch persists or recurs.

Prognosis

- related to the severity of the SL branch lesion:
  - fair for one branch
  - poor for both branches.
- Sesamoiditis does not appear to affect the prognosis.
- dystrophic mineralisation associated with a poorer outcome.
- reinjury has a poor prognosis.

Complete breakdown of the suspensory ligament

Clinical presentation

- extremely rare.
- most common in Thoroughbred racehorses, particularly those racing over fences.
- often involves fracture of both proximal sesamoid bones (PSBs) and/ or breakdown of the distal sesamoidean ligaments with loss of suspensory support to the fetlock.

- acute, sudden-onset lameness with decreased weight bearing, marked dropping of the fetlock and diffuse oedema.
- progressive weakening/breakdown of the hindlimb suspensory:
  - occasional complication of chronic proximal SL desmopathy.
  - degenerative disease in old broodmares.
  - progressive hyperextension of fetlock, usually with a straight hock conformation.

Diagnosis

- **Radiography**
  - fetlock views to rule out concurrent phalangeal, metacarpal/tarsal or proximal sesamoid bone fractures.
  - rule out avulsion at the attachments of the suspensory sesamoidean ligaments (see Figs. 4.25, 4.26).
- **Ultrasonography**
  - complete loss of definition of the SL in both transverse and longitudinal views in acute rupture (Figs. 8.91, 8.92).
  - heterogeneous haemorrhagic tissue between the frayed ends of the rupture.
  - body and/or both branches affected (Fig. 8.93).

Management

- goal is salvage for breeding or pasture soundness.
- conservative treatment:
  - fibreglass casts, followed by Robert Jones bandaging and splints for several months.

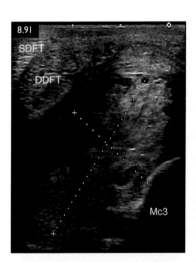

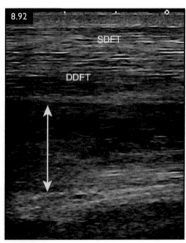

FIGS. 8.91, 8.92 Transverse (8.91) (from a palmaromedial approach) and longitudinal (8.92) (sagittal) ultrasound scans of the palmar metacarpus in zone II. The SL is very enlarged, hypoechogenic and with complete loss of visible fibres. The margins are ill defined (dotted lines and double yellow arrow). There is a mixture of slightly echogenic but homogeneous fibrinous tissue and anechogenic fluid. The periligamentous tissues are thickened due to oedema (red arrows).

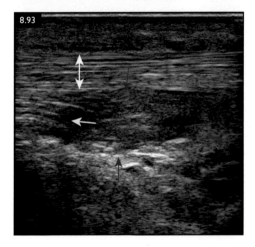

FIG. 8.93 Longitudinal scan from the medial aspect of the metacarpus (frontal plane zone IIB/IIIA) at the level of the SL bifurcation. There is partial rupture of the SL body and branch. Part of the SL remains intact with regular fibre alignment (white double arrow). The ruptured distal end (yellow arrow) floats in amorphous, hypoechogenic tissue (haematoma) (red arrows). Note the marked, hypoechogenic periligamentous and subcutaneous tissue thickening.

- successful in some horses, although severe pressure sores, fetlock OA and contralateral laminitis are common complications.
- surgical arthrodesis of the fetlock is the preferred treatment method for salvage.

Prognosis

- hopeless for sports activities.
- poor for pasture soundness with conservative management.
- fair for pasture and breeding soundness after surgical arthrodesis.

Distal sesamoidean ligament desmopathy

Definition/overview

- Short (deep) and cruciate distal sesamoidean ligament (DDSL) injury:
  - rare but probably underdiagnosed as these ligaments are difficult to image (Fig. 8.94).
  - may be a component of plantar fragmentation of the proximal phalanx or basilar fracture of the PSBs.
- Oblique distal sesamoidean ligament (ODSL) and straight distal sesamoidean ligament (SDSL) injury:
  - part of suspensory apparatus breakdown injuries:
    - includes SL desmitis and PSB transverse/basilar fractures.

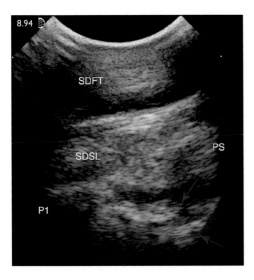

- rare but prevalence increased with MRI diagnosis.
- may result from overreach injuries.
- severe injury can lead to PIP joint subluxation.

Aetiology/pathophysiology

- unclear:
  - trauma (overreach injury).
  - single strain injury, possibly due to asymmetric overextension of the fetlock.
  - chronic, recurrent strain.
- occasionally concurrent SL branch injuries.
- occasionally with avulsion fractures of attachment sites:
  - base of the PSBs, proximal phalanx, or middle scutum.

Clinical presentation

- lameness is usually marked to moderate depending on chronicity and severity.
- distension of digital sheath or fetlock joint are rare.
- ODSL injury sometimes with focal thickening on the palmarolateral or palmaromedial aspect of the proximal phalanx.
- often few or no localising signs.
- pain on passive digital flexion and hyperextension.

Differential diagnosis

- digital tenosynovitis
- deep digital flexor tendonitis in the pastern.
- tendonitis of the branches of the SDFT
- basilar fracture of a PSB.
- digital palmar/plantar annular ligament injuries.

FIG. 8.94 Long-axis, parasagittal scan of the palmaromedial fetlock. A microconvex probe is placed between the axial border of the base of the medial PSB and P1, just medial to the ergot. The normal deep distal sesamoidean ligaments (arrows) are visualised deep to the straight distal sesamoidean ligaments (SDSL). The latter originates from the proximal scutum (PS) and base of the PSBs. The scan plane catches the medial edge of the SDFT.

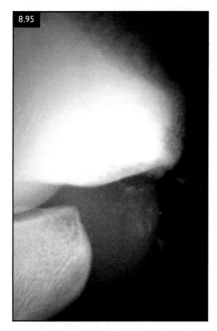

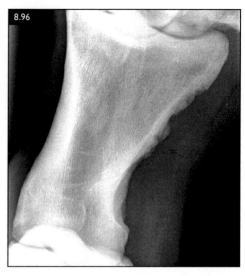

FIG. 8.96 Oblique radiograph of the pastern region. New bone production along the palmaroabaxial aspect of P1 represents entheseous new bone production at the insertion of the ODSLs and is generally considered to be an incidental radiographic finding (arrows).

FIG. 8.95 Oblique radiograph of the fetlock region of a horse with chronic oblique distal sesamoidean ligament desmopathy showing entheseophytic new bone production at the distal border of the medial proximal sesamoid bone (arrow) and ectopic mineralisation within the soft tissues in the area of projection of the ODSL (arrowhead).

Diagnosis

- diagnostic analgesia:
  - improvement following palmar digital nerve block or intrathecal anaesthesia of digital sheath.
  - abolished by abaxial sesamoid nerve block.
- **Radiography**
  - remodelling (entheseopathy) at the ligaments' attachment sites:
    - particularly at the base of the PSBs (Fig. 8.95).
  - rule out other associated proximal sesamoid bone lesions.
    - fractures, displacement, and osteitis.
  - new bone palmar aspect of the diaphysis of the proximal phalanx:
    - entheseopathy of the ODSLs (Fig. 8.96).
    - incidental finding often without lameness.
  - PIP joint subluxation may imply damage to:
    - middle scutum
    - palmar/plantar capsule of the PIP joint.
    - straight DSL
    - SDFT insertion branches.
  - rarely avulsion fracture of the palmar/plantar proximal aspect of middle phalanx.
  - proximal displacement of both PSBs when complete disruption of the DSLs (Fig. 8.97).
  - basilar fragmentation of the PSBs may represent avulsion of the DSLs (Fig. 8.98).
- **Ultrasonography**
  - difficult anatomical area to scan – requires experience.
  - short and cruciate DSLs:
    - microconvex array transducer (Fig. 8.94).
    - marked enlargement.
    - hypoechogenic
    - heterogenous echogenicity.
    - bone remodelling at the attachment sites.
    - complete rupture rare (Fig. 8.99).
  - ODSL desmitis:
    - obvious thickening

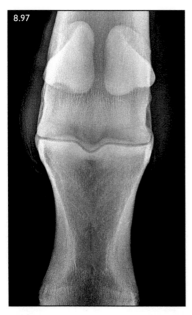

FIG. 8.97 Dorsopalmar non-weight-bearing radiograph showing proximal displacement of the sesamoid bones associated with complete rupture of the DSLs.

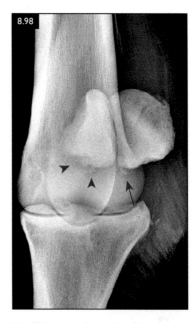

FIG. 8.98 Oblique radiograph of the same horse as in 8.97. Note the multiple fragmentation at the base of both PSBs (arrows).

- ♦ heterogeneous decrease in echogenicity.
- ♦ diffuse or focal lesions
- ♦ most often near the sesamoidean origin.
- ♦ one or both ODSLs affected.
- ♦ marked roughening and irregularity of bone surface at attachment sites.
- ♦ loss of normal fibre pattern and mineralisation in chronic injuries (Fig. 8.100).
  - ○ Straight DSL injuries:
    - ♦ most often localised in the middle portion of the ligament.
    - ♦ discrete, hypo- to anechogenic core lesion.

- ♦ focal or diffuse lesions (Figs. 8.101, 8.102).
- ♦ rupture is uncommon.
- ♦ chronic lesions characterised by:
  - – marked increase in CSA.
  - – loss of normal fibre alignment.
  - – abnormal heterogeneity and hyperechogenic foci.
  - – focal, ectopic mineralisation possible.

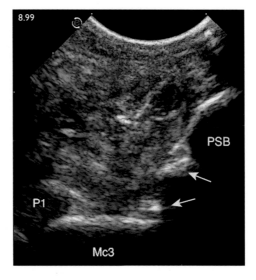

FIG. 8.99 Similar view as in 8.94. This horse sustained partial breakdown of the DSLs. The short (deep) DSLs are ruptured with frayed proximal fragments visible (yellow arrows). The straight DSL was partially torn from the proximal scutum, with amorphous tissue filling the space between the scutum and ligament (red arrow). This was associated with fragmentation of the base of the PSB (not visible in this image).

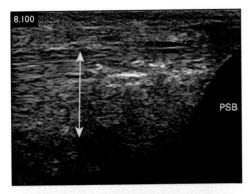

FIG. 8.100 Longitudinal scan over the base of the medial PSB showing a thickened ODSL (yellow double arrow) with loss of fibre alignment in the proximal portion and hyperechogenic foci (red arrows), representing chronic fibrosis and mineralisation.

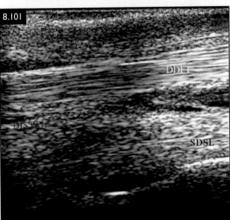 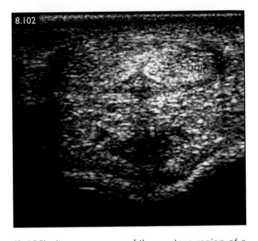

FIGS. 8.101, 8.102 Longitudinal (8.101) and transverse (8.102) ultrasonograms of the pastern region of a riding horse which sustained an acute injury of the SSL. Note the enlarged ligament in both planes and central area of disrupted fibres.

- **MRI and CT**
  - identify small focal or complex DSL injuries, overlooked with ultrasonography.
  - determine true significance of false-positive sonographic findings.
  - useful to look for proximal DSL injuries.

Management

- conservative treatment with rest and controlled exercise as for SL branch desmitis.

Prognosis

- results from recent retrospective study:
  - 55% of 51 horses sound.
  - 33% returned to previous level of performance.
  - 45% persistent or recurrent lameness.

Intersesamoidean ligament desmopathy

Definition/overview

- Intersesamoidean ligament (ISL):
  - transversely orientated fibres join the two PSBs.
  - dorsal surface continuous with the fetlock joint.
  - palmar surface continuous with proximal scutum and digital flexor tendon sheath.
- injury rare but can lead to severe, persistent lameness.
- post-mortem prevalence of 26% in one study, mostly without clinical signs.

Aetiology/pathophysiology

- unclear but injury probably due to mechanical overload.

- Friesian horses appear to be particularly predisposed.
- septic desmopathy associated with osteitis of the axial border of the PSB, infectious arthritis or tenosynovitis.

Clinical presentation

- lameness usually acute and severe with painful flexion.
- diagnostic analgesia with marked improvement following:
  - low 4-point nerve block.
  - intrathecal analgesia of digital sheath.
  - intra-articular anaesthesia of the fetlock joint.

Differential diagnosis

- any cause of lameness in the fetlock joint or the digital flexor tendon sheath.

Diagnosis

- **Radiography**
  - remodelling of axial borders of the PSBs with irregular erosions, fragmentation, and mineralisation within the intersesamoidean space (see Figs. 4.23, 4.24).
  - several dorsopalmar/plantar projections at different angles.
  - avulsion fractures are usually associated with a longitudinal radiolucent line separating a thin fragment on the axial surface of the PSB.
  - occasionally visible widening of the space between the two PSBs in severe cases.
  - septic osteitis is characterised by marked, irregular lysis of the axial border of one or both PSBs, which may extend to the flexor surface of the bone.
  - discrete entheseopathy of axial borders of the PSBs occasionally encountered as an incidental finding.
- **Ultrasonography**
  - examination from a palmar/plantar approach.
  - alterations of the ISL include:
    - ♦ decreased and heterogeneous echogenicity.
    - ♦ irregular bone-to-ligament interfaces representing either fragmentation or entheseophytosis (Figs. 8.103, 8.104).
    - ♦ dorsopalmar ligament thickening.

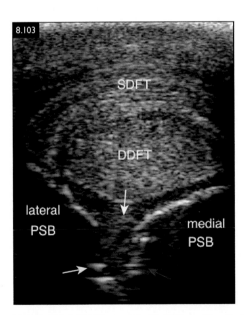
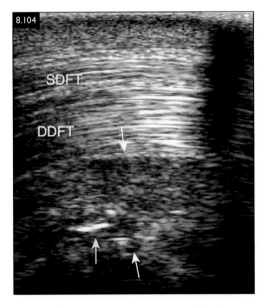

FIGS. 8.103, 8.104 Transverse (8.103) and longitudinal (8.104) scans over the palmar aspect of the fetlock. The axial border of the PSBs is irregular (red arrow) with a bone fragment visible just off the axial border of the lateral PSB (yellow arrow). The ISL(white arrows) is heterogeneous with loss of transversely arranged fibre pattern on the transverse image.

- heterogeneous appearance, mineralisation of the fibrocartilage and palmar axial surface of the PSBs (erosions).
  - loss of transversely arranged linear pattern.
  - rarely, complete rupture with penetration of hypoechogenic synovial fluid.
- **MRI and CT**
  - structural detail of the ISL and presence of bone oedema (MRI).
  - identify subtle axial margin resorption (CT).
  - contrast enhancement of the ligament (CT).
- **Scintigraphy**
  - occult lameness to identify increased radionuclide uptake in the ISL and or PSBs.

Management and Prognosis

- arthroscopic debridement of necrotic/ septic part of the ligament and the axial margins of the PSBs with a fair prognosis for return to function in recent cases.
- poor prognosis with conservative treatment, or in chronic cases.
- guarded for specific condition in Friesians.

Inferior check ligament (accessory ligament of the deep digital flexor tendon) desmitis

Definition/overview

- similar to SDF tendinopathy.
- partial to complete fibre tearing in the inferior check ligament.
- occasionally only superficial fibres of the accessory ligament are involved.
- all types of horses may be affected.
- most common in the forelimb, rarely in the hindlimb.

Aetiology/pathophysiology

- aetiology of spontaneous tears is unclear:
  - probably involves gradual, age-related accumulation of microdamage and degeneration, predisposing to an acute tear.

- differential stretching of the SDFT and the inferior check ligament causing tearing of the superficial accessory ligament fibres.
- common in general purpose or retired ponies and in older animals used for jumping:
  - occasionally in young animals, in association with a sudden-onset distal interphalangeal (DIP) joint flexural deformity.
- adhesions between the flexor tendons and the check ligament:
  - may result from focal trauma or chronic tendinopathy and desmopathy.
  - may predispose to check ligament tears or to recurrence of injury.

Clinical presentation

- swelling over the middle or proximal palmar metacarpus/metatarsal area:
  - centred between flexor tendons and the metacarpus, and more prominent laterally.
  - pain on palpation over swelling.
  - variable size of swelling:
    - mild, laterally localised swelling when tearing of superficial accessory ligament fibres.
    - severe diffuse swelling when tearing in the ligament.
- lameness variable:
  - swelling without lameness is common.
  - acute lameness that occurs during, and immediately after, exercise.
  - chronic progression possible with recurrent bouts of swelling and lameness.
- occasionally, flexural deformity of the DIP joint with persistent lameness and postural limb changes (Fig. 8.105), especially in hindlimbs.

Differential diagnosis

- all causes of diffuse soft tissue swelling of the proximal palmar metacarpal region.

Diagnosis

- **Clinical examination**
  - swelling
  - pain on palpation and flexion of limb.
- **Ultrasonography**

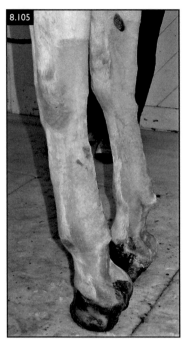

FIG. 8.105 Chronic inferior check ligament injury associated with severe, acquired loss of extension of the DIP joint ('flexural deformity').

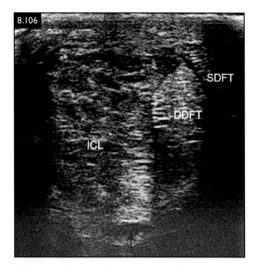

FIG. 8.106 A transverse scan from a lateral approach in the mid-metacarpal region. Subacute ICL desmitis with a markedly enlarged, ICL with a mottled, heterogeneous echogenicity. (arrows denote margin of ICL).

- ○ palmar/plantar and lateral approaches to assess the ligament.
- ○ dynamic evaluation (flexing/extending the limb) to detect adhesions and complete rupture.
- ○ ultrasonographic signs include:
  - ◆ markedly decreased echogenicity throughout ligament (mostly diffuse and heterogenous) (Fig. 8.106).
  - ◆ increased CSA of the AL-DDFT (Figs. 8.107, 8.108).
  - ◆ near-complete rupture with complete loss of parenchyma is rare (Fig. 8.109).
  - ◆ core lesions generally in the middle section of the ligament.
  - ◆ occasionally only tearing of the superficial accessory ligament fibres.
  - ◆ scarring with increased echogenicity with persistent and marked thickening (Fig. 8.110).
  - ◆ lateral adhesions with the deep and superficial digital flexor tendons are common in chronic cases (Figs. 8.109, 8.110).

- ◆ secondary ('sympathetic') effusion within the carpal flexor tendon sheath is common.

Management

- rest and conservative treatment as for tendonitis are usually effective:
  - ○ healing is faster than for the SDFT.
- physiotherapy with controlled exercise and passive manipulations to decrease restrictive adhesion formation.
- intralesional injections of biological products (PRP, stem cells) are probably not indicated.
- corrective farriery:
  - ○ trimming: shorten the toe and leave slightly long heels.
  - ○ shoeing: rolled toe to facilitate breakover.
  - ○ avoid heel wedges long term as may cause retraction of the ligament and flexural deformity.
- surgical resection (inferior check ligament desmotomy or desmectomy):
  - ○ chronic recurrent cases.
  - ○ secondary flexural deformity of the DIP joint (with toe extension shoes).

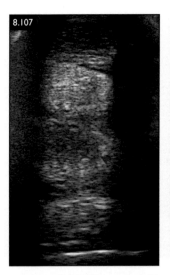
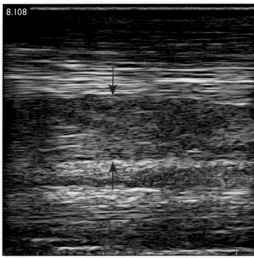

FIGS. 8.107, 8.108 (8.107) Acute desmitis of the accessory ligament of the DDFT (inferior check ligament). A large, diffuse and poorly delineated, hypoechogenic lesion is visible in the body of the ICL (arrows). (8.108) Same horse as 8.107, longitudinal (sagittal) scan.

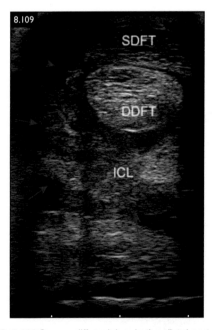

FIG. 8.109 Severe diffuse injury in the distal part of the ICL with near-complete loss of normal structure and loss of the ligament margins (arrows). Note the mottled tissue extending laterally to the edge of the SDFT, with associated loss of definition of its lateral border (arrowhead).

FIG. 8.110 A chronic /healed injury where the echogenicity is near-normal, although heterogeneous, and the ligament remains markedly enlarged with poorly defined margins (arrows). Note the poorly defined interface between the AL-DDFT, the dorsal border of the DDFT and the lateral border of the SDFT (arrowheads).

Prognosis

- overall fair prognosis with two-thirds of cases making a complete recovery.
- more guarded in chronic recurrent cases.
- poor if secondary flexural deformity, especially in hindlimbs.

Deep Digital Flexor tendinopathy (outside sheathed areas)

Overview

- injury to unsheathed portions of the DDFT is rare.
- occasionally seen as:
  ○ distal extension of inferior check ligament lesions.
  ○ lesion observed in young Thoroughbreds in training.
- cause unclear.
- treated conservatively with success.

Paratendonitis

Definition/overview

- Paratenon is connective tissue layer that ensheaths tendons outside synovial sheaths and bursae.
- adheres to the tendon parenchyma:
  ○ made of non-aligned, loose connective tissue.
  ○ continuous with the endotenon.
  ○ contains network of blood vessels that supply the tendon.
- organised in a coiled manner to allow for tendon elongation.
- Peritendinopathy, paratendonitis or bandage bow represents injury of the paratenon, with or without tendinopathy, with possible consequences of:
  ○ thickening and fibrosis   ○ adhesions.
  ○ decreased vascular supply to the tendon.
  ○ persistent pain   ○ exercise intolerance.

Aetiology/pathophysiology

- inflammation and/or haemorrhage in the contiguous tendon parenchyma.
- significant component of tendon lesions that extend to the peripheral margins of the tendon or ligament:
  ○ injuries of SL branches, SDFT, and AL-DDFT.
- direct trauma from kicks, interference injuries, hitting fixed objects, or pressure from tight or slipped bandages (Fig. 8.111).

Clinical presentation

- most common sites:
  ○ digital flexor and extensor tendons.
  ○ AL-DDFT, SL, and the common calcaneal tendon.
- marked swelling with oedema and pain on palpation.
- appearance often indistinguishable from tendinopathy/desmitis (Fig. 8.112).
- plantar aspect of the tarsal region typical site for SDFT paratendonitis:
  ○ Curb deformity (Fig. 8.113).
- variable lameness depending on duration, and involvement of tendon or neurovascular bundles (palmar or plantar nerves, dorsal metacarpal nerves etc.).

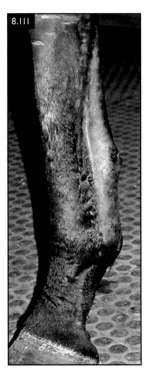

FIG. 8.111 Self-incurred traumatic injury to the palmar aspect of the metacarpus. Note the associated swelling ('bowed leg') due to contusion, oedema and haemorrhage, in and around the paratenon.

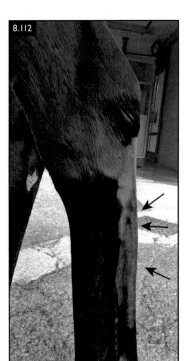

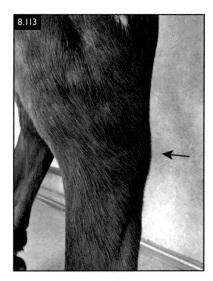

FIG. 8.113 Focal deformity of the plantar aspect of the distal tarsus is often referred to as 'curb'. The most common cause of this swelling is thickening of the paratenon of the SDFT secondary to trauma.

FIG. 8.112 Diffuse deformity of the plantar aspect of the proximal metatarsus, giving a bow-legged appearance. This was due to paratendonitis without tendon parenchymal involvement; however, the clinical appearance is indistinguishable from tendinopathy.

- chronic cases often have minimal thickening but persistent pain on palpation.

Differential diagnosis
- primary tendinopathy or desmitis
- oedema from other causes
- lymphangitis.

Diagnosis
- based on a history of trauma, slipped bandage or acute edema, and ultrasonography.
- **Ultrasonography**
  - check for primary tendon or ligament damage.
  - diffuse thickening and hypoechogenicity of the paratenon (Figs. 8.114, 8.115).
  - haemorrhage may separate the paratenon from the underlying tendon

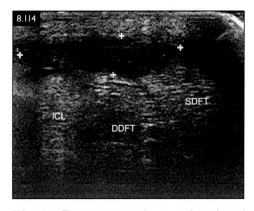

FIG. 8.114 Transverse scan from a palmarolateral approach at mid-metacarpus. A poorly defined hypoechogenic area (calipers) is present over the lateral border of the SDFT and extends along the DDFT and ICL. The border of the SDFT is irregular and poorly delineated. The overlying subcutaneous tissue is very thickened.

tissue (either hypoechogenic or hyperechogenic line) (Fig. 8.115).
  - spread of fluid:
    ♦ around tendon or neurovascular bundles, within confines of fascia.
      – typical crescent or triangular shape in cross-sectional images.

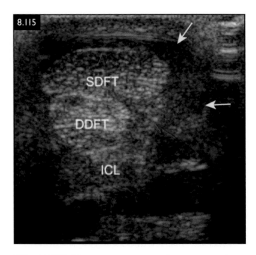

FIG. 8.115 Transverse scan from a palmar approach, mid-metacarpus. Severe, hypoechogenic thickening of the SDFT paratenon is visible along the palmar aspect of the tendon (arrowhead). It extends around the medial aspect of the SDFT and DDFT, displacing the fascia and neurovascular bundle (yellow arrows). The medial border of the SDFT is irregular and poorly defined. Echogenic material is present within the medial angle between the SDFT and DDFT (red arrow).

- o hyperechogenic paratenon layer in chronic cases (Fig. 8.116, 8.117).
- o adhesions characterised by:
  - ◆ heterogeneous, hyperechogenic tissue blurring borders of the normal tendon.
  - ◆ continuous tissue between the tendon or ligament and adjacent structures (periosteum, other tendons etc.) (Figs. 8.118, 8.119).

- o secondary tendon damage caused by collagenase release and inflammation (Fig. 8.120).
- o pressure damage to the neurovascular bundles.
- o Colour Doppler imaging confirms vascular compression or intravascular thrombosis.

Management

- acute cases:
  - o limit haemorrhage by strict box rest and cryotherapy:
    - ◆ icepacks or cold-water hosing/ bathing.
  - o pressure bandages to decrease swelling.
  - o lesion should resorb rapidly (haematoma several weeks).
  - o pain may persist for up to 6 weeks.
  - o damage to the associated tendon/ ligament and surrounding tissues (bone, neurovascular structures) treated specifically.
- adhesion prevention:
  - o early aggressive anti-inflammatory treatment,
  - o in-hand walking as early as possible (when oedema has receded).
    - ◆ within 2–3 weeks of injury.
  - o cryotherapy and bandaging continued for several weeks.
- chronic adhesions:
  - o can cause recurrence of the lesion, persistent lameness, and overt tendinopathy.
  - o surgical adhesiolysis.

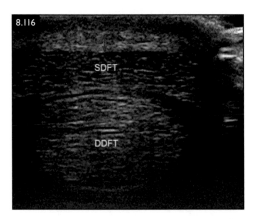

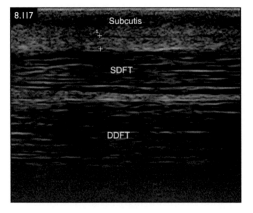

FIGS. 8.116, 8.117 Chronic thickening of the paratenon with a markedly hyperechogenic appearance in tranverse (8.116) and longitudinal (sagittal) scan (8.117) (double arrow and caliper respectively).

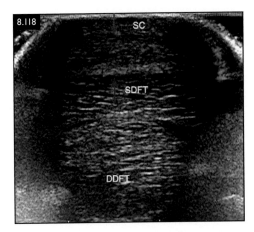

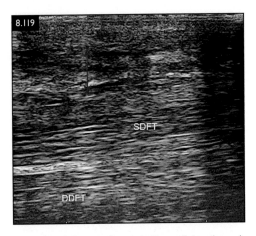

FIGS. 8.118, 8.119 (8.118) Chronic thickening of the paratenon with hyperechogenic tissue lining the palmar aspect of the SDFT (between arrows). This fibrous tissue is continuous with the SDFT parenchyma and the subcutaneous tissue (SC). (8.119) Longitudinal (sagittal) scan over the palmar metacarpus. Note the diffuse, chronic loss of fibre alignment in the SDFT and complete loss of the interface between the SDFT, paratenon and subcutis (arrow).

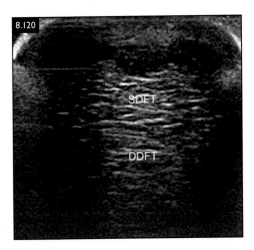

FIG. 8.120 Transverse scan over the palmar aspect of the metacarpus (zone IIb). The SDFT paratenon is very hypoechogenic and thickened (between arrows). The palmar border of the SDFT is irregular and ill-defined and there is a focal, encroaching lesion in the palmar sagittal portion of the tendon (arrowhead). This may be due to the initial trauma and/or due to focal destruction of the parenchyma due to platelet and inflammatory cell-derived mediators and collagenases.

Prognosis

- good in the absence of complications.
- guarded with restrictive adhesions and/or tendinopathy.

Extensor tendon injuries and extensor tenosynovitis

Definition/overview

- common in horses of all breeds and ages.
- most commonly involves:
  - extensor carpi radialis (ECR), lateral digital extensor (LaDE), and common digital extensor (CDE) tendons in the forelimb.
  - tibialis cranialis, fibularis (peroneus) tertius (FT), lateral digital extensor (LaDE) and long digital extensor (LDE) tendons in the hindlimb.
- mostly caused by direct trauma to the dorsal aspect of the limb.
- occasional spontaneous injury of the CDE in foals.
- tenosynovitis of surrounding sheaths also caused by direct trauma, with or without damage to the tendon.
- idiopathic tenosynovitis with increased synovial fluid but without pain or inflammation.
- septic tenosynovitis of the extensor sheaths following open or penetrating wounds.

Aetiology/pathophysiology

- trauma from kicks or hitting jumps, gates, hedges or wire fences.

- most common where tendons and tendon sheaths are only protected by skin:
  - dorsal aspect of the distal limb, particularly in the metacarpus/metatarsus.
  - dorsal distal radius, carpus, or dorsal tarsal regions.
- direct damage to the sheath and subsequent haemorrhage induces a severe inflammation.
- chronic tenosynovitis may lead to:
  - chronic synovial thickening and fibrosis.
  - restrictive adhesion formation.
  - eventually partial carpal or tarsal joint motion restriction.

Clinical presentation

- dorsal soft tissue swelling in the metacarpal/metatarsal and carpal/tarsal regions.
- mild to moderate lameness if partial tendon damage.
- mechanical gait deficit if complete rupture:
  - unable to extend the carpus, tarsus, fetlock and/or digit.
  - knuckling over at the fetlock and dragging the toe if ruptured LDE or CDE.
  - compensatory behaviour by flicking the limb forwards:
    - ♦ speedy recovery of extensor tone if LaDE tendon is not affected.
- trauma without open wound:
  - painful fluctuant swelling and thickening.
  - tendon sheath swelling over the dorsal carpal or tarsal regions characterised by:
    - ♦ longitudinal vertical fluid distension interrupted over the carpal/tarsal bones because of compression of the sheaths by transverse retinacula (Fig. 8.121).
  - marked lameness except in idiopathic tenosynovitis.
  - chronic cases:
    - ♦ firm and pain-free thickening.
    - ♦ painful and restricted carpal or tarsal passive flexion.
- trauma with open skin wound:
  - tendon and/or periosteum exposed.
  - frayed tendon ends may be visible at wound edges or retracted for several centimetres proximally/distally.

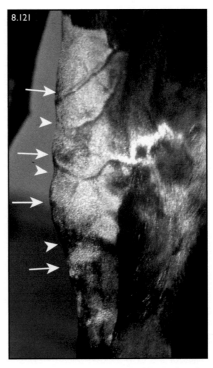

FIG. 8.121 Tenosynovitis of the ECR tendon sheath has a characteristic appearance. The enlarged sheath forms a sausage-like swelling along the longitudinal path of the tendon over the dorsal carpus (arrows), with transverse notches induced by the retinacula focally containing the distension (arrowheads).

- periosteum may be exposed or stripped.
- associated septic or inflammatory open tenosynovitis.
- occasionally penetrating wound into the carpal/tarsal joints.

Differential diagnosis

- causes of dorsal soft tissue swelling in the metacarpal/metatarsal and carpal/tarsal regions.

Diagnosis

- **Clinical examination:**
  - dorsal soft tissue swelling in the metacarpal/metatarsal and carpal/tarsal regions.
  - palpation and sterile examination of any wound.
- **Radiography**

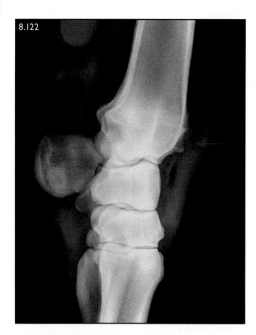

FIG. 8.122 Lateromedial radiograph of the carpus and distal radius of a horse with persistent tenosynovitis of the ECR tendon sheath subsequent to chronic trauma. Note the extensive roughened new bone on the cranial distal aspect of the radius (arrow).

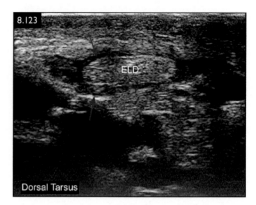

FIG. 8.123 Transverse plane scan over the dorsal aspect of the tarsus. The synovial sheath of the long digital extensor tendon (ELD) is slightly distended with mild thickening of the synovial lining membrane (arrows). The tendon has a normal ultrasonographic appearance.

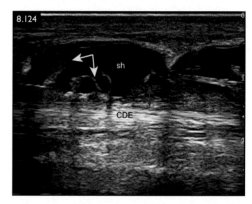

FIG. 8.124 Longitudinal scan over the dorsal aspect of the carpus. The synovial sheath of the CDE tendon is distended with anechogenic fluid (sh). The synovial membrane is markedly thickened by inflammatory changes (double arrows), with enlarged villi protruding into the sheath cavity (yellow arrows). Note the focal constriction of the swollen sheath by a small carpal retinaculum (arrowhead).

- rule out concurrent fractures, bone lysis or new bone formation.
- periosteal new bone formation on the distal radius is common sequel of chronic tenosynovitis of the ECR or CDE tendon sheaths (Fig. 8.122).
- extensive new bone may occur from trauma originally outwith the sheaths but subsequently interferes with the extensor tendons and causes secondary tendonitis.
- contrast tenography or arthrography may help define:
  ♦ communication with pentetrating wounds.
  ♦ abnormal traumatic communications between synovial sheaths and/or joints.
- **Ultrasonography**
  - diagnostic method of choice
  - findings of tenosynovitis may include:
    ♦ thickened synovial membrane
    ♦ anechogenic fluid distension.
    ♦ synovial mass formation due to synovial thickening and/or pannus accumulation (Figs. 8.123–8.125).
  - **important to identify any extensor tendinopathy:**
    ♦ diffuse damage with loss of fibrillar pattern, diffusely decreased echogenicity and increased cross-sectional area (Figs. 8.126, 8.127).
    ♦ frequently longitudinal hypoechogenic, sagittal clefts along the tendon:
      – partial loss of organisation of parenchyma and signs of severe synovitis (Fig. 8.128).

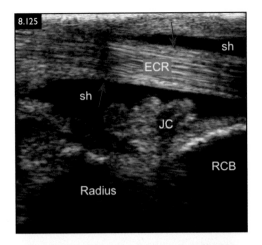

FIG. 8.125 Chronic tenosynovitis can lead to marked distension of the tendon sheaths without associated tendon changes. On this longitudinal scan over the dorsal proximal aspect of the carpus, the sheath of the ECR tendon is severely distended (sh), revealing the underlying, normally curled up capsule of the antebrachiocarpal joint (JC). The tendon has a normal appearance with a regular, longitudinal fibre pattern, although the visceral synovial sheath lining is moderately thickened. RCB = radial carpal bone.

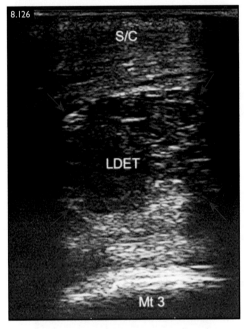

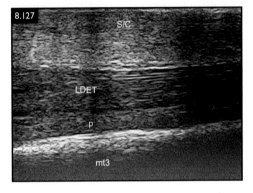

FIGS. 8.126, 8.127 (8.126) Transverse scan showing diffuse tendonitis of the long digital extensor tendon (LDET) in the proximal metatarsal area. The tendon is enlarged (arrows), heterogeneous and diffusely hypoechogenic. The surrounding soft tissues (subcutaneous and periosteal) are oedematous and thickened (S/C). Mt3 = third metatarsal bone, dorsal aspect. (8.127) Longitudinal scan image. Note the enlargement, loss of fibre alignment, and loss of definition of the tendon contours. p = periosteum; mt3 = third metatarsal bone.

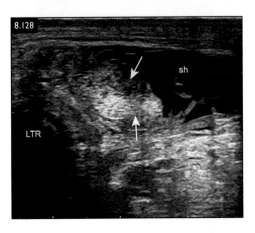

FIG. 8.128 Transverse scan over the dorsolateral aspect of the tarsus at the level of the lateral trochlear ridge of the talus (LTR). The long digital extensor tendon is relatively echogenic but contains a sagittal, hypoechogenic tear (white arrow). Granulation tissue and synovial proliferation extends from the tear into the sheath cavity (yellow arrow). The tendon sheath (sh) is markedly distended with thickening and villous hypertrophy of the synovial membrane (red arrows).

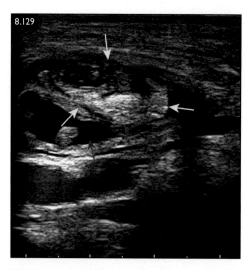

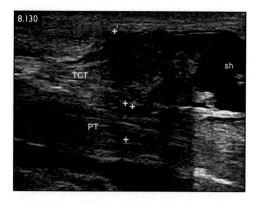

FIGS. 8.129, 8.130 (8.129) Transverse scan at the level of the crurotarsal joint, dorsal aspect. The tibialis cranialis tendon has been severed over the distal tarsus, the proximal end has retracted proximally, leaving an abnormal void, filled with fluid, within the tendon sheath (red arrow). The synovial membrane is thickened, forming villous masses. Some amorphous tissue remains (yellow arrows) in contact with the intact, though heterogeneous, peroneus tertius tendon (arrowhead). (8.130) Longitudinal scan (proximal to the left) at the distal crus level. The proximal end of the torn tibialis cranialis tendon (TCT) is displaced proximally, leaving a large, fluid-filled void within the tarsal tendon sheath (sh). The frayed end of the tendon is enlarged like cauliflower (arrow). The underlying peroneus tertius (PT) tendon is very heterogeneous.

- o complete rupture:
  - ♦ enlarged, hypoechogenic, and markedly thickened tendon stumps.
  - ♦ visible several centimetres apart (Figs. 8.129, 8.130).
  - ♦ haematoma formation between the stumps.
  - ♦ granulation tissue develops with gradual increase in echogenicity as fibrosis ensues.
- o **Septic tenosynovitis:**
  - ♦ severe synovial changes.
  - ♦ heterogeneous lesions which may extend into the tendon.
  - ♦ distension of the sheath cavity with heterogeneous 'cellular' material.
    - – represents exudate, fibrin, and debris (Fig. 8.131).
  - ♦ fistula extending from the wound into the sheath may be visible.
    - – associated gas and debris.
  - ♦ septic tendinopathy from primary laceration of the tendon:
    - – secondary invasion into the tendon tissue.

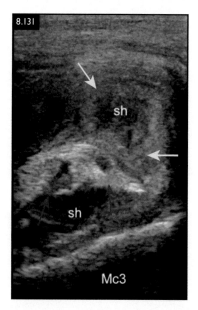

FIG. 8.131 Transverse scan over the dorsal proximal aspect of the metacarpus. The CDE tendon has very irregular contours with two irregular and hypoechogenic clefts (red arrows), the sheath membrane is thickened (yellow arrows) and fluid distension has an echogenic, 'cellular' appearance. This was due to a septic tenosynovitis secondary to a wound over the dorsal tarsus.

Management

- conservative treatment:
  - box rest
  - systemic and topical anti-inflammatory drugs.
  - regular in-hand exercise.
  - hosing for 15–20 minutes with cold water 2–4 times daily (or ice packs applied).
  - healing typically rapid and most horses can return to work within 2–12 weeks.
  - wounds treated by appropriate wound management and bandaging.
  - casts or splints to limit movement of the wound edges and tendon stumps.
    - or if there is severe impairment of motion.
  - passive manipulation and physiotherapy to aid in regaining full joint motion.
  - intrathecal hyaluronic acid and/ or short-acting corticosteroids in appropriate cases.
- complete rupture:
  - strict stall rest is necessary.
  - limb support with dorsal or palmar/ plantar splints:
    - elbow or mid-crus to the foot over a padded bandage.
    - aluminium or resin splint system that encloses the foot or is attached to the toe of the shoe to prevent hock flexion.
    - maintain splinted for 3–4 weeks, followed by a Robert Jones bandage for a further 2–6 weeks depending on the clinical appearance of the wound.
- chronic, adhesive tenosynovitis with restricted range of motion:
  - complete surgical removal of the sheath synovium via open incisional approach with sharp dissection, or tenoscopy with synovial resector.
  - complete tenectomy of septic necrotic tendons has been performed for salvage.

Prognosis

- good to fair for most extensor tendon injuries.
  - persistent synovial distension without lameness is a common sequel:

- mechanical impairment may affect performance.
  - severed tendons heal through fibrosis and scar tissue formation.
    - many affected horses adapt and regain full function over time.
- guarded for septic tenosynovitis following aggressive surgical debridement or occasional complete tenectomy/ tenosynovectomy.

Ruptured extensor tendon in foals (see page 32)

Rupture of the extensor carpi radialis tendon in adult horses

Definition/overview

- spontaneous partial to complete rupture of the ECR tendon is a rare condition.
- affects adult horses, particularly those used for showjumping.

Aetiology/pathophysiology

- single, sudden trauma on a tense tendon during carpal flexion.
- repeated trauma to the dorsal carpus.
- repetitive strain injuries.

Clinical presentation

- partial tear:
  - swelling of the ECR synovial sheath.
  - mild to moderate lameness.
- complete tear:
  - sudden-onset lameness.
  - marked sheath distension:
    - over the cranial aspect of the distal antebrachium and dorsal carpus.
  - exaggerated, stringhalt-like flexion of the carpus during the stride:
    - caused by lack of counter-resistance to the flexor muscle action.

Diagnosis

- typical swelling of ECR sheath dorsal carpus.
- ultrasonography to identify partial or complete tendon rupture (Fig. 8.132).

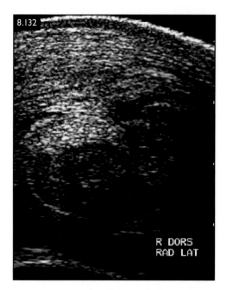

FIG. 8.132 Transverse ultrasound image of the distal dorsal radius just proximal to the carpus showing a severely damaged ECR within a distended sheath dorsal to the radius. Note the subcutaneous fibrosis and thickening.

Management

- conservative treatment:
  - support bandaging
  - splint or a tube cast from elbow to fetlock.
  - gait abnormality and sheath distension usually persist.
- surgical treatment:
  - tenoscopic debridement of the sheath and frayed tendon ends to resolve synovitis.
  - repair of the ruptured tendon is not indicated as not effective in most cases.

Prognosis

- fair for soundness.
- poor for return to athletic activities.

Digital flexor tendon sheath (DFTS) tenosynovitis

Definition/overview

- inflammation, and usually synovial effusion, of the DFTS (tendinous windgalls).

Aetiology/pathophysiology

- Idiopathic synovitis without lameness (cold windgalls):
  - common especially in hindlimbs of older, larger horses.
  - may be more prominent during periods of inactivity.
  - may be caused by chronic low-grade inflammation associated with repetitive microtrauma of daily exercise.
- Primary traumatic tenosynovitis:
  - direct trauma to the area of the sheath.
  - tearing of the synovial membrane through overextension or overuse.
  - intrasynovial haemorrhage and synovitis.
- Secondary traumatic tenosynovitis:
  - synovial inflammation caused by a primary tendinous, ligamental or bony lesion within the sheath or sheath wall.
    - injury to the DDFT, manica flexoria or the branches of the SDFT.
    - fractures of the PSBs or phalanges.
    - primary injury to the palmar annular ligament (PAL) of the fetlock, or the intersesamoidean ligament, or the distal sesamoidean ligaments.
- Sympathetic tenosynovitis refers to non-inflammatory, transitory synovial fluid distension due to inflammation elsewhere in the distal limb.
- chronic inflammation may result in:
  - permanent thickening of the visceral and parietal layers of the synovial membrane.
  - fibrous adhesions between the parietal and visceral sheath layers.
  - permanently changed permeability of the synovial membrane.
  - fibrosis of the fibrous layer of the digital sheath wall, including the PAL of the fetlock.
  - propensity for chronic subcutaneous fibrosis secondary to chronic tenosynovitis in some types of ponies and horse (native breeds, cobs and draught types).
  - palmar/plantar constriction of the fetlock canal by the thickened soft tissue (PAL and subcutaneous fibrosis).

Clinical presentation

- synovial effusion of the DFTS:
  - lateral and medial protrusion of the proximal pouches of digital sheath along the digital flexor tendons and proximally to the PAL of the fetlock (Fig. 8.133).
  - generally (but not always) accompanied by palmar protrusion of the distal recess of the sheath on the palmar/plantar aspect in the distal pastern area:
    - between both branches of the SDFT and proximal to the proximal margin of the distal digital annular ligament.
- Idiopathic tenosynovitis (windgalls):
  - synovial effusion, often soft, without obvious heat, pain or thickening of the sheath.
  - usually bilateral
  - no associated lameness.
  - flexion of the fetlock may occasionally elicit pain through increased pressure and mechanical impairment.
- Traumatic primary and secondary tenosynovitis:
  - synovial effusion with variable heat and pain.

- pain on deep palpation of the palmar/plantar pastern and over proximal pouch of sheath.
  - diffuse oedema possible in acute cases.
  - pain elicited by distal limb flexion.
  - lameness variable:
    - moderate to severe lameness with decreased digital flexion and reduced foot flight arc possible following acute injury.
    - mild to moderate intermittent lameness more commonly associated with repetitive low-grade injury associated with exercise (e.g. jumping).
    - decreased fetlock sinking during weight bearing.
- Chronic tenosynovitis:
  - more diffuse and firm swelling with less obvious pain on palpation.
  - lameness less obvious but decreased amplitude of flexion and extension of the fetlock.
  - pain on forced flexion.

Differential diagnosis

- any cause of distal limb oedema.
- SL branch injuries.
- pronounced effusion of the palmar/plantar pouches of the fetlock joint.

Diagnosis

- **Clinical examination:**
  - clinical findings.
  - lameness partially or totally abolished with:
    - abaxial sesamoid nerve block.
    - low 4-point nerve block.
    - intrathecal injection of 5–7 ml of local anaesthetic solution into the sheath.
- **Radiography:**
  - contrast tenography of the digital sheath:
    - following injection of 5–10 ml radiocontrast medium.
    - diagnose primary tendinous pathology (tears of manica flexoria and longitudinal margin of the DDFT).
- **Ultrasonography:**
  - longitudinal, oblique transverse and non-weight-bearing scanning

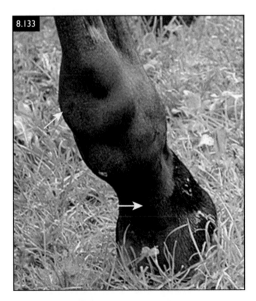

FIG. 8.133 Swelling of the digital sheath (tendinous 'windgalls') is obvious proximal to the PAL of the metacarpophalangeal joint (top arrow) and over the palmar mid-pastern (bottom arrow).

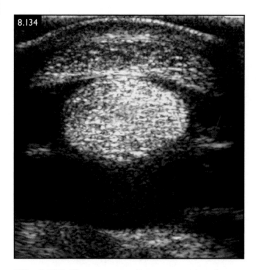

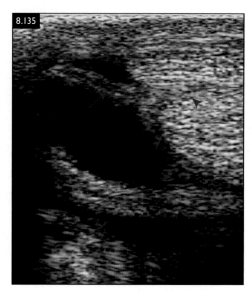

FIG. 8.134 Transverse ultrasonography image over the distal metacarpus of a horse with idiopathic distension of the digital tendon sheath. The sheath is distended by anechogenic fluid. The flexor tendons are well delineated, but no thickened synovial membrane is visible. The proximal vinculae of the DDFT are visible within the proximal recesses of the sheath (arrow).

FIG. 8.135 Subacute, traumatic tenosynovitis. The vinculae are thickened (arrow) and there is a halo of hypoechogenic tissue around the tendons (arrowheads). The sheath is distended by anechogenic fluid.

techniques used in addition to the standard transverse weight-bearing technique to increase the possibility of detecting tendinous abnormalities.

- Idiopathic asymptomatic tenosynovitis:
  - ♦ anechogenic fluid distension without synovial membrane thickening.
  - ♦ no thickening of intrasynovial structures such as plicae, vincula, mesotenons and the manica flexoria (less than 1 or 2 mm in thickness) (Fig. 8.134).
- Acute tenosynovitis:
  - ♦ distension with anechogenic fluid.
  - ♦ early haemorrhage may cause strands of hypoechogenic material (fibrin strands and pannus) in an echogenic and grainy fluid ('cellular appearance').
  - ♦ diffusely thickened and hypoechogenic synovium.
    - – 2–4 mm thick halo around the tendons (Fig. 8.135).
    - – hypoechogenic interface between flexor tendons in the fetlock canal.

- Chronic tenosynovitis:
  - ♦ partial to complete obliteration of the sheath by a thickened, echogenic synovium (Figs. 8.136, 8.137).
  - ♦ lateral and medial plicae of the DDFT proximal to the fetlock 3–10 mm thick.
  - ♦ dynamic examination (flexing and extending the limb) may show restricted tendon gliding due to adherence to the sheath wall.
- Primary tendon injury inside the digital sheath:
  - ♦ longitudinal marginal tears of the SDFT, DDFT or both.
  - ♦ may be difficult to detect ultrasonographically.
  - ♦ may be visible as a hypoechogenic cleft, wedge-shaped area or as diffuse lesions, extending into the centre or peripheral portion of the tendon (Fig. 8.138, 8.139).
  - ♦ extending proximodistally.
  - ♦ occasionally complete split into two or more independent longitudinal tendon portions.

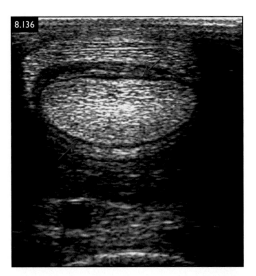

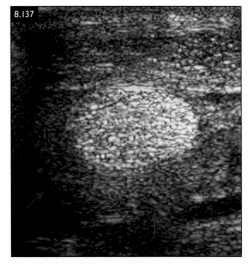

FIGS. 8.136, 8.137 Chronic, traumatic tenosynovitis. (8.136) There is a halo of echogenic synovial tissue around the tendons, separating the DDFT from the SDFT and its manica flexoria (arrows). (8.137) In extreme cases, thickened synovium can totally obliterate the sheath (arrows), enclosing the tendons and vinculae in solid tissue and thus causing restriction of movement in the affected part of the limb.

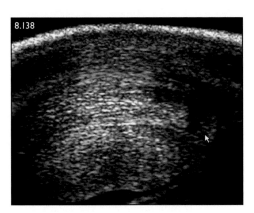

FIG. 8.138, 8.139 (8.138) A transverse ultrasound image of the palmaromedial aspect of the DFTS just at the level of the PSB showing an obvious hypoechoic cleft in the medial edge of the SDFT plus subcutaneous fibrous thickening, mild thickening of the AL, and a possible adhesion from the palmar aspect of the tendon to the sheath wall. (8.139) Surgical dissection at an open tenosynoviotomy, which clearly confirms the full thickness split in the medial SDFT.

○ tears of the manica flexoria may be detected, often with difficulty:
  ◆ partial tears as a focal, hypoechogenic thickening near the attachment of the manica, often forming an amorphous mass (Fig. 8.140).
  ◆ complete tears may result in curling of manica or floating manica in one of the proximal recesses of the sheath or lateral to the SDFT:
    – may be difficult to differentiate from hypertrophic synovial masses.
○ echogenic, fibrous adhesions between tendons or between tendon(s) and parietal sheath (Fig. 8.141) may be present:
  ◆ may be obscured by synovial thickening and the tight space.
  ◆ strong suspicion based on restricted tendon movement during dynamic ultrasound.

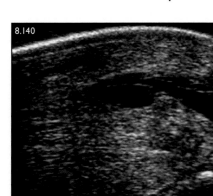

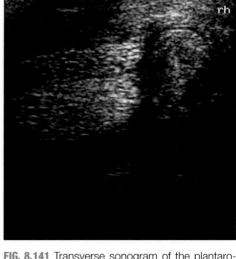

FIG. 8.140 Transverse sonogram of the plantaro-medial aspect of the DFTS, clearly showing a mature and large adhesion from the medial edge of the SDFT to the plantar sheath wall. Note also the damaged medial edge of the SDFT, the absence of the origin of the manica flexoria, and the poor delineration of the medial DDFT.

FIG. 8.141 Transverse sonogram of the plantaro-lateral aspect of the DFTS. The lateral edge of the SDFT is abnormal and the manica flexoria is not visible but a coiled-up mass of damaged tissue is visible laterally. This was confirmed at tenoscopy as a severely damaged manica.

- ♦ fetlock flexion should cause both flexor tendons to move together in relation to the sheath wall.
    - ♦ DIP joint flexion should cause the DDFT to move independently relative to the SDFT.
  - ○ other lesions may include:
    - ♦ erosions of the proximal scutum and the intersesamoidean ligament.
    - ♦ entheseous new bone at the insertions of the PALs on the sesamoid bones and phalanges.
    - ♦ thickening of the annular ligaments and subcutaneous tissue.
- **MRI**
  - ○ High-field MRI is particularly sensitive to detect lesions within the flexor tendons, distal sesamoidean ligaments, and to identify adhesions.
  - ○ very useful for lesions in the distal fetlock and proximal pastern region where the ergot and the sharp angle make ultrasonographic assessment difficult.
  - ○ Low-field MRI is less sensitive due to motion interference but can still identify tendinous lesions (marginal

tears and manica tearing) that may be undetectable with ultrasonography.
- **Tenoscopy**
  - ○ provides excellent visualisation of the sheath cavity and synovial surfaces.
  - ○ technique of choice to confirm:
    - ♦ adhesions (Fig. 8.142) and tendon surface lesions (Fig. 8.143) that communicate with the sheath.
    - ♦ unable to explore the tendon parenchyma or structures outside the sheath.
  - ○ complementary to ultrasonography and MRI.
  - ○ treatment of many conditions.

Management

- Idiopathic distension:
  - ○ usually not treated but may respond to rest and pressure bandages.
  - ○ intrathecal corticosteroid injections may provide temporary resolution of the distension, but in most cases the swelling recurs within a few weeks or months.
- Acute tenosynovitis:

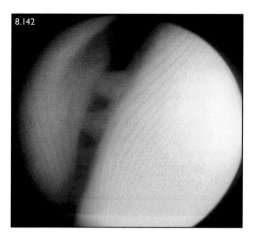

FIG. 8.142 Tenoscopic image of multiple mature adhesions between the plantar aspect of the SDFT and the plantar aspect of the sheath wall.

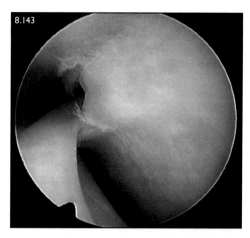

FIG. 8.143 Tenoscopic image of the DDFT as it emerges from beneath the manica flexoria of the SDFT. Note the tearing of the lateral edge of the DDFT.

- rest with controlled exercise
- systemic NSAIDs.
- cold hosing and/or application of ice.
- pressure bandages for 2–4 weeks.
- intrathecal IRAP, sodium hyaluronate and/or short-acting corticosteroid injections.
- physiotherapy to decrease fibrosis, resolve inflammation and distension, and encourage healing with improved limb mobility.
- Primary tendinous injury:
  - tenoscopic intervention as a first stage treatment:
    - resection of torn manica flexoria.
    - removal of space-occupying mass-like lesions.
    - debridement of longitudinal tendon tears.
  - recurrence is common for margin DDFT tears.
- Chronic or recurrent tenosynovitis:
  - **difficult to treat.**
  - intrathecal hyaluronate, corticosteroids or IRAP.
  - tenoscopy to perform:
    - debridement of adhesions.
    - annular ligament desmotomy.
    - removal of synovial masses.

Prognosis

- good for idiopathic asymptomatic distension (windgalls).

- fair for acute tenosynovitis without lesions to the tendons or scutums.
- fair for manica flexoria tears and synovial masses.
- guarded for lesions of the DDFT, especially in forelimbs.
- poor for large or deep tendon tears.
- generally poorer in the forelimb than in the hindlimb.

Palmar/plantar annular ligament syndrome

Definition/overview

- palmar/plantar annular ligament of the fetlock (PAL):
  - 1 mm thin ligamentous band with transversely orientated fibres extending from the periosteum on the abaxial aspect of the PSBs.
  - part of the fibrous wall of the DFTS.
  - restrains the digital flexor tendons on the palmar/plantar aspect of the fetlock during flexion.
- PAL syndrome:
  - desmitis of the PAL.
  - lameness associated with tenosynovitis of the digital sheath:
    - characteristic notched appearance on the palmar/plantar aspect of the fetlock.
    - thickening of the PAL.

- may induce constriction of the
flexor tendons in the fetlock canal:
  - leading to pain and lameness.
  ○ notched appearance more often
  associated with:
    - bulging of proximal sheath –
    distension proximal to the inelastic
    PAL.
    - regardless of whether the ligament
    is affected or not (Fig. 8.144).

Aetiology/pathophysiology

- Primary desmitis of the PAL:
  ○ caused by direct trauma to the palmar
  aspect of the fetlock or repetitive
  overextension of the digit with
  overstretching of the PAL.
  ○ rare condition.
- Secondary desmitis of the PAL:
  ○ chronic thickening of the PAL
  associated with thickening of the wall
  of a chronically inflamed digital sheath
  (chronic tenosynovitis).
  ○ chronic stretching due to the presence
  of chronically enlarged superficial
  digital flexor tendonitis within the
  digital sheath.

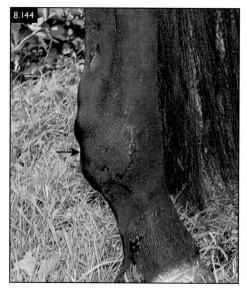

FIG. 8.144 Notched appearance over the palmar
fetlock (arrow), sometimes wrongly attributed to
annular ligament thickening but actually due to
distension of the proximal recesses of the digital sheath proximal to the non-elastic, PAL of the
fetlock.

○ most common presentation of PAL
syndrome.
- thickening and fibrosis of subcutaneous
tissues over the palmar/plantar aspect:
  ○ may cause similar notched or
  thickened appearance without desmitis
  of the PAL.
  ○ some types of ponies and horse (native
  breeds, cobs and draft types).
- concept of a true compressive syndrome
like human carpal tunnel syndrome is
questionable as no major neurovascular
structures are contained within the
fetlock canal.

Clinical presentation

- characteristic notching and/or soft tissue
thickening at the palmar/plantar aspect
with bulging of the sheath proximal and
palmar/plantar to the ligament.
- some cases may have diffuse thickening
over the whole palmar/plantar aspect of
the fetlock.
- local pain on palpation in acute cases.
- lameness with reduced sinking of the
fetlock during weight bearing.
- additional signs related any tenosynovitis.

Differential diagnosis

- tenosynovitis • subcutaneous fibrosis.

Diagnosis

- Clinical examination:
  ○ palpation and lameness examination.
- Radiography:
  ○ eliminate other causes of swelling
  in palmar fetlock (e.g. fractures of
  sesamoid bones).
  ○ identify entheseous new bone on the
  palmar aspect of the PSBs.
- Ultrasonography:
  ○ diagnostic technique of choice.
  ○ normal PAL is thin band of tissue with
  a transversely orientated fibre pattern.
    - extends between palmar margins of
    the PSB.
    - measures <2 mm in thickness (Fig.
    8.145).
  ○ haematoma may be visible in acute
  cases in and around PAL.
  ○ thickening visible >3 mm in association
  with desmitis due to chronic

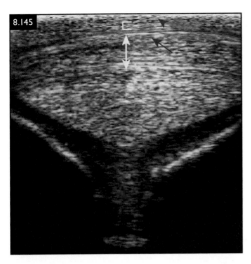

FIG. 8.145 Transverse sonogram over the palmar aspect of the fetlock. The normal PAL is a thin, fibrous band (yellow bracket) with fibrillar organisation, lying directly over the palmar aspect of the SDFT (yellow double arrow). Note the barely visible synovial tissue between the tendon and the ligament (red arrow) and the thin subcutaneous tissue (arrowhead).

tenosynovitis or chronic trauma (Figs. 8.146, 8.147).
- ○ diffuse fibrosis in chronic disease of cob-type horses may cause marked thickening between the skin and the SDFT (Figs. 8.148, 8.149).

- important to distinguish subcutaneous fibrosis from thickening of the visceral synovium of the SDFT and parietal synovium of the sheath (Fig. 8.150).
  - ♦ identify the attachments of PAL to PSBs first and follow the ligament from there.
  - ○ entheseopathy on PSB with local thickening and irregular bony insertion (Fig. 8.151).

Management

- acute case:
  - ○ rest, local anti-inflammatory treatment (e.g. cold hosing, ice packs) and bandaging.
  - ○ tenosynovitis treatment as described earlier (see page 327).
- chronic case with ultrasonographic evidence of thickening of the PAL:
  - ○ tenoscopic transection of the PAL (PAL desmotomy) with curved tenotomy knife ('hook blade'), electrosurgical hook blades, or radiofrequency (Coblation®) hook probe.
  - ○ tenoscopic inspection of DFTS for primary tendinous injuries.
  - ○ early passive manipulations of the fetlock and walking exercise to reduce adhesions following surgery.

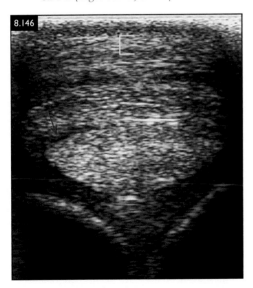

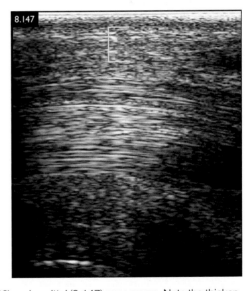

FIGS. 8.146, 8.147 PAL desmopathy: transverse (8.146) and sagittal (8.147) sonograms. Note the thickening and increased echogenicity of the PAL (yellow bracket), associated with mild synovial thickening (red arrows). The arrowhead shows the mesotenon of the SDFT, outlined by the synovitis.

○ open surgery through a larger skin incision can be successful but is associated with a higher risk of wound breakdown, fibrosis and adhesions.

Prognosis

• good in acute and subacute cases with conservative management.

• good to fair after desmotomy in cases of primary PAL desmitis with ultrasonographic evidence of PAL thickening without tendinous injuries in the sheath.
• guarded for cases of chronic tenosynovitis.

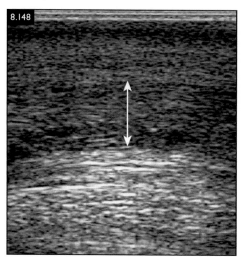

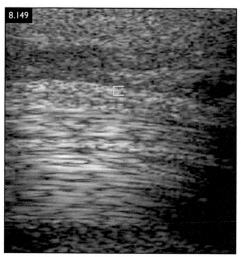

FIGS. 8.148, 8.149 Subcutaneous tissue thickening palmar to the annular ligament is unlikely to cause a stenosing syndrome. (8.148) Transverse ultrasound scan showing a normal annular ligament and synovial sheath. Hypoechogenic thickening of the subcutaneous tissues is probably due to focal contusion/haematoma formation (double arrow). (8.149) Longitudinal scan of the same area with the annular ligament indicated by the yellow bracket.

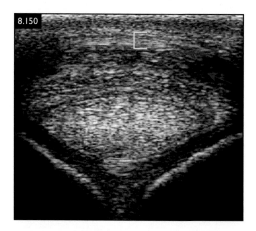

FIG. 8.150 Thickening of the synovial membrane due to septic tenosynovitis. The annular ligament is normal (yellow bracket), but the inflamed, space-occupying synovial tissue causes stenosis of the fetlock canal (arrows).

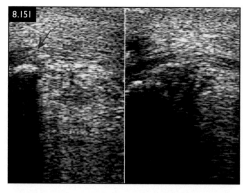

FIG. 8.151 Entheseopathy of the lateral insertion of the annular ligament (left image). Note the thickening in comparison with the medial insertion (right), irregular bony contour at the insertion (entheseophytes), hypoechogenicity of the ligament, and loss of fibre pattern (desmopathy) (arrow).

Deep digital flexor tendonitis in the digital flexor tendon sheath

Definition/overview
- spontaneous injury to the DDFT nearly always occurs in sheathed areas:
  - DFTS – digital tenosynovitis.
  - podotrochlear bursa – palmar foot pain.
  - carpal flexor tendon sheath (see carpal canal syndrome, page 143) associated with:
    - protruding osteochondromas.
    - physeal scar spikes.
    - rarely injury of the inferior check ligament.
  - Tarsal sheath (see Thoroughpin, page 333).

Aetiology/pathophysiology
- spontaneous strain injury due to repeated trauma (cyclic injury) or overextension injury.
- direct trauma and puncture of the tendon through the palmar fetlock or pastern area without contamination of the sheath.
- injuries may be diffuse, focal (either central or peripheral) or in the form of longitudinal marginal tears.
  - central core lesions:
    - may/may not extend to the periphery of the tendon.
    - most commonly found in the fetlock region.
    - relatively short (1–3 cm in length).
  - longitudinal marginal tears:
    - may be superficial, most frequently on the lateral aspect.
    - distal fetlock region or immediately distal to the ergot.
    - may spread through the parenchyma in a reticulated, star-shaped pattern (with multiple tears).
- healing of intrathecal tendon lesions:
  - through fibrosis and fibrocartilaginous metaplasia.
  - invasion by synovial fluid of tendon lesions delays or impairs healing, causing lesions to persist in the long term.

- central core lesions without breach of the peripheral margins heal better:
  - lesions often remain hypoechogenic despite resolution of the lameness because of fibrocartilage formation (metaplasia).
  - mineralisation can occur within the metaplastic foci.
- associated tenosynovitis is often severe, especially in communicating lesions:
  - forms a major component of the disease.

Clinical presentation
- tenosynovitis:
  - marked distension and inflammation of the synovial sheath.
  - more common in the hindlimbs, especially in ponies and cobs.
- lameness variable from mild to severe and is usually unilateral and acute in onset.
- pain on forced flexion of the fetlock and increases lameness markedly:
  - pain on deep digital pressure over the margins of the tendon is more variable.
- lameness improved or resolved by low 4-point nerve block or after intrathecal analgesia of the digital sheath, though this may only be partial.
- puncture wounds:
  - may be difficult to diagnose if the wound is not obvious.
  - localised palmar soft tissue swelling and marked focal pain on pressure.
  - clipping the hair for better identification of puncture site is recommended.

Differential diagnosis
- all causes of digital tenosynovitis and tendonitis of the SDFT.

Diagnosis
- **Clinical findings:**
- **Radiography:**
  - contrast tenography after injection of 5–7 ml sodium meglumine diatrozoate with or without 10 ml mepivacaine.
  - accurate technique for diagnosis of longitudinal marginal tears of the DDFT and manica flexoria tears.
  - identify dystrophic mineralisation in the tendon and/or sheath boundaries.

- **Ultrasonography:**
  - four types of lesions of the DDFT have been identified in the digital sheath:
    - diffuse enlargement and change in the shape of the tendon, with asymmetric thickening of one or both lobes (Fig. 8.152).
    - focal hypoechoic lesions within the tendon (core lesions) or on its margins (Figs. 8.152, 8.153).

- dystrophic mineralisation within the DDFT especially if chronic (Fig. 8.154).
- marginal tears identified as a focal cleft or superficial hypoechogenic area on the tendon periphery (Fig. 8.155)
- may be difficult to visualise.
  - associated synovial masses or tendon granuloma possible in the sheath cavity.

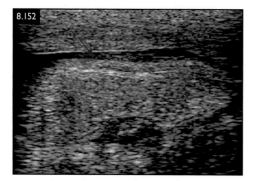

FIG. 8.152 Transverse ultrasonographic image of the palmar aspect of the pastern at the level of the PIP joint. The lateral part of the DDFT is enlarged and irregular in shape (left side). There is an associated thickening of the synovial membrane of the digital sheath (arrow).

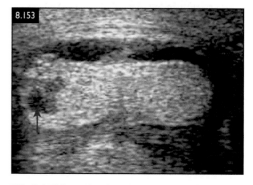

FIG. 8.153 Irregular, longitudinal lesion within the lateral part of the DDFT in the pastern region (arrow). The lesion extends to the periphery and synovial sheath cavity, which appears distended. There is, however, no obvious synovial thickening.

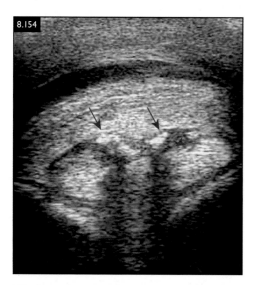

FIG. 8.154 Longitudinal tears of the DDFT in the fetlock region (arrowhead) associated with mineralisation within the lesion (arrows), visible as hyperechogenic surfaces casting acoustic shadows.

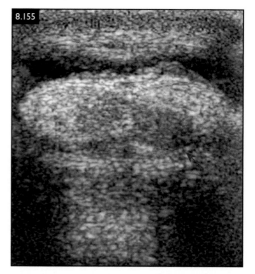

FIG. 8.155 Transverse ultrasonographic image of the palmar aspect of the pastern at the level of the PIP joint. The medial part of the DDFT (right side) is slightly enlarged. There is an hypoechogenic lesion opening at the dorsomedial aspect, suggesting a superficial tear (arrow).

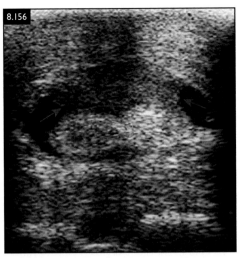

FIG. 8.156 A large adhesion is present on the palmar aspect of the DDFT in the pastern region (arrows). There is mild distension of the digital sheath. The synovial membrane is thickened and irregular, indicating chronic synovitis.

- adhesions possible in more chronic cases (Fig. 8.156).
- lesion may be reticular in pattern or combine several of the above characteristics.
- acute or chronic digital sheath synovitis is usually evident and severe.
- **MRI:**
  - useful if ultrasonography inconclusive.
    - thick skin in cob-type horses.
    - distal fetlock region with change of angle and the presence of the ergot.
  - High-field MRI more sensitive than ultrasonography to detect:
    - superficial tendon lesions.
    - adhesions.
  - Low-field MRI useful to detect marginal tears of the dorsal margin of the DDFT in the pastern.
- **Tenoscopy:**
  - indicated if:
    - tenosynovitis unresponsive to medical/conservative management.
    - marginal SDFT or DDFT lesion identified on ultrasound or MRI.
    - marginal DDFT tears often only identified by tenoscopy (see Fig. 8.143).
  - digital sheath tenoscopy is difficult and may cause iatrogenic injury to the tendons.

Management

- core lesions and lesions that do not extend to the tendon margins:
  - rest, cryotherapy and anti-inflammatory therapy.
  - followed by graduated, controlled exercise.
  - intralesional injection of biological (IRAP®, PRP) drugs under ultrasonographic or tenoscopic guidance:
    - PRP may increase the risk of post-adhesion formation.
    - IRAP® lacks objective evidence of beneficial effects.
    - MSCs.
    - risk of leakage of the injectate into the sheath lumen.
  - tenosynovitis is dealt with as described in the DFTS section (see pages 326–327).
- surgical treatment justified whenever a lesion extends through the tendon margins or if there is concurrent pathology in the sheath:
  - tenoscopic exploration of the sheath.
  - debridement of torn fibres, synovial masses, adhesions, annular ligament desmotomy.
- corrective foot care:
  - adequate trimming to shorten the toe and encourage heel growth,
  - shoes with rolled toes and heel support (e.g. egg bar shoes).

Prognosis

- fair for core lesions with adequate healing after 6–9 months of rest and controlled exercise.
- fair for short longitudinal margin tears.
- guarded to poor for longer marginal tears and those that communicate with the lumen.
- worse in the forelimb than in the hindlimb.

Tenosynovitis of the tarsal sheath (thoroughpin)

Definition/overview

- synovial sheath of the lateral digital flexor tendon (LDF).

tendon runs over the medial aspect of the hock and sustentaculum tali of the calcaneus:
- joins medial digital flexor tendon (MDF) in the proximal metatarsus to form the DDFT.
- Thoroughpin refers to distension of the sheath.

Aetiology/pathophysiology

- idiopathic asymptomatic tarsal sheath effusion.
- spontaneous or degenerative injury to the tendon (LDF tendinopathy) or other sheath components (tears of the mesotenon, erosions of the sustentaculum tali):
  - superficial marginal tears of the LDF cause tenosynovitis.
  - erosion of fibrocartilage covering the sustentaculum tali leads to contact/ shearing lesions of the overlying tendon.
- direct trauma (e.g. kick injuries, interference with the opposite limb).
  - fragmentation of the plantaromedial edge of the sustentaculum tali of the calcaneus.
  - septic osteitis and/or septic tenosynovitis following penetrating injury.
- adhesion formation possible in chronic or recurrent tarsal sheath inflammation.

Clinical presentation

- swelling of the proximal sheath pouch in the distal crus (Fig. 8.157):
  - particularly laterally and medially between the tibia and common calcaneal tendon
- lesser swelling medial to the DDFT in the proximal third of the metatarsus.
- lameness variable, from no lameness (idiopathic synovitis) to severe lameness in septic tenosynovitis and osteitis of the sustentaculum.
- often positive limb flexion.

Differential diagnosis

- other swellings on the caudal distal crus, including:
  - crurotarsal joint effusion
  - deep capped hock (calcaneal bursitis).

- false thoroughpin (deep gastrocnemius bursitis)
- sympathetic effusion.
- Synoviocoele of the tarsal sheath.

Diagnosis

- **Clinical examination:**
  - careful palpation helps differentiate the swelling from other causes.
- intrathecal injection of local anaesthetic solution (5–8 ml):
  - confirms pain is arising from the tarsal sheath.
- **Radiography:**
  - four standard projections of the tarsus and a flexed caudoproximal plantarodistal ('skyline') view of the calcaneus.
  - fragmentation, osteolysis or new bone formation of the sustentaculum tali mainly visible on dorsolateral– plantaromedial oblique and skyline projections (Fig. 8.158).

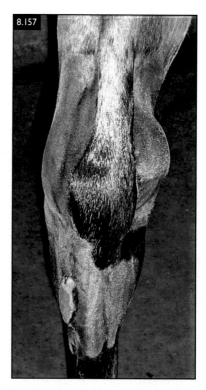

FIG. 8.157 Substantial tarsal sheath distension clearly visible as a fluid-filled pouch laterally (right), and less obviously medially and distomedially.

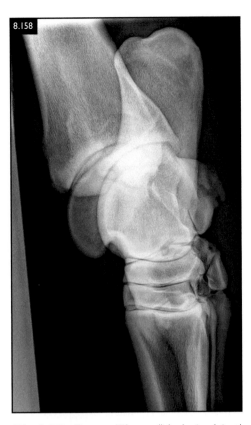

FIG. 8.158 Dorso 45° medial–plantarolateral oblique radiograph of the tarsus of a horse that had sustained a kick injury to the medial aspect of the hock. There is an obvious defect and fragmentation of the edge of the sustentaculum tali of the calcaneus.

- ○ entheseophytes along the medial edge of the sustentaculum tali and on the axial insertion of the sheath on the plantaromedial calcaneus in chronic cases.
  - ○ contrast tenography of the tarsal sheath to identify communication between the synovial cavity and a penetrating wound.
- **Ultrasonography:**
  - ○ tarsal sheath identified cranial to the flexor tendons along the caudal aspect of the tibia and then around the LDFT over the plantaromedial aspect of the hock:
    - ♦ extends distally to proximal ¼ to ⅓ of the metatarsus, between the DDFT and tarsal check ligament.
  - ○ signs of tenosynovitis include:
    - ♦ synovial membrane thickening:
      - – hypoechogenic in acute stages, then increasing in echogenicity.
    - ♦ fluid distension
    - ♦ adhesions (Figs. 8.159, 8.160).
    - ♦ chronic cases – large synovial masses that obliterate proximal pouch (Fig. 8.161).
  - ○ LDF tendon lesions with variable appearance:
    - ♦ diffuse and longitudinally arranged (large clefts in tendon) (Fig. 8.162).
    - ♦ thin, superficial longitudinal tears with adherent, echogenic pannus on the tendon surface can be difficult to identify.

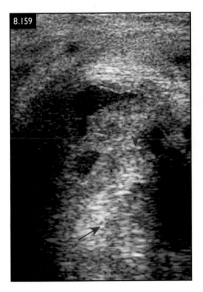

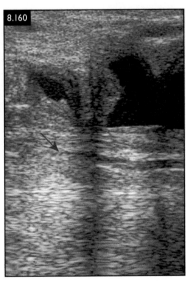

FIGS. 8.159, 8.160 Transverse (8.159) and longitudinal (8.160) ultra sonographic image over the plantaromedial aspect of the proximal hock. The LDF tendon is deformed and frayed (arrows) and there is a large, fibrous adhesion (arrowheads) forming a restrictive link to the thickened parietal sheath.

- fibrocartilage erosion of the sustentaculum tali with irregular, focal defects (Fig. 8.163).
- fragmentation of the edge of the sustentaculum tali with thickening and decreased echogenicity of the plantar retinaculum (Fig. 8.164).
- careful examination of wounds for a contiguous fistula.

Management

- acute cases: rest, cryotherapy, bandaging and intrathecal medication.
- chronic cases are often non-responsive and require surgery:
  - tenoscopic exploration, debridement of erosive lesions, and partial synovectomy.

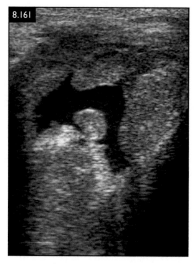

FIG. 8.161 Chronic tarsal sheath tenosynovitis. There are several echogenic synovial tissue masses protruding into the sheath cavity.

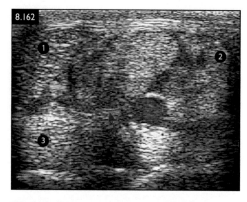

FIG. 8.162 Transverse ultrasonographic image over the plantaromedial aspect of the distal tarsus. The LDF tendon contains a large, irregular, hypoechogenic lesion at its dorsolateral border (arrow). This was a longitudinal tear forming a deep cleft on the surface of the tendon. 1 = SDFT; 2 = medial digital flexor tendon; 3 = LPL lateral plantar ligament.

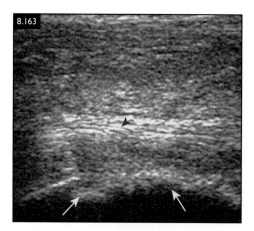

FIG. 8.163 Longitudinal sonogram over the sustentaculum tali. There is a large, irregular defect at the surface of the latter (yellow arrows). The overlying LDF tendon is hypoechogenic and has lost its normal fibre pattern (arrowhead). There is a tremendous amount of synovial thickening in this chronically inflamed tarsal tendon sheath (red arrow).

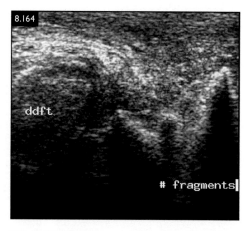

FIG. 8.164 Transverse sonogram over the plantaromedial aspect of the distal tarsus which had sustained a kick to this area. Note the bony fragmentation of the medial rim of the sustentaculum tali and associated soft tissue fibrous reaction.

- septic osteitis of the sustentaculum tali with septic tenosynovitis:
  - aggressive local debridement, tenoscopic lavage of tarsal sheath with synovectomy.
- septic tendonitis:
  - resection of tarsal portion of LDF tendon has been described as salvage procedure.

Prognosis

- good to fair in acute cases but decreasing with chronicity.
- guarded for severe tendon tears and chronic tenosynovitis.
- good for early aggressive treatment of infectious tenosynovitis.
- fragmentation of the sustentaculum tali carries a good prognosis, unless a significant proportion of the gliding surface for the tendon has been damaged.

Luxation of the superficial flexor tendon from the tuber calcis

Definition/overview

- lateral and medial margins of the fibrocartilaginous cap of the SDFT are attached to the tuber calcis of the calcaneus with a lateral and a medial retinaculum.
- retinacula prevent the SDFT slipping sideways during flexion of the hock.
- rupture of a retinaculum (medial more commonly than lateral) results in the tendon slipping in the opposite direction during hock flexion.
- usually unilateral but occasionally bilateral.

Aetiology/pathophysiology

- traumatic either by direct injury or during fast exercise.
- luxation may be complete or partial (subluxation):
  - lateral luxation much more common than medial.
  - intermittent (luxation during flexion, repositioning during extension) or permanent.

- luxation may also result from a parasagittal longitudinal tear of fibrocartilage cap of SDFT:
  - with or without tearing of a retinaculum.

Clinical presentation

- acute injury during exercise, causing sudden-onset lameness.
- severe lameness and acute distress as luxations are unstable (intermittent) initially:
  - SDFT comes off the tuber calcis during hock flexion.
  - returns to normal position during limb extension.
  - can be readily palpated during hock manipulation.
- associated diffuse soft tissue swelling and distension of the calcaneal bursa.
- tendon eventually settles in a permanently dislocated position in many cases (Fig. 8.165).

Differential diagnosis

- Capped hock
- Calcaneal bursitis.

Diagnosis

- Clinical examination:
  - movement of tendon
  - instability of tendon on palpation.

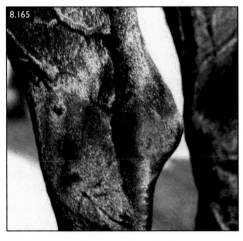

FIG. 8.165 Permanent lateral luxation of the SDFT off the tuber calcis. The tendon is visible under the skin, following a straight path over the lateral aspect of the calcaneus.

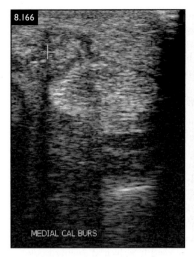 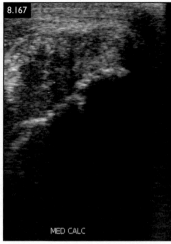

FIGS. 8.166, 8.167 Transverse sonograms of the calcaneal region of the plantar hock of a horse with an acute lateral subluxation of the SDFT. (8.166) shows the damaged medial retinaculum (+) and the slight lateral position of the SDFT. (8.167) is a slightly more medially obliqued scan showing the damaged ligament adjacent to an area of calcaneal bone roughening caused by the avulsion of the ligament from its insertion.

- **Radiography:**
  - detect possible concurrent fractures of the tuber calcis of the calcaneus.
- **Ultrasonography:**
  - confirm luxation of the SDFT.
  - identify tearing of the lateral or medial retinaculum (Figs. 8.166, 8.167).
    - torn retinaculum usually replaced by hypoechogenic tissue:
      - haematoma, oedema and frayed ligament initially.
      - subsequently echogenic fibrous scar tissue.
  - identify intact, opposite retinaculum:
    - hypoechogenic, mass-like structure, often retracted like an accordion.
  - detect possible parasagittal longitudinal tear of the fibrocartilage cap of the SDFT.

Management

- acute management:
  - stall rest, with various forms of support (bandage cast, Robert Jones bandage, plantar splints) and anti-inflammatory medication.
  - reassess after 1–2 months to determine if luxation permanent or intermittent without bandage support.
- conservative management:
  - indicated for horses with stable complete luxation.
  - progressive programme of in-hand walking after 6–8 weeks for a period of 6–8 weeks.

- small paddock turnout can be resumed after 3–4 months.
- normal exercise resumed gradually from 6 months.
- tendon becomes stabilised in its dislocated position in most cases.
- mild mechanical lameness persists in all cases.
- horses are pain free and can often return to their previous level of performance.
- surgical intervention may be useful if persistent severe swelling and lameness or persistence of intermittent luxation:
  - calcaneal bursoscopy:
    - remove torn tissues that may perpetuate calcaneal bursitis and pain.
    - remove persistent attachments of partially torn retinaculum and stabilise luxation.
    - remove abaxial margin of fibrocartilage cap (with its retinaculum) in the presence of a parasagittal longitudinal tear of the cap of the SDFT and stabilise luxation.
  - surgical repair of the torn retinaculum:
    - polypropylene mesh and anchoring screws in the calcaneus.
    - full limb cast support for at least 6 weeks.
    - complications possible (laminitis, cast sores, failure of repair, infection).
    - long rehabilitation.

Prognosis

- good for stable lateral luxation.
- guarded for medial luxation.
- poor for intermittent luxation without surgical intervention.

Calcaneal bursitis

Definition/overview

- calcaneal bursa is a synovial pouch located deep to the SDFT as it traverses the calcaneus:
  - extends from the distal crus to the distal tarsus.
- located between the cranial surface of the SDFT and the caudal surface of the gastrocnemius tendon and between the dorsal surface of the SDFT and the plantar fibrocartilage layer covering the tuber calcis.
- communicates with a smaller synovial pouch located cranial to the insertion of the gastrocnemius tendon to the tuber calcis.

Aetiology

- idiopathic asymptomatic effusion possible but rare.
- occasionally accompanies gastrocnemius tendonitis.
- commonly caused by trauma to the point of the hock.
- septic bursitis most common:
  - associated with penetrating injury to the plantar hock region.
  - often with puncture of the SDFT.
  - occasionally with septic osteitis and sequestration of the tuber calcis.

Clinical presentation

- characteristic soft swelling on both sides of the point of the hock (Fig. 8.168):
  - extends proximally along the common calcaneal tendon.
  - distally along the margins of the SDFT.
- mild to moderate lameness in the presence of gastrocnemius tendonitis.
- severe lameness and subcutaneous swelling if septic bursitis.
- diffuse thickening with loss of landmarks in chronic cases.

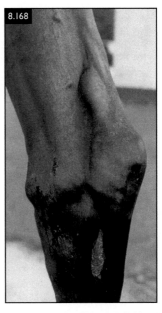

FIG. 8.168 This picture of the medial hock shows swelling of the calcaneal bursa just dorsomedial to the common calcaneal tendon. This was associated with injury to the gastrocnemius tendon. (Photo courtesy S J Dyson)

Diagnosis

- **Clinical examination:**
  - fluid swelling on the plantar aspect of the tuber calcis, with lateral and medial bulging.
  - wounds on the plantar aspect of the calcaneus (Fig. 8.169).
  - synoviocentesis and fluid analysis to confirm the presence of infection.
- **Radiography:**
  - rule out associated bone disorders, including septic osteitis of the calcaneus, avulsion or fractures (Fig. 8.170).
  - contrast bursography to confirm communication between the bursa and penetrating skin wound.
- **Ultrasonography:**
  - confirm inflammation (thickening of the synovial membrane).
  - assess for injury to the SDFT and plantar surface of the calcaneus.
  - presence of gastronemius tendonitis.
  - signs of sepsis include:
    - severe synovial hypertrophy, echogenic and heterogeneous fluid with fibrin clots (Fig. 8.171).

 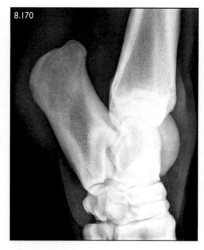 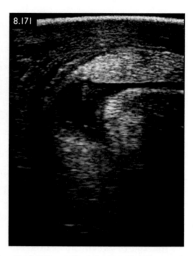

FIGS. 8.169–8.171 (8.169) A puncture wound of some days duration is visible on the plantar distal aspect of the point of the hock (calcaneus). This enters the calcaneal bursa and has led to septic synovitis. Note the swelling around the wound and proximally either side of the calcaneal tendons. (8.170) Lateromedial radiograph of the hock of 8.169. Note the soft tissue swelling at the point of the hock and small defect in the plantar calcaneus due to impact of the wounding trauma. (8.171) Transverse ultrasound image 4 cm proximal to the point of the hock of the horse in 8.169. Note the hypoechogenic distended lateral pouch of the calcaneal bursa with synovial thickening and proliferation lining the bursa and tendons. The increased fluid in the bursa is speckled (suggesting increased cellular content) and separates the SDFT and gastrocnemius tendons.

- ◆ irregular defect and/or new bone production of the calcaneus.

Management

- acute management:
  - ○ rest, bandage, cryotherapy, intrasynovial medication.
- chronic cases with adhesions and thickening:
  - ○ partial synovectomy and adhesiolysis via bursoscopy.
- septic bursitis:
  - ○ aggressive bursoscopic lavage and debridement.
  - ○ sequestrum removal and debridement.
  - ○ intrabursal or intravenous regional antimicrobial therapy.

Prognosis

- good for idiopathic asymptomatic swelling.
- guarded for gastrocnemius tendonitis.
- guarded for septic bursitis:
  - ○ early bursoscopic lavage and debridement improve prognosis.
  - ○ poor if septic SDFT tendonitis and/or septic calcaneal osteitis.

Rupture of the fibularis ('peroneus') tertius tendon

Definition/overview

- fibularis tertius has evolved into a tendinous band without muscle fibres in the horse.
- extends from its origin on lateral femoral epicondyle to the dorsal proximal aspect of the 3rd and 4th metatarsal bones.
- major component of the reciprocal apparatus counteracted by the combined Achilles tendon.
- forces the tarsus to flex passively when the stifle is flexed.

Aetiology/pathophysiology

- sudden strain injury causing the tendon to overstretch to rupture through overextension:
  - ○ fall with hyperextension of the hock while the stifle is flexed.
    - ◆ distal limb is caught over a fence, door, jump or in a farrier's hold.
  - ○ during recovery from anaesthesia with a full limb cast.

o rarely in foals, avulsion fracture of the origin of the tendon from the extensor fossa on the lateral femoral epicondyle.
- transection of the distal insertion from a penetrating wound to dorsal aspect of the distal tarsal/proximal metatarsal region.

Clinical presentation

- normal stance posture.
- during the swing phase of the limb:
  o hock lags behind and remains extended while the stifle is flexed and brought forwards.
- lameness is variable.
- hock can be extended manually while the stifle is flexed (Fig. 8.172).
- relaxation of the achilles tendon during concurrent hock extension/stifle flexion causes a characteristic dimple in the tendon.

Diagnosis

- **Clinical examination**
- **Ultrasonography:**
  o not necessary for the diagnosis.
  o identify the level of the rupture
  o monitor the healing process.

Management

- stall rest for 2–3 months followed by gradual resumption of exercise over further 2 months.
- full functional recovery is expected.

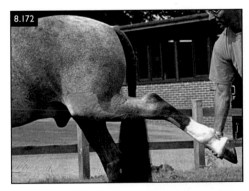

FIG. 8.172 Rupture of the peroneus tertius tendon. The hock is extended by pulling the limb backward while the stifle remains flexed. Note the slight dimpling of the common calcaneal tendon due to loss of reciprocal tension from the peroneus tertius tendon. (Photo courtesy Roger Smith)

Prognosis

- good for all soft tissue ruptures.
- poor for avulsion fracture of the origin from the lateral epicondyle of the femur.

Injuries to the common calcaneal ('Achilles') tendon

Definition/overview

- common calcaneal ('Achilles') tendon comprises:
  o tendons of insertion of the gastrocnemius muscles on the tuber calcis of the calcaneus.
  o SDFT.
  o deep tarsal tendons, two ligamentous branches that receive fibres from the:
    ♦ soleus, biceps femoris, gracilis, semitendinosus, and semimembranosus muscles.
    ♦ insert on the tuber calcis, dorsal to the gastrocnemius tendons.
- SDFT lies deep to the gastrocnemius tendon proximally but twists around its medial aspect and becomes more superficial in the distal crus.
- Achilles tendon is essential for weight bearing as part of the reciprocal apparatus:
  o locks the hock into extension when the stifle is extended:
    ♦ actively during locomotion.
    ♦ passively through locking of the patella on the medial femoral trochlear ridge.
- strain injuries are rare:
  o gastrocnemius injury may occur at origin on the femur or at the insertion to the calcaneus.
  o tendonitis of the deep tarsal tendons is very rare.

Aetiology/pathophysiology

- tendonitis due to repeated strain or to a single event causing hyperflexion of the hock while the stifle is extended.
- rupture associated with a sharp wound or blunt injury to the caudal aspect of the crus.

Clinical presentation

- Gastrocnemius tendons most affected:
  - enlargement of the distal part of the common calcaneal tendon.
    - sometimes difficult to palpate.
  - distension of the calcaneal bursa.
  - occasionally diffuse enlargement of area proximal to the point of the hock (see Fig. 8.168).
  - hindlimb lameness ○ mild to moderate.
  - specific gait abnormality with lateral rotation of the tuber calcis during the stance phase of the stride, shortened caudal phase, and reduced foot flight arc, has been described.
- injury at the proximal attachment of the gastrocnemius muscles to the caudal femur difficult to diagnose as no typical signs.
- complete rupture:
  - laceration usually obvious.
  - partial collapse of the hock during stance.
    - characteristic posture with stifle extended and hock flexed.
  - inability to bear weight on the limb when laceration complete:
    - confirmed by ability to flex the hock manually without flexing the stifle.

Differential diagnosis

- Calcaneal bursitis   • Capped hock

Diagnosis

- **Clinical examination:**
  - swelling in distal crus or calcaneal region.
  - characteristic stance/gait.
- **Radiography:**
  - sometimes entheseous new bone over the caudal distal femoral metaphysis at the proximal attachment of the gastrocnemius muscle.
- **Scintigraphy:**
  - increased radionuclide uptake at caudal aspect of distal femoral metaphysis associated with entheseopathy of proximal attachment of the gastrocnemius muscle.
- **Ultrasonography:**
  - necessary to confirm tendonitis of the gastrocnemius tendon or the deep tarsal tendons (Figs. 8.173, 8.174).
  - gastrocnemius tendonitis lesions similar to digital flexor tendon injury:
    - diffuse lesions with mottled pattern rather than discrete hypoechogenic lesion.
    - associated synovial thickening and distension of the gastrocnemius bursa.
    - entheseopathy/avulsion at the femoral origin difficult to diagnose.
  - confirm partial or complete rupture of the SDFT.

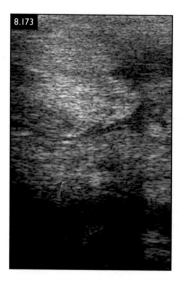
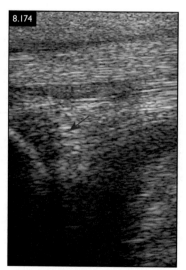

FIGS. 8.173, 8.174 Transverse (8.173) and sagittal (8.174) ultrasonographic images of the common calcanean tendon (arrows) showing severe increase in size and a heterogeneously decreased echogenicity of the deep part of the tendon. These images are typical of deep tarsal tendonitis.

Management

- Gastrocnemius tendonitis:
  - box rest with controlled exercise for up to 12 months to full work.
  - recurrence very common.
- SDFT rupture:
  - box rest
  - full limb cast or plantar splints for up to 3 months.
  - fibrosis may allow partial to complete functional repair.
- complete Achilles tendon rupture:
  - surgical repair and application of a full limb cast for 12 weeks.
    - ♦ often fails on recovery from anaesthetic.
  - bandage and splint support for a further 12 weeks.

Prognosis

- good for deep tarsal tendonitis.
- guarded for distal gastrocnemius tendonitis.
- fair for proximal entheseopathy of the gastrocnemius muscles.
- guarded for SDFT rupture.
- grim for complete Achilles tendon rupture.

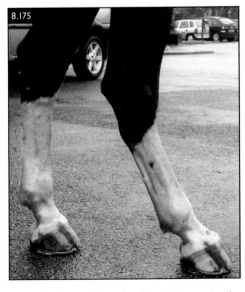

FIG. 8.175 Curb deformity due to trauma to the plantar distal aspect of the right hind tarsus, giving a bowed appearance to this region.

Curb

Definition/overview

- convex swelling over the distal plantar aspect of the hock (Fig. 8.175).
- injury sites:
  - long plantar tarsal ligament.
  - SDFT.
  - peritendinous subcutaneous tissues and fascia.

Aetiology/pathophysiology

- direct trauma and contusion (e.g. kicking stable door or manger) of subcutaneous tissues.
- direct trauma of the SDFT and associated haematoma of peritendinous tissue.
- predisposition to plantar ligament strain, especially horses with a sickle hock conformation.
- curb-like appearance in young foals with dorsal collapse of cuboidal bones.

Clinical presentation

- typical convex deformity over the plantar aspect of the hock.
- soft swelling initially, but firm in chronic cases.
- lameness is variable:
  - marked in SDFT tendonitis.
  - low-grade and recurrent in plantar ligament desmitis.
  - often no lameness.

Differential diagnosis

- long plantar tarsal ligament desmitis.
- SDFT tendonitis or paratendinitis.
- subcutaneous and peritendinous haematoma.
- dorsal collapse of cuboidal bones.

Diagnosis

- **Clinical examination.**
- **Radiography:**
  - rule out other causes or damage to the underlying bones:
    - ♦ calcaneus and heads of the 2nd and 4th metatarsal bones.
    - ♦ cuboidal bone collapse.
- **Ultrasonography:**
  - subcutaneous and peritendinous tissue thickening is obvious (Fig. 8.176).

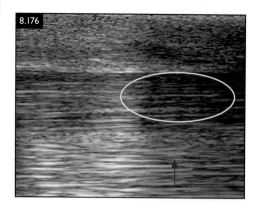

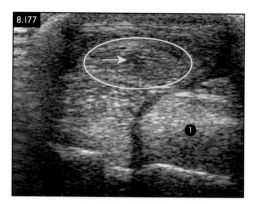

FIG. 8.176 Longitudinal sonogram over the plantar aspect of the calcaneus. The SDFT (circled) and LPL (arrow) are normal, but there is marked thickening of the subcutaneous tissues (arrowhead), probably as a result of repeated, mild trauma.

FIG. 8.177 Transverse ultrasound scan over the plantar aspect of the calcaneus of a horse presenting with curb. The SDFT (circled) is enlarged, with a focal, central, and hypoechogenic lesion (yellow arrow), as well as a peritendinous, hypoechogenic lesion suggesting paratenon thickening (red arrow). The LPL is normal. 1 = DDFT.

- ♦ haematomas visible as hypoechogenic tissue forming crescent-shaped halo plantar to SDFT.
- ♦ sometimes fills space between SDFT and other underlying structures (plantar ligament, DDFT).
- ♦ vessels may be compressed or damaged.
- ○ SDFT lesions:
  - ♦ typically focal hypoechogenic and associated marked tendon enlargement (Fig. 8.177).
  - ♦ paratendonitis and/or haematoma associated with focal lesions on the plantar or abaxial aspects of the tendon.
- ○ Plantar ligament desmitis associated with:
  - ♦ ligament thickening (Fig. 8.178)
    - – compare with the opposite limb.
  - ♦ diffuse or focal hypoechogenic lesions within the ligament.
  - ♦ bone remodelling, particularly at distal insertion on fourth metatarsal bone.

Management

- conservative treatment with box rest, in-hand exercise, cryotherapy and anti-inflammatory treatments.

Prognosis

- good for lameness, although the swelling may persist.
- variable for SDFT injuries depending on the severity and extent of the lesion.

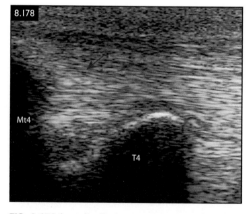

FIG. 8.178 Longitudinal sonogram over the plantarolateral aspect of the distal tarsus. The long branch of the LPL (arrow) runs distally from the calcaneus, over the plantar aspect of the fourth tarsal bone (T4), to insert over the proximal aspect of the head of the fourth metatarsal bone (Mt4). There is a mild, hypoechogenic thickening of the LPL and overlying subcutaneous tissue. This is a very rare cause of curb-like deformity.

Cunean tendonitis/bursitis

Overview

- cunean tendon is medial branch of the distal tendon of the tibialis cranialis muscle:
  - attaches to the head of the second metatarsal bone.
- small, subtendinous bursa exists between the surface of the second and central tarsal bones, and the cunean tendon.
- Cunean tendonitis and bursitis have been described as a cause of lameness and poor performance, especially in racing Standardbreds.
- **existence of the condition is disputed.**
- ultrasonography useful to identify injury from direct trauma (Fig. 8.179).
- treatment with rest, application of cold, and intrabursal corticosteroids.

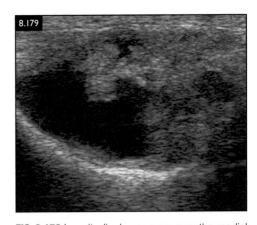

FIG. 8.179 Longitudinal sonogram over the medial aspect of a tarsus with cunean tendonitis. The tendon (arrow) is enlarged, hypoechogenic, and heterogeneous. The underlying bursa is distended with anechogenic fluid (arrowhead).

Index